Strong Approximations
in Probability and Statistics

Strong Approximations in Probability and Statistics

M. Csörgő and P. Révész

Akadémiai Kiadó · Budapest 1981

Copyright © 1981 by Akadémiai Kiadó, Budapest

All rights reserved. No part of this publication may be reproduced or transmitted in any form or by any means, electronic or mechanical, including photocopy, recording, or any information storage and retrieval system, without permission in writing from the publishers.

Joint edition published by Akadémiai Kiadó, Budapest and
Academic Press, New York—San Francisco—London

ISBN 963 05 2441 4

Printed in Hungary

Contents

Preface 9

Introduction 11

1. **Wiener and some Related Gaussian Processes** 21

   1.0 On the notion of a Wiener process 21
   1.1 Definition and existence of a Wiener process 21
   1.2 How big are the increments of a Wiener process? 29
   1.3 The law of iterated logarithm for the Wiener process 36
   1.4 Brownian bridges 41
   1.5 The distributions of some functionals of the Wiener and Brownian bridge processes 43
   1.6 The modulus of non-differentiability of the Wiener process 44
   1.7 How small are the increments of a Wiener process? 47
   1.8 Infinite series representations of the Wiener process and Brownian bridge 53
   1.9 The Ornstein–Uhlenbeck process 55
   1.10 On the notion of a two-parameter Wiener process 57
   1.11 Definition and existence of a two-parameter Wiener process 58
   1.12 How big are the increments of a two-parameter Wiener process? ... 61
   1.13 A continuity modulus of $W(x, y)$ 74
   1.14 The limit points of $W(x, y)$ as $y \to \infty$ 75
   1.15 The Kiefer process 80
   Supplementary remarks 82

2. **Strong Approximations of Partial Sums of I.I.D.R.V. by Wiener Processes** 88

   2.0 Notations 88
   2.1 A proof of Donsker's theorem with Skorohod's embedding scheme 88
   2.2 The strong invariance principle appears 91
   2.3 The stochastic Geyser problem as a lower limit to the strong invariance problem ... 95
   2.4 The longest runs of pure heads and the stochastic Geyser problem 97
   2.5 Improving the upper limit 101
   2.6 The best rates emerge 106
   Supplementary remarks 112

3. A Study of Partial Sums with the Help of Strong Approximation Methods 115

3.0 Introduction 115
3.1 How big are the increments of partial sums of I.I.D.R.V. when the moment generating function exists? 115
3.2 How big are the increments of partial sums of I.I.D.R.V. when the moment generating function does not exist? 117
3.3 How small are the increments of partial sums of I.I.D.R.V.? 120
3.4 A summary 122
    Supplementary remarks 122

4. Strong Approximations of Empirical Processes by Gaussian Processes 127

4.1 Some classical results 127
4.2 Why should the empirical process behave like a Brownian bridge? 129
4.3 The first strong approximations of the empirical process 130
4.4 Best strong approximations of the empirical process 133
4.5 Strong approximation of the quantile process 143
    Supplementary remarks 154

5. A Study of Empirical and Quantile Processes with the Help of Strong Approximation Methods 156

5.0 Introduction 156
5.1 The law of iterated logarithm for the empirical process 156
5.2 The distance between the empirical and the quantile processes 160
5.3 The law of iterated logarithm for the quantile process 162
5.4 Asymptotic distribution results for some classical functionals of the empirical process 163
5.5 Asymptotic distribution results for some classical functionals of the quantile process 171
5.6 Asymptotic distribution results for some classical functionals of some k-sample empirical and quantile processes 181
5.7 Approximations of the empirical process when parameters are estimated .. 188
5.8 Asymptotic quadratic quantile tests for composite goodness-of-fit 202
5.9 On testing for exponentiality 213
    Supplementary remarks 216

6. A Study of Further Empirical Processes with the Help of Strong Approximation Methods 219

6.0 Introduction 219
6.1 Strong invariance principles and limit distributions for empirical densities 219
6.2 Strong theorems for empirical densities 230
6.3 Empirical regression 237

6.4 Empirical characteristic functions 242
     Supplementary remarks 248

7. Random Limit Theorems via Strong Invariance Principles 250

7.0 Introduction and some historical remarks 250
7.1 Laws of large numbers for randomly selected sequences 252
7.2 Invariance (strong and weak) principles for random-sum limit theorems 255
7.3 Invariance (strong and weak) principles for random size empirical processes 261

References 263

Author Index 277

Subject Index 281

Summary of Notations and Abbreviations 283

Preface

Without knowing that both of us were there, the authors of this volume were random-walking on the streets of London in 1966 when, due to a theorem of Pólya, they met. Obviously this occasion called for a beer and a chat about mathematics. The beer turned out to be good enough to suggest that we should work together, and the idea of writing this book was born then. We are deeply indebted to the innkeeper for his hospitality on this occasion.

At that time we did not really know each other, though we had a common root in that both of us were students of Alfréd Rényi. The first named author actually studied mathematics at McGill University in Montreal and never took any courses from him. It was the papers and book of Rényi at that time, however, which influenced him most, and moulded his interest in doing research in probability-statistics. This also led to meeting him several times personally, thus directly benefiting from his most stimulating and unique way of thinking about mathematics. The second named author was a student of Rényi, indeed taking his courses in Budapest, and learning the secrets of doing research in probability directly from him. Rényi's great enthusiasm for the beauty of doing mathematics has inspired him to also try his hands at it. Both of us are deeply convinced that, without his lasting influence and help while we were young, we could have never written this book.

Our real collaboration began in 1972. During these past years we were fortunate enough to be able to visit each other several times, working in Ottawa where M. Csörgő is located and in Budapest where P. Révész is. This intensive collaboration would have been impossible without the understanding and support of our respective home institutions, the Department of Mathematics at Carleton University and the Mathematical Institute of the Hungarian Academy of Sciences.

Generous financial support was received in Canada: National Research Council Operating Grants throughout these years, Canada Council Leave Fellowship (1976–1977), The Carleton University Norman Paterson Centre (1976–1977, 1979), Canada Council Killam Senior Research Scholarship (1978–1980) and in Hungary: The Bolyai János Mathematical Society throughout these years, the Institute of Cultural Relations (1976–1977). We are deeply indebted to all these institutions and can only say that without their support our work together could have never taken place.

In various stages of development, preprints of the different chapters of this book have been distributed to a number of prominent mathematicians who have commented on a large number of topics involved. Many of their valuable remarks were incorporated in the final version. We express our best thanks to all of them, and we especially appreciate the help of M. D. Burke (University of Calgary), A. H. C. Chan (Ontario Hydro), S. Csörgő (University of Szeged), I. A. Ibragimov (University of Leningrad) and R. J. Tomkins (University of Regina).

The thankless task of reading the semi-final version of all the chapters fell to the referees of our book, I. Berkes (Mathematical Institute of the Hungarian Academy of Sciences) and K. Tandori (University of Szeged). Their expert, inquisitive reading of our manuscript in a very short time, resulted in their listing a large number of misprints, oversights and mistakes in our text. We are more than grateful to them and our sincere thanks are recorded here with much appreciation. We also express our gratitude to Mrs. Gill S. Murray of Ottawa for her expert and patient typing of the many versions of our manuscript. Similar thanks are due to the Hungarian printers of our book.

While trying to correct all the mistakes of our manuscript which we have noticed and/or had been pointed out to us, we must have also left a few in and introduced some further ones. We should be very happy to learn of any found by the reader together with whatever his or her comments might be.

Ottawa, October 13, 1979

M. Csörgő
Department of Mathematics
Carleton University
Colonel By Drive
Ottawa
Canada

P. Révész
Mathematical Institute of the
Hungarian Academy of Sciences
H–1053 Budapest
Reáltanoda u. 13–15
Hungary

Introduction

Let $X_1, X_2, \ldots$ be i.i.d.r.v. with $EX_1 = 0$, $EX_1^2 = 1$ and let F be their distribution function. Let $Y_1, Y_2, \ldots$ be i.i.d. normal r.v. with mean zero and variance one $(Y_1 \in \mathcal{N}(0, 1))$ and put $S_n = \sum_{i=1}^{n} X_i$, $T_n = \sum_{i=1}^{n} Y_i$ with $S_0 = T_0 = 0$. The classical central limit theorem states

(0.1) $$P\{n^{-1/2} S_n \leq y\} \to \Phi(y) = \frac{1}{\sqrt{2\pi}} \int_{-\infty}^{y} e^{-u^2/2} \, du$$

for any real y as $n \to \infty$. Since

$$P\{n^{-1/2} T_n \leq y\} = \Phi(y) \quad (n = 1, 2, \ldots),$$

the central limit theorem can also be stated as follows:

(0.2) $$P\{n^{-1/2} S_n \leq y\} - P\{n^{-1/2} T_n \leq y\} \to 0,$$

which, roughly speaking, means that the limiting behaviour of S_n and T_n is the same. In other words, as time goes on, S_n forgets about the distribution function F where it has come from. However, it is also true that observing the sequence $S_1, S_2, \ldots$ (or, only $S_n, S_{n+1}, \ldots$ from any fixed n on), one can determine F with probability one via the Glivenko–Cantelli theorem.

Thus one can say that each individual S_n forgets about F but the complete sequence $\{S_n; n = 1, 2 \ldots\}$ (or a tail of it) remembers F. One of the main goals of this book is to investigate to what extent can S_n remember F and to what extent can it forget about it.

The first questions of this type were formulated by Erdős and Kac (1946) (cf. also Kac (1946)). They wanted to evaluate the limit distributions

(i) $$G_1(y) = \lim_{n \to \infty} P(n^{-1/2} \max_{1 \leq k \leq n} S_k \leq y),$$

(ii) $$G_2(y) = \lim_{n \to \infty} P(n^{-1/2} \max_{1 \leq k \leq n} |S_k| \leq y),$$

(iii) $$G_3(y) = \lim_{n \to \infty} P\left(n^{-2} \sum_{k=1}^n S_k^2 \leq y\right),$$

(iv) $$G_4(y) = \lim_{n \to \infty} P\left(n^{-3/2} \sum_{k=1}^n |S_k| \leq y\right).$$

Erdős and Kac realized that these limit distributions can be easily evaluated for some special distributions F. For example (i) and (ii) can be immediately evaluated if F is the distribution $P(X_1 = +1) = P(X_1 = -1) = \frac{1}{2}$, while for (iii) and (iv) the normal law turns out to be a good starting point. Hence a program for finding the limits (i)–(iv) may be carried out in two steps. First, they should be evaluated for any specific distribution function F, and then one should show that the functionals of $\{S_n\}$, in any one of the cases (i)–(iv), do not remember the initially taken distribution.

Indeed, Erdős and Kac proved that the ability of S_n to forget is strong enough for the above program, that is, they proved that the limit distributions (i)–(iv) exist and they do not depend on the initial distribution of X_1. They called this method of proof the invariance principle, and their paper has initiated a new methodology for proving limit laws in probability theory.

The first step in this general development was taken by Donsker (1951). Donsker's idea was that from the partial sums $S_0, S_1, \ldots, S_n$ one should construct a sequence of stochastic processes $\{S_n(t); 0 \leq t \leq 1\}$ on $C(0, 1)$ as follows:

(0.3) $$S_n(t) = n^{-1/2}\{S_{[nt]} + X_{[nt]+1}(nt - [nt])\}.$$

Clearly $S_n\left(\dfrac{k}{n}\right) = S_k/\sqrt{n}$, and $S_n(t)$ is the linear interpolation of the latter for $\dfrac{k}{n} < t < \dfrac{k+1}{n}$. The advantage of the map (0.3) is that one can study the limiting behaviour of the sequence $S_k/\sqrt{n}$ via that of $S_n(t)$ on $C(0,1)$. Indeed, using a multivariate version of the central limit theorem, one can immediately say that

(0.4) $$(S_n(t_1), S_n(t_2), \ldots, S_n(t_k)) \xrightarrow{\mathcal{D}} (W(t_1), W(t_2), \ldots, W(t_k))$$

for any fixed sequence $0 \leq t_1 < t_2 < \ldots < t_k \leq 1$ as $n \to \infty$, where $W(t)$ is a standard Wiener process. This would then suggest that the distributional properties of $\{S_n(t); 0 \leq t \leq 1\}$ should coincide[1] with those of $\{W(t); 0 \leq t \leq 1\}$ as $n \to \infty$. One possible way of saying this precisely is:

Theorem 0.1. (Donsker 1951). *We have*

$$(0.5) \qquad h(S_n(t)) \xrightarrow{\mathscr{D}} h(W(t))$$

for every continuous functional $h: C(0, 1) \to R^1$.

We note here that (0.4) only suggests that (0.5) should also be true and a precise proof of it was not at all easy to produce. Indeed, if $\{X_n(t)\}_{n=0}^{\infty}$ is a sequence of stochastic processes taking values from a function space M endowed with a metric ϱ, and

$$(0.6) \qquad (X_n(t_1), X_n(t_2), \ldots, X_n(t_k)) \xrightarrow{\mathscr{D}} (X_0(t_1), X_0(t_2), \ldots, X_0(t_k))$$

for any fixed sequence $t_1 < t_2 < \ldots < t_k$, then the statement that

$$(0.7) \qquad h(X_n(t)) \xrightarrow{\mathscr{D}} h(X_0(t))$$

should hold for every continuous functional $h: M \to R^1$, is not necessarily true. A complete methodology for proving (0.7), assuming that (0.6) is true, was worked out by Prohorov (1956) and Skorohod (1956).

In fact they proved a stronger statement to the effect that, under some conditions, the sequence of probability measures generated by $\{X_n(t)\}$ converges (in the so-called weak topology) to the measure generated by $X_0(t)$. An excellent summary and further development of these ideas and techniques can be found in the books of Billingsley (1968) and Parthasarathy (1967).

Replacing the functional h in Theorem 0.1 by

$$h_1(f) = \sup_{0 \leq t \leq 1} f(t), \qquad h_2(f) = \sup_{0 \leq t \leq 1} |f(t)|,$$

$$h_3(f) = \int_0^1 f^2(t)\, dt, \qquad h_4(f) = \int_0^1 |f(t)|\, dt,$$

and taking into account that these functionals are continuous with respect to the topology of $C(0, 1)$, Theorem 0.1, in particular, also implies that

---

[1] In this connection we should also mention that Kolmogorov (1931, 1933a) and Khinchine (1933) investigated the problem of evaluating the asymptotic probability of the event $f_1(t) < S_n(t) \leq f_2(t)$ for two functions $f_1(t) < 0 < f_2(t)$, and proved that under some conditions on these functions the latter probability is equal to $P\{f_1(t) < W(t) \leq \leq f_2(t)\}$. Their approach is based on the heat equation.

$G_i(x)$ ($i=1, 2, 3, 4$) of (i)–(iv) do not depend on F. That is to say the invariance principle of Erdős and Kac follows from Donsker's theorem and, at the same time, the latter can also be applied for any other continuous functional.

After the development of the theory of weak convergence of probability measures on metric spaces, a completely new form of the invariance principle was introduced by Strassen (1964). He proposed to construct a Wiener process $W(t)$ on the very same probability space where the r.v. $\{X_i\}$ live in such a way that $|S_n - W(n)|$ would be small in the sense that the relation

$$(0.8) \qquad \frac{|S_n - W(n)|}{g(n)} \xrightarrow{\text{a.s.}} 0$$

should hold for a suitably increasing function g. In fact the possibility of such a construction depends not only on the distribution F but also on the structure of the basic space. Hence the question in a more adequate form is the following:

Given a distribution function F with $\int x dF = 0$, $\int x^2 dF = 1$, can we construct a probability space $\{\Omega, \mathcal{A}, P\}$, a sequence $\{X_i\}$ of i.i.d.r.v. with $P(X_1 \leq y) = F(y)$ living on Ω, and a Wiener process $W(t)$ also defined on Ω, such that (0.8) should hold?

Answering this question Strassen (1964) proved the following

Theorem 0.2.

$$(0.9) \qquad \frac{|S_n - W(n)|}{\sqrt{n \log \log n}} \xrightarrow{\text{a.s.}} 0.$$

That is to say for any F with $\int x dF = 0$, $\int x^2 dF = 1$, one can construct a probability space where the i.i.d. sequence $\{X_i\}$ and a Wiener process $W(t)$ can be realized such that (0.9) holds.

In order to get a form of Theorem 0.2 resembling that of Theorem 0.1, we give the following reformulation of the former.

Theorem 0.2*.

$$(0.9^*) \qquad \sup_{0 \leq t \leq 1} \frac{|S_n(t) - n^{-1/2} W(nt)|}{\sqrt{\log \log n}} \xrightarrow{\text{a.s.}} 0.$$

Comparing Theorems 0.1 and 0.2 (or 0.2*), a great advantage of the latter is that is speaks about almost sure convergence instead of convergence in distribution.

Strassen used his strong invariance principle (Theorem 0.2) to prove the law of iterated logarithm for i.i.d.r.v. with finite second moment (the Hartman–Wintner theorem (1941)) via first proving such a theorem for the Wiener process. In fact studying the sequence $\left\{\dfrac{n^{-1/2}W(nt)}{\sqrt{2\log\log n}}; 0\leq t\leq 1\right\}$ of stochastic processes, Strassen also obtained a deeper insight into the properties of the sequence $\left\{\dfrac{S_n(t)}{\sqrt{2\log\log n}}; 0\leq t\leq 1\right\}$ (cf. Theorem 1.3.2).

In this spirit then Theorem 0.2 is like Theorem 0.1, the latter being applicable to prove weak convergence theorems for i.i.d.r.v. using distributional properties of the Wiener process, while the former is useful for proving strong theorems via similar properties of the Wiener process.

Theorem 0.2, however, does not imply Theorem 0.1, and this is because the rate of convergence in (0.9) is not strong enough. Should one be able to prove (0.8) with $g(n)=o(n^{1/2})$, then clearly we could also get (0.5) as a consequence of such a strong invariance principle. Chapter 2 of this book is mainly devoted to the question of the best possible rate in (0.8).

The precise connection between weak and strong invariance principles was established by Strassen (1965a) (cf. also Dudley (1968) and Wichura (1970)) via the so-called Prohorov distance of probability measures. In fact these results state a kind of equivalence between these two forms of invariance.

Our book is mainly devoted to the overall question of strong invariance theorems.

Our reason for concentrating on strong invariance methodology (instead of the weak one) can, perhaps, be justified by the fact that this approach has developed so much in recent years that it was capable of producing a number of results in probability and statistics which, in spite of the above mentioned equivalence of the two principles, would have been quite difficult to produce by the usual weak convergence methodology.

When talking about the origin of the invariance principle, another, independent source should be also mentioned besides the 1946 paper of Erdős and Kac. It is the paper of Doob (1949), entitled "Heuristic approach to the Kolmogorov–Smirnov theorems". The idea of this paper is the following: Let $U_1, U_2, \ldots$ be a sequence of i.i.d.r.v., coming from the uniform $U(0, 1)$ law. Let

$$E_n(x) = n^{-1} \sum_{k=1}^{n} I_{(0,x]}(U_k)$$

be the empirical distribution function, and let

$$\alpha_n(x) = n^{1/2}(E_n(x) - x)$$

be the empirical process. Observe that the limit of the joint distribution of $\alpha_n(x_1), \alpha_n(x_2), \ldots, \alpha_n(x_k)$ $(0 \leq x_1 < x_2 < \ldots < x_k \leq 1; \ k = 1, 2, \ldots)$ is the corresponding finite dimensional distribution of a Brownian bridge; that is to say

(0.10) $\quad \{\alpha_n(x_1), \alpha_n(x_2), \ldots, \alpha_n(x_k)\} \xrightarrow{\mathscr{D}} \{B(x_1), B(x_2), \ldots, B(x_k)\}$

as $n \to \infty$, where $B(x)$ is a Brownian bridge. This then suggests that the limit properties of the empirical process $\alpha_n(x)$ should agree with the corresponding properties of a Brownian bridge. For example, the limit distribution of $\sup_x \alpha_n(x)$ (resp. $\sup_x |\alpha_n(x)|$) should agree with the distribution of $\sup_x B(x)$ (resp. $\sup_x B|(x)|$). Since the direct evaluation of the limit distribution of $\sup_x \alpha_n(x)$ (resp. $\sup_x |\alpha_n(x)|$) is rather complicated, while the evaluation of the distribution of $\sup_x B(x)$ (resp. $\sup_x |B(x)|$) is easier, the above sketched approach is obviously useful. Indeed, besides posing the above invariance argument, Doob (1949) proceeded to evaluate the distribution of these latter functionals of $B(x)$, leaving the problem of justification of his approach open. Donsker (1952) was the first one again who attacked this latter problem and succeeded in justifying and extending Doob's heuristic approach.

Comparing this problem to that of Theorem 0.1, we can see that a difficulty is coming from the fact that the sample functions of $\alpha_n(x)$ do not belong to $C(0, 1)$. This difficulty was again solved by Prohorov (1956) and Skorohod (1956), while working on the so-called $D(0, 1)$ function space. Naturally, an analogue of Theorem 0.1 is also true for a continuous approximation of $\alpha_n(x)$ on $C(0, 1)$.

In the light of Strassen's strong invariance principle, it was only natural to look for analogous approximations also for the empirical process $\alpha_n(x)$. This task turned out to be quite difficult and it took a bit of time to get results. The first one of them is due to Brillinger (1969), and reads as follows:

Theorem 0.3. *Given independent* $U(0, 1)$ *r.v.* $U_1, U_2, \ldots,$ *there exists a probability space with sequences of Brownian bridges* $\{B_n(x); 0 \leq x \leq 1\}$ *and empirical processes* $\{\tilde{\alpha}_n(x); 0 \leq x \leq 1\}$ *such that*

(0.11) $\quad \{\tilde{\alpha}_n(x); 0 \leq x \leq 1\} \stackrel{\mathscr{D}}{=} \{\alpha_n(x); 0 \leq x \leq 1\}$ *for each* $n = 1, 2, \ldots,$

and

(0.12) $\sup_{0\leq x\leq 1} |\tilde{\alpha}_n(x) - B_n(x)| \stackrel{\text{a.s.}}{=} O(n^{-1/4}(\log n)^{1/2}(\log\log n)^{1/4}).$

This theorem immediately implies the above mentioned analogue of Theorem 0.1. Namely, in terms of weak convergence, we have

(0.13) $\alpha_n(\cdot) \stackrel{\mathcal{D}}{\longrightarrow} B(\cdot).$

On the other hand, in spite of the indicated a.s. convergence in (0.12), Theorem 0.3 is not really a strong approximation theorem like Theorem 0.2 is. The reason for this is that in (0.12) we only have an approximation for each n, and only for a version $\tilde{\alpha}_n(x)$ of $\alpha_n(x)$. More precisely then, while Theorem 0.3 is a good first step in the right direction, it does not succeed in bringing together the stochastic processes $\{\alpha_n(x); 0\leq x\leq 1, n=1, 2, ...\}$ and $\{B_n(x); 0\leq x\leq 1, n=1, 2, ...\}$. Consequently, no strong law type behaviour of the process $\alpha_n(x)$, say like the law of iterated logarithm, can be deduced from (0.12).

Kiefer (1969b) was the first one to call attention to the desirability of viewing the empirical process $\alpha_n(x)$ as a two parameter process and that a strong approximation theorem for $\alpha_n(x)$ should be given in terms of an appropriate two dimensional Gaussian process. He also succeeded in giving a first solution to this problem (Kiefer 1972; cf. Theorem 4.3.1). Preceding this work, Müller (1970) proved a corresponding two dimensional weak convergence of $\alpha_n(x)$, using Rényi's (1953) exponential representation of the empirical process.

In the present book we intend to summarize and elaborate on a number of recent strong invariance type results for partial sums and empirical processes of i.i.d.r.v., putting an emphasis on the applicability of strong approximation methodology to a variety of problems in probability and statistics. This is why, in the title, we use the expression "strong approximations" instead of "strong invariance principles".

In Chapter 1 we study the Wiener process together with some further Gaussian processes derived from it. In fact, in this Chapter we have intended to collect mostly those theorems for Gaussian processes which can be extended to partial sums and empirical processes of i.i.d.r.v. via strong approximation methods.

Chapter 2 is addressed to the problem of best possible strong approximations of partial sums of i.i.d.r.v. by a Wiener process, and it contains those theorems which tell us a complete story of this problem.

The content of Chapter 3 can be summarized in one sentence: Take "almost" any theorem of Chapter 1 concerning the one-time parameter Wiener process, then it can be extended to partial sums of i.i.d.r.v. via the results of Chapter 2. In most of the cases when the approximation methods do not work we can also conclude that the corresponding results cannot be extended at all. This Chapter does not intend to give a full systematic treatment of the asymptotic behaviour of partial sum processes and we concentrate only on those properties which can be deduced from invariance principles. For a detailed discussion of sums of random variables we refer to Petrov (1975) and Stout (1974).

Chapter 4 contains strong approximation theorems (in terms of suitable Gaussian processes) for the empirical and quantile processes based on i.i.d.r.v.

The role of Chapter 5 in the theory of empirical and quantile processes is similar to that of Chapter 3 in the theory of partial sums of i.i.d.r.v. Namely, in this Chapter we show that by applying the results of Chapter 4, the theorems of Chapter 1 concerning Brownian bridges and the so-called Kiefer process are also valid for empirical and quantile processes. This phenomenon of inheriting properties from appropriate Gaussian processes is not so complete here as in the case of partial sums of i.i.d.r.v. and, to some extent, we also touch upon the problem of similar and non-similar behaviour beyond invariance (cf. Remark 5.1.1). For a recent and more detailed discussion of this topic we refer to the survey paper of Gaenssler and Stute (1979).

In Chapter 6 we show that suitably defined sequences of empirical density, regression and characteristic functions can be approximated by appropriate Gaussian processes. Here it will be seen that some results on Gaussian processes can be extended also to these by strong approximation methods.

The aim of Chapter 7 is to demonstrate that strong approximation methodology can also be applied to study weak and strong convergence properties of random size partial sum and empirical processes.

A common property of Chapters 3, 5, 6 and 7 is that their respective topics are treated only so far as one can see them via strong approximation methods, and we did not aim at completeness at all in treating them.

The subject of this book is restricted to i.i.d.r.v. when the time and state parameters belong to the real line. There is an exception in Chapter 1, when we also study certain properties of two-time parameter Wiener and Kiefer processes. Our reason for this is due to the fact that certain properties of the empirical process $\alpha_n(x)$ can only be described and handled via viewing it as a two-time parameter process in x and n.

We intend to study the problems of strong approximation of multi-time parameter partial sum and empirical processes by appropriate multi-time parameter Gaussian processes in the second volume of this book.

The case when the state space is also a higher dimensional Euclidean space (or a Banach space) has been investigated by several authors (cf. e.g. J. Kuelbs 1973, J. Hoffman–Jørgensen–G. Pisier 1976, Garling 1976) and it should be the subject of a third volume. The subject of a fourth volume should be the case of non-independent and/or non-identically distributed r.v. (for a preliminary version we refer to W. Philipp and W. Stout (1975), an excellent survey of the present situation of this topic). However, the authors have realized that the lifetime of a human being is not only a one-dimensional but also a strictly bounded r.v. Hence, they do not intend to write the mentioned third and fourth volumes, though they would be glad to live long enough to read these by someone else.

1. Wiener and some Related Gaussian Processes

1.0. On the notion of a Wiener process

The English botanist Brown observed in 1826 that microscopic particles suspended in a liquid are subject to continual molecular impacts and execute zigzag movements (Brownian motion). Einstein found (1905) that these movements can be analysed by laws of probability. One of the simplest models for a one dimensional Brownian motion can be given in terms of the coin tossing or random walk model. Suppose that a particle is moving on the real line, starting from the origin. In each time unit it can only move one step to the right, or to the left, with probability one half, and these steps are assumed to be independent. Say the ith step of the particle is X_i; then $X_1, X_2, \ldots$ are i.i.d.r.v. with

$$(1.0.1) \qquad P(X_1 = +1) = P(X_1 = -1) = 1/2,$$

and after n steps the particle will be located at $S_n = X_1 + X_2 + \ldots + X_n$. The thus created path $S_1, S_2, \ldots$ imitates Brownian motion quite well if the time unit and steps are short enough.

In a more realistic model of a Brownian motion the particle makes instantaneous steps to the right or to the left, that is a continuous time scale is used instead of a discrete one, and the lengths X_i of steps are normally distributed instead of the distribution (1.0.1). In the next section the definition of a Wiener process takes into account the just sketched definition of a Brownian motion.

1.1. Definition and existence of a Wiener process

A stochastic process $\{W(t; \omega) = W(t); 0 \leq t < \infty\}$, where $\omega \in \Omega$, and $\{\Omega, \mathcal{A}, P\}$ is a probability space, is called a Wiener process if
 (i) $W(t) - W(s) \in \mathcal{N}(0, t-s)$ for all $0 \leq s < t < \infty$ and $W(0) = 0$,
 (ii) $W(t)$ is an independent increment process, that is $W(t_2) - W(t_1)$,

$W(t_4)-W(t_3), \ldots, W(t_{2i})-W(t_{2i-1})$ are independent r.v. for all $0 \leq t_1 < t_2 \leq t_3 < t_4 \leq \ldots \leq t_{2i-1} < t_{2i} < \infty$ $(i=2, 3, \ldots)$,

(iii) the sample path function $W(t, \omega)$ is continuous in t with probability one.

We note that (i) and (ii) imply that the covariance function of a Wiener process is
$$R(s, t) = EW(s)W(t) = s \wedge t.$$

Remark 1.1.1. Conversely, a Gaussian process having the latter covariance function must also satisfy properties (i) and (ii), that is to say, a continuous or at least separable Gaussian process with the above covariance function is a Wiener process.

The aim of this section is to give a constructive proof for the existence of this process. Towards this end, let $\{r_n\}$ be the sequence of positive dyadic rational numbers (i.e., numbers of the form $k/2^n$, $k=1, 3, \ldots$, $n=1, 2, \ldots$) and let $\{X_{r_n}\}$ be independent $\mathcal{N}(0, 1)$ r.v. defined on a probability space $\{\Omega, \mathcal{A}, P\}$. On this probability space we construct a Wiener process as follows:

For any positive integer k, let
$$W(k) = X_1 + X_2 + \ldots + X_k$$
and
$$W(k+\tfrac{1}{2}) = \frac{W(k)+W(k+1)}{2} + \frac{X_{k+1/2}}{\sqrt{4}}.$$

Now we wish to define $W(k/2^n)$ for $k=1, 3, \ldots$ and $n=1, 2, \ldots$. Assume that it is already defined for $k=1, 2, \ldots$ and $n=1, 2, \ldots, n_0$. Then, for $k=1, 2, \ldots$ and $n=n_0+1$, let
$$W\left(\frac{2k+1}{2^n}\right) = \frac{W\left(\frac{2k}{2^n}\right)+W\left(\frac{2k+2}{2^n}\right)}{2} + \frac{X_{(2k+1)2^{-n}}}{\sqrt{2^{n+1}}}.$$

Whence, by induction, we have defined our Wiener process at every dyadic rational point r_n. For an arbitrary $0 < t = \sum_{k=0}^{\infty} \frac{\varepsilon_k(t)}{2^k}$ ($\varepsilon_0(t)=0, 1, 2, \ldots$; $\varepsilon_k(t)=0, 1$; $k=1, 2, \ldots$ and $\varepsilon_k(t)$ should not be identically 1 from some k on) we define
$$W(t) \stackrel{a.s.}{=} \lim_{n \to \infty} W([2^n t]/2^n) = \lim_{n \to \infty} W(t_n)$$
$$= W(\varepsilon_0(t)) + \lim_{n \to \infty} \sum_{k=1}^{n} (W(t_k)-W(t_{k-1})),$$

with $t_n = \sum_{j=0}^{n} \varepsilon_j(t)/2^j$. The existence of the above limit follows immediately from Kolmogorov's Three Series Theorem for every fixed $t>0$. However, the exceptional set of probability zero, where this latter a.s.–convergence might not hold, can depend on the particular fixed t. This, however, presents no problems, because there exists a set $\Omega_0 \subset \Omega$ of probability zero such that the series $\sum_{k=1}^{\infty} (W(t_k) - W(t_{k-1}))$ converges for every t whenever $\omega \in \Omega - \Omega_0$. In fact, we are going to prove the stronger statement that the above limit representation of $W(t)$ holds uniformly in t with probability one. In order to see this, it suffices to show

$$\sum_{k=1}^{\infty} \sup_{0 \leq t \leq 1} |W(t_k) - W(t_{k-1})| < \infty \quad \text{a.s.,}$$

which, in turn, is simplied by the well known estimation (cf. Feller 1968, p. 175)

(1.1.1) $\quad \dfrac{1}{\sqrt{2\pi}} \left(\dfrac{1}{x} - \dfrac{1}{x^3} \right) e^{-x^2/2} \leq 1 - \Phi(x) \leq \dfrac{1}{\sqrt{2\pi}\, x} e^{-x^2/2}, \quad x > 0,$

as follows. First, we have

$$P\left\{ \sup_{0 \leq t \leq 1} |W(t_k) - W(t_{k-1})| \geq u_k \frac{1}{\sqrt{K}} \right\} \leq 2K e^{-u_k^2/2},$$

where $K = 2^k$ and $u_k = C\sqrt{2 \log K}$, $C = \text{const.} > 1$. Consequently, with $L = C \sum_{k=1}^{\infty} \sqrt{\dfrac{2 \log 2^k}{2^k}}$,

$$P\left\{ \sum_{k=1}^{\infty} \sup_{0 \leq t \leq 1} |W(t_k) - W(t_{k-1})| \geq CL \right\} \leq \sum_{k=1}^{\infty} \frac{2}{2^{k(C^2-1)}}$$

$$= \frac{2}{2^{C^2-1} - 1} \to 0 \quad \text{as} \quad C \to \infty,$$

gives the desired a.s. convergence.

A little calculation now shows that the thus defined process $\{W(t); 0 \leq t < \infty\}$ satisfies conditions (i) and (ii). Condition (iii), however, is not immediate at all. The rest of this section is devoted to proving and further elaborating on condition (iii) for the above constructed process $\{W(t); 0 \leq t < \infty\}$. The following lemma plays a key role in doing this.

Lemma 1.1.1. For any $\varepsilon > 0$ there exists a constant $C = C(\varepsilon) > 0$ such that the inequality

$$(1.1.2) \quad P\left\{\sup_{0 \leq s \leq 1-h} \sup_{0 < t \leq h} |W(s+t) - W(s)| \geq v\sqrt{h}\right\} \leq \frac{C}{h} e^{-\frac{v^2}{2+\varepsilon}}$$

holds for every positive v and $h < 1$.

Proof. Using again the above notation, for a positive real number s and integer r, let $s_r = [2^r s]/2^r = \sum_{j=0}^{r} \varepsilon_j(s)/2^j$. Also write $R = 2^r$. Clearly, for each $\omega \in \Omega$ and s, t, r fixed, we have

$$|W(s+t) - W(s)| \leq |W((s+t)_r) - W(s_r)| + |W(s+t) - W((s+t)_r)|$$
$$+ |W(s_r) - W(s)| \leq |W((s+t)_r) - W(s_r)|$$
$$+ \sum_{j=0}^{\infty} |W((s+t)_{r+j+1}) - W((s+t)_{r+j})|$$
$$+ \sum_{j=0}^{\infty} |W(s_{r+j+1}) - W(s_{r+j})|.$$

Since $\sup_{0 < t \leq h} |(s+t)_r - s_r| \leq h + R^{-1}$, $\sup_{0 < t \leq h} |(s+t)_{r+j+1} - (s+t)_{r+j}| \leq 2^{-(r+j+1)}$ and $W((s+t)_r) - W(s_r) \in \mathcal{N}(0, (s+t)_r - s_r)$, for any positive h, u, x_j and integers r, j we have

$$P\left\{\sup_{0 \leq s \leq 1-h} \sup_{0 < t \leq h} |W((s+t)_r) - W(s_r)| \geq u\sqrt{h + 1/R}\right\}$$
$$\leq 2e^{-u^2/2} R(Rh+1),$$
$$P\left\{\sup_{0 \leq s \leq 1-h} \sup_{0 < t \leq h} |W((s+t)_{r+j+1}) - W((s+t)_{r+j})| \geq x_j \frac{1}{\sqrt{2^{r+j+1}}}\right\}$$
$$\leq 2e^{-x_j^2/2} 2^{r+j+1},$$

and similarly

$$P\left\{\sup_{0 \leq s \leq 1-h} \sup_{0 \leq t \leq h} |W(s_{r+j+1}) - W(s_{r+j})| \geq x_j \frac{1}{\sqrt{2^{r+j+1}}}\right\}$$
$$\leq 2e^{-x_j^2/2} 2^{r+j+1}.$$

Whence

$$(1.1.3) \quad P\left\{\sup_{0 \leq s \leq 1-h} \sup_{0 < t \leq h} |W(s+t) - W(s)| \geq u\sqrt{h + 1/R} + 2\sum_{j=0}^{\infty} \frac{x_j}{\sqrt{2^{r+j+1}}}\right\}$$
$$\leq 2R(Rh+1)e^{-u^2/2} + 8R \sum_{j=0}^{\infty} 2^j e^{-x_j^2/2}.$$

Put $x_j=\sqrt{2j+u^2}$ and R such that $2R>K/h\geq R$, where K is a positive constant and will be specified later on. Then

$$8R\sum_{j=0}^{\infty}2^je^{-x_j^2/2}\leq\frac{8K}{h}\sum_{j=0}^{\infty}(2/e)^je^{-u^2/2}=\frac{AK}{h}e^{-u^2/2},$$

where

$$A=8\sum_{j=0}^{\infty}(2/e)^j$$

and

$$u\sqrt{h+1/R}+2\sum_{j=0}^{\infty}\frac{x_j}{\sqrt{2^{r+j+1}}}\leq u\sqrt{h+1/R}+2\sqrt{\frac{h}{K}}\left[\sum_{j=0}^{\infty}\sqrt{\frac{2j}{2^j}}+u\sum_{j=0}^{\infty}\frac{1}{\sqrt{2^j}}\right]$$

$$\leq u\sqrt{h+\frac{2h}{K}}+2\sqrt{\frac{h}{K}}[B+uG]=\sqrt{h}\left\{u\left[\sqrt{1+\frac{2}{K}}+2G\sqrt{\frac{1}{K}}\right]+2B\sqrt{\frac{1}{K}}\right\},$$

where

$$B=\sum_{j=0}^{\infty}\sqrt{\frac{2j}{2^j}}\quad\text{and}\quad G=\sum_{j=0}^{\infty}\frac{1}{\sqrt{2^j}}.$$

Letting now $v=u\left[\sqrt{1+2/K}+2G\sqrt{1/K}\right]+2B\sqrt{1/K}$ we get by (1.1.3) that

$$P\{\sup_{0\leq s\leq 1-h}\sup_{0<t\leq h}|W(s+t)-W(s)|\geq\sqrt{h}v\}$$

$$\leq\frac{2K}{h}(K+1)e^{-u^2/2}+\frac{AK}{h}e^{-u^2/2}=\frac{1}{h}e^{-u^2/2}[2K(K+1)+AK]\leq\frac{C_1}{h}e^{-v^2/(2+\varepsilon)}$$

where the last inequality follows from the inequality

$$u=\frac{v-2B\sqrt{1/K}}{\sqrt{1+2/K}+2G\sqrt{1/K}}\geq\frac{v}{\sqrt{1+\varepsilon/2}},$$

which, in turn, is true for all $v\geq 1$ and any given $\varepsilon>0$ upon taking K large enough. This proves our lemma with $v\geq 1$, while it is trivially true for $v\in(0,1)$, since, in the latter case, the right hand side of (1.1.2) is larger than one for C big enough.

With the help of our Lemma 1.1.1 we can also prove now that the above constructed $\{W(t);0\leq t<\infty\}$ is continuous in t with probability one, that is condition (iii) is also satisfied. This will immediately follow from the next theorem, which also gives more, namely the modulus of continuity of the Wiener process.

Theorem 1.1.1 (P. Lévy 1937, 1948). *We have*

(1.1.4) $$\lim_{h \to 0} \frac{\sup_{0 \leq s \leq 1-h} \sup_{0 < t \leq h} |W(s+t)-W(s)|}{\sqrt{2h \log 1/h}} \stackrel{a.s.}{=} 1$$

and

(1.1.5) $$\lim_{h \to 0} \frac{\sup_{0 \leq s \leq 1-h} |W(s+h)-W(s)|}{\sqrt{2h \log 1/h}} \stackrel{a.s.}{=} 1.$$

Proof. Let

(1.1.6) $$A_h = \sup_{0 \leq s \leq 1-h} \sup_{0 < t \leq h} |W(s+t)-W(s)|.$$

First we prove

(1.1.7) $$\varlimsup_{h \to 0} \frac{A_h}{\sqrt{2h \log 1/h}} \leq 1 \quad \text{a.s.}$$

We apply the inequality of (1.1.2) with $v=(1+\varepsilon)\sqrt{2 \log 1/h}$, $\varepsilon > 0$. Then

$$P\left\{\frac{A_h}{\sqrt{2h \log 1/h}} \geq 1+\varepsilon\right\} \leq \frac{C}{h} \exp\left\{-\frac{2\left(\log \frac{1}{h}\right)(1+\varepsilon)^2}{2+\varepsilon}\right\} \leq Ch^{\varepsilon}.$$

Take $T > 1/\varepsilon$ and let $h = h_n = n^{-T}$. Then

$$\sum_{n=1}^{\infty} P\left\{\frac{A_{h_n}}{\sqrt{2h_n \log 1/h_n}} \geq 1+\varepsilon\right\} \leq \sum_{n=1}^{\infty} Cn^{-T\varepsilon} < \infty$$

and the Borel–Cantelli lemma implies that

$$\varlimsup_{n \to \infty} \frac{A_{h_n}}{\sqrt{2h_n \log 1/h_n}} \leq 1+\varepsilon \quad \text{a.s.}$$

Let us take $h_{n+1} < h < h_n$. Then, for each $\omega \in \Omega$, we have

$$\varlimsup_{h \to 0} \frac{A_h}{\sqrt{2h \log 1/h}} \leq \varlimsup_{n \to \infty} \frac{A_{h_n}}{\sqrt{2h_{n+1} \log 1/h_{n+1}}}$$

$$= \varlimsup_{n \to \infty} \frac{A_{h_n}}{\sqrt{2h_n \log 1/h_n}} \frac{\sqrt{2h_n \log 1/h_n}}{\sqrt{2h_{n+1} \log 1/h_{n+1}}} \leq 1+\varepsilon, \quad \text{a.s.},$$

for all $\varepsilon > 0$, and whence we have (1.1.7).

Next we show

(1.1.8) $$\varlimsup_{h \to 0} \sup_{0 \leq s \leq 1-h} \frac{|W(s+h)-W(s)|}{\sqrt{2h \log 1/h}} \geq 1 \quad \text{a.s.}$$

We have by (1.1.1)

$$P\left\{\left|W\left(\frac{k+1}{n}\right)-W\left(\frac{k}{n}\right)\right|<(1-\varepsilon)\sqrt{\frac{2\log n}{n}}\right\}$$

$$\leq 1-\frac{1}{n^{1-\varepsilon}}\frac{1}{\sqrt{8\pi\log n}}.$$

Consequently, by property (ii)

(1.1.9) $\quad \sum_{n=1}^{\infty} P\left\{\max_{0\leq k\leq n-1}\left|W\left(\frac{k+1}{n}\right)-W\left(\frac{k}{n}\right)\right|<(1-\varepsilon)\sqrt{\frac{2\log n}{n}}\right\}$

$$\leq \sum_{n=1}^{\infty}\left(1-\frac{1}{n^{1-\varepsilon}}\frac{1}{\sqrt{8\pi\log n}}\right)^n \leq \sum_{n=1}^{\infty}\exp\left(-\frac{n^\varepsilon}{\sqrt{8\pi\log n}}\right)<\infty.$$

Also, for almost all $\omega \in \Omega$

(1.1.10) $\quad \lim_{n\to\infty}\sup_{0\leq s\leq 1-\frac{1}{n}}\frac{|W(s+1/n)-W(s)|}{\sqrt{2(1/n)\log n}}$

$$\geq \lim_{n\to\infty}\sup_{0\leq k\leq n-1}\frac{\left|W\left(\frac{k+1}{n}\right)-W\left(\frac{k}{n}\right)\right|}{\sqrt{2(1/n)\log n}} \geq 1,$$

by (1.1.9).

Considering now $h_{n+1}<h<h_n$ with $h_n=1/n$, we get

$$\lim_{h\to 0}\sup_{0\leq s\leq 1-h}\frac{|W(s+h)-W(s)|}{\sqrt{2h\log 1/h}}$$

$$\geq \lim_{n\to\infty}\sup_{0<s<1-\frac{1}{n+1}}\frac{\left|W\left(s+\frac{1}{n+1}\right)-W(s)\right|}{\sqrt{2\frac{1}{n+1}\log(n+1)}}\frac{\sqrt{2\frac{1}{n+1}\log(n+1)}}{\sqrt{2\frac{1}{n}\log n}}$$

$$-\overline{\lim}_{n\to\infty}\sup_{0<s<1-\frac{1}{n+1}}\sup_{0\leq t\leq \frac{1}{n(n+1)}}\frac{|W(s+t)-W(s)|}{\sqrt{2\frac{1}{n}\log n}}$$

where the latter r.v. is $\overset{\text{a.s.}}{=} o(1)$ by (1.1.7), and the first one is a.s. ≥ 1 by (1.1.10). Hence we get (1.1.8).

Remark 1.1.2. The following trivial generalizations of Theorem 1.1.1 are easily obtained:

$$\lim_{h\to 0} \frac{\sup_{a\leq s\leq b} \sup_{0<t\leq h} (W(s+t)-W(s))}{(2h\log 1/h)^{1/2}}$$

$$\stackrel{a.s.}{=} \lim_{h\to 0} \frac{\sup_{a\leq s\leq b} \sup_{0<t\leq h} |W(s+t)-W(s)|}{(2h\log 1/h)^{1/2}}$$

$$\stackrel{a.s.}{=} \lim_{h\to 0} \frac{\sup_{a\leq s\leq b} (W(s+h)-W(s))}{(2h\log 1/h)^{1/2}}$$

$$\stackrel{a.s.}{=} \lim_{h\to 0} \frac{\sup_{a\leq s\leq b} |W(s+h)-W(s)|}{(2h\log 1/h)^{1/2}} \stackrel{a.s.}{=} 1$$

for any $0 \leq a < b < \infty$.

In fact, in the above relations the statements with two sups can be obtained directly from the corresponding statements with one sup. For example, (1.1.4) can be obtained as a consequence of (1.1.5). In order to see this, it is enough to prove that (1.1.5) implies

(1.1.11) $$\varlimsup_{h\to 0} \frac{A_h}{(2h\log 1/h)^{1/2}} \leq 1 \quad \text{a.s.,}$$

where A_h is defined by (1.1.6). Moreover, the following stronger statement is also true:

(1.1.12) $$\varlimsup_{h\to 0} \sup_{0\leq s\leq 1-h} \sup_{0<t\leq h} \frac{|W(s+t)-W(s)|}{(2t\log 1/t)^{1/2}} \leq 1 \quad \text{a.s.}$$

By (1.1.5) for all $\varepsilon > 0$ and for almost all $\omega \in \Omega$ there exists an $h_0 = h_0(\omega, \varepsilon) > 0$ such that

$$|W(s+\chi; \omega) - W(s; \omega)| \leq (1+\varepsilon)(2\chi \log 1/\chi)^{1/2}$$

for all $0 \leq s \leq 1$, and for all $0 < \chi \leq h_0$; that is to say

$$\sup_{0\leq s\leq 1-h} \sup_{0<\chi\leq h} \frac{|W(s+\chi; \omega) - W(s; \omega)|}{(2\chi \log 1/\chi)^{1/2}} \leq 1+\varepsilon,$$

provided $h \leq h_0$. This clearly implies (1.1.12) which, in turn, implies (1.1.11).

Throughout this Chapter we will several times formulate similar statements with two sups and with one sup. The above sketched idea, which is saying that a statement with two sups follows from the corresponding statement with one sup, can be applied also in those cases. Our only reason for spelling out also statements with two sups is the fact that the proofs of these are generally simpler if we already have the inequality (1.1.2). We emphasize however that the inequality (1.1.2) itself does not follow from the corresponding inequality with one sup (cf. also Supplementary Remarks to Section 1.1).

1.2. How big are the increments of a Wiener process?

In Theorem 1.1.1 we saw how large the increments of a Wiener process over subintervals of length h of the unit interval can be when h is small. In this section we are going to study the similar problem of how large the increments of a Wiener process over subintervals of length a_T of the interval $[0, T]$ can be when $T \to \infty$ and a_T is a non-decreasing function of T. These two problems are closely related to each other and can be studied from the same source of information, namely from Lemma 1.1.1. Towards this end we first extend the statement of the latter from the unit interval to any finite interval of the positive half-line (Lemma 1.2.1). From this latter lemma the main result (Theorem 1.2.1) of this section will follow just like Theorem 1.1.1 did from Lemma 1.1.1. This then shows that Theorems 1.1.1 and 1.2.1 are closely linked. They do not seem to follow directly from each other though (cf., however, Supplementary Remarks, Section 1.2).

The above mentioned immediate analogue of Lemma 1.1.1 is

Lemma 1.2.1. *For any $\varepsilon > 0$ there exists a constant $C = C(\varepsilon) > 0$ such that the inequality*

$$(1.2.1) \quad P\{\sup_{0 \leq s \leq T-h} \sup_{0 \leq t \leq h} |W(s+t) - W(s)| \geq v\sqrt{h}\} \leq \frac{CT}{h} e^{-\frac{v^2}{2+\varepsilon}}$$

holds for every positive v, T and $0 < h < T$.

Proof. This lemma follows from (1.1.1) and from the following

Observation. For any fixed $T > 0$ we have

$$\{W(s); 0 \leq s \leq T\} \stackrel{\mathcal{D}}{=} \left\{\sqrt{T} W\left(\frac{s}{T}\right); 0 \leq s \leq T\right\}.$$

Theorem 1.2.1 (Csörgő, Révész 1979b). *Let a_T ($T \geq 0$) be a monotonically non-decreasing function of T for which*
  (i) $0 < a_T \leq T$,
  (ii) T/a_T *is monotonically non-decreasing.*
Then

(1.2.2) $$\overline{\lim_{T \to \infty}} \sup_{0 \leq t \leq T-a_T} \beta_T |W(t+a_T) - W(t)| \stackrel{a.s.}{=} 1,$$

(1.2.3) $$\overline{\lim_{T \to \infty}} \beta_T |W(T+a_T) - W(T)|$$
$$\stackrel{a.s.}{=} \overline{\lim_{T \to \infty}} \sup_{0 \leq s \leq a_T} \beta_T |W(T+s) - W(T)| \stackrel{a.s.}{=} 1$$

and

(1.2.4) $$\overline{\lim_{T \to \infty}} \sup_{0 \leq t \leq T-a_T} \sup_{0 \leq s \leq a_T} \beta_T |W(t+s) - W(t)| \stackrel{a.s.}{=} 1,$$

where

$$\beta_T = \left(2a_T \left[\log \frac{T}{a_T} + \log \log T\right]\right)^{-1/2}.$$

If we have also
  (iii) $\lim_{T \to \infty} (\log T/a_T)(\log \log T)^{-1} = \infty$,

then

(1.2.5) $$\lim_{T \to \infty} \sup_{0 \leq t \leq T-a_T} \beta_T |W(t+a_T) - W(t)| \stackrel{a.s.}{=} 1$$

and

(1.2.6) $$\lim_{T \to \infty} \sup_{0 \leq t \leq T-a_T} \sup_{0 \leq s \leq a_T} \beta_T |W(t+s) - W(t)| \stackrel{a.s.}{=} 1.$$

Remark 1.2.1. The proof of this theorem will show that statements (1.2.2)–(1.2.6) remain true if any one, two or all of T, t and s are running over all the integers, or if we omit the absolute value signs in these statements. Also, because of the symmetry of W, if we replace the lim sup by lim inf and sup by inf in (1.2.2)–(1.2.6), then the above results will be true with -1 instead of $+1$, when also omitting the absolute value signs. For example

(1.2.2*) $$\underline{\lim_{T \to \infty}} \inf_{0 \leq t \leq T-a_T} \beta_T (W(t+a_T) - W(t)) \stackrel{a.s.}{=} -1$$

if conditions (i)–(ii) hold true and

(1.2.5*) $$\lim_{T \to \infty} \inf_{0 \leq t \leq T-a_T} \beta_T (W(t+a_T) - W(t)) \stackrel{a.s.}{=} -1$$

if conditions (i)–(iii) hold true.

Choosing a_T as $c \log T$, cT and 1 respectively, the following corollaries are immediate.

Corollary 1.2.1. *For any* $c>0$ *we have*

$$(1.2.7) \qquad \lim_{T\to\infty} \sup_{0\leq t\leq T-c\log T} \frac{|W(t+c\log T)-W(t)|}{c\log T} \stackrel{a.s.}{=} \sqrt{\frac{2}{c}}.$$

This latter statement is the Erdős–Rényi (1970) law of large numbers for the Wiener process (cf. also Theorem 2.4.3).

Corollary 1.2.2. *For* $0<c\leq 1$ *we have*

$$(1.2.8) \qquad \varlimsup_{T\to\infty} \sup_{0\leq t\leq T-cT} \frac{|W(t+cT)-W(t)|}{\sqrt{2cT\log\log T}} \stackrel{a.s.}{=} 1,$$

$$(1.2.8^*) \qquad \varlimsup_{T\to\infty} \sup_{0\leq t\leq T-cT} \sup_{0\leq s\leq cT} \frac{|W(t+s)-W(t)|}{(2cT\log\log T)^{1/2}} \stackrel{a.s.}{=} 1.$$

(1.2.8) and (1.2.8*) also follow from Strassen's law of iterated logarithm (1964). In Section 3 of this Chapter we will, however, follow the opposite road, proving Strassen's law via (1.2.8*).

Corollary 1.2.3. *We have*

$$(1.2.9) \qquad \varlimsup_{T\to\infty} \sup_{0\leq t\leq T-1} \frac{|W(t+1)-W(t)|}{\sqrt{2\log T}} \stackrel{a.s.}{=} 1.$$

This is a well-known result which (when T and t run over the integers; cf. Remark 1.2.1) in terms of the order statistics $X_i^{(n)}$ ($i=1, 2, ..., n$) of n independent $\mathcal{N}(0, 1)$ r.v. reads

$$(1.2.10) \qquad \lim_{n\to\infty} \frac{X_n^{(n)}}{\sqrt{2\log n}} \stackrel{a.s.}{=} 1.$$

Proof of Theorem 1.2.1. The proof is formulated in three steps, which together will imply our statements.

Step 1. Let

$$A(T) = \sup_{0\leq t\leq T-a_T} \sup_{0\leq s\leq a_T} \beta_T |W(t+s)-W(t)|.$$

Suppose that conditions (i), (ii) *of Theorem 1.2.1 are fulfilled. Then*

$$(1.2.11) \qquad \varlimsup_{T\to\infty} A(T) \leq 1 \quad a.s.$$

Proof. By Lemma 1.2.1 we have for any $\varepsilon > 0$

$$P(A(T) \geq \sqrt{1+\varepsilon}) \leq C \frac{T}{a_T} \exp\left\{-(1+\varepsilon)\left[\log \frac{T}{a_T} + \log\log T\right]\right\}$$

$$= C\left(\frac{a_T}{T}\right)^{\varepsilon} \frac{1}{(\log T)^{1+\varepsilon}}.$$

Let $T_k = \theta^k$ ($\theta > 1$). Then

$$\sum_{k=1}^{\infty} P(A(T_k) \geq \sqrt{1+\varepsilon}) < \infty$$

for every $\varepsilon > 0$, $\theta > 1$. Hence by the Borel–Cantelli lemma

(1.2.12) $$\varlimsup_{k \to \infty} A(T_k) \leq 1.$$

We also have

(1.2.13) $$1 \leq \frac{\beta_{T_k}}{\beta_{T_{k+1}}} \leq \theta$$

if k is big enough.

Now choosing θ near enough to one, (1.2.11) follows from (1.2.12) and (1.2.13), because $\beta_T^{-1} A(T)$ is non-decreasing and β_T is non-increasing in T.

Step 2. Let

$$B(T) = \beta_T |W(T) - W(T - a_T)|.$$

Suppose that the conditions (i), (ii) *of Theorem 1.2.1 are fulfilled. Then*

(1.2.14) $$\varlimsup_{T \to \infty} B(T) \geq 1.$$

Proof. For any $\varepsilon > 0$, by (1.1.1), we have

(1.2.15) $$P(B(T) \geq 1 - \varepsilon) \geq \frac{\exp\left\{-(1-\varepsilon)^2\left[\log \frac{T}{a_T} + \log\log T\right]\right\}}{\sqrt{2\pi}\left[2\left(\log \frac{T}{a_T} + \log\log T\right)\right]^{1/2}}$$

$$\geq \left(\frac{a_T}{T \log T}\right)^{1-\varepsilon}$$

if T is big enough. Let $T_1 = 1$ and define T_{k+1} by

$$T_{k+1} - a_{T_{k+1}} = T_k \quad \text{if} \quad \varrho < 1$$

and

$$T_{k+1} = \theta^{k+1} \quad \text{if} \quad \varrho = 1,$$

where $\theta > 1$ and $\lim a_T/T = \varrho$. (We note that our conditions (i) and (ii) imply that a_T is a continuous function of T and that $T - a_T$ is a strictly increasing function if $\varrho < 1$.)

In case of $\varrho < 1$, (1.2.14) follows from the simple fact that

$$\sum_{k=2}^{\infty} \left(\frac{a_{T_k}}{T_k \log T_k} \right)^{1-\varepsilon} = \infty$$

and that the r.v. $B(T_k)$ ($k = 1, 2, \ldots$) are independent.

In order to see the divergence of the above series, we have

$$\sum_{k=2}^{n} \left(\frac{a_{T_k}}{T_k \log T_k} \right)^{1-\varepsilon} \geq \sum_{k=2}^{n} \frac{a_{T_k}}{T_k} \frac{1}{(\log T_k)^{1-\varepsilon}} \geq \frac{1}{(\log T_n)^{1-\varepsilon}} \sum_{k=2}^{n} \frac{a_{T_k}}{T_k},$$

and, because $-\log(1-x) \leq K_\delta x$ for all $x \in (0, 1-\delta)$ and some $K_\delta > 0$,

$$\log T_n = \sum_{k=2}^{n} \log \frac{T_k}{T_{k-1}} = -\sum_{k=2}^{n} \log \left(1 - \frac{a_{T_k}}{T_k} \right) \leq K_\delta \sum_{k=2}^{n} \frac{a_{T_k}}{T_k}.$$

These two statements combined give the stated divergence.

In case of $\varrho = 1$, $a_{T_{k+1}} \geq T_{k+1} - T_k$ (if k is big enough), hence

$$B(T_{k+1}) \geq \beta_{T_{k+1}} |W(T_{k+1}) - W(T_k)| - \beta_{T_{k+1}} \sup_{0 \leq u \leq v \leq T_k} |W(v) - W(u)|.$$

By Step 1,

(1.2.16) $\qquad \overline{\lim}_{k \to \infty} \beta_{T_{k+1}} \sup_{0 \leq u \leq v \leq T_k} |W(v) - W(u)| \leq 2\theta^{-1/2}.$

We also have by (1.1.1)

$$P\{\beta_{T_{k+1}} |W(T_{k+1}) - W(T_k)| \geq 1 - \varepsilon\} = O\left(k^{-(1-\varepsilon)^2 \frac{\theta}{\theta - 1}} \right).$$

Now for any given $\varepsilon > 0$, choosing θ big enough and applying the Borel–Cantelli lemma we get

(1.2.17) $\qquad \overline{\lim}_{k \to \infty} \beta_{T_{k+1}} |W(T_{k+1}) - W(T_k)| \geq 1 - \varepsilon$ a.s.

Whence combining (1.2.17) and (1.2.16), and choosing again θ big, we get (1.2.14).

We note that $\varrho = 1$ if and only if $a_T = T$, i.e., in this case (1.2.2) and (1.2.4) reduce to the well-known laws of the iterated logarithm.

Step 3. Let

$$C(T) = \sup_{0 \leq t \leq T - a_T} \beta_T |W(t + a_T) - W(t)|.$$

Suppose the conditions (i)–(iii) *of Theorem* 1.2.1 *are fulfilled. Then*

(1.2.18) $\qquad \lim_{T \to \infty} C(T) \geq 1$ a.s.

Proof. Since the r.v.

$$\beta_T |W((k+1)a_T) - W(ka_T)| \quad (k = 0, 1, 2, \ldots, [T/a_T]-1)$$

are independent, by (1.2.15) we have

$$P\{\max_{0 \leq k \leq [T/a_T]-1} \beta_T |W((k+1)a_T) - W(ka_T)| \leq 1-\varepsilon\}$$

$$\leq \left(1 - \left(\frac{a_T}{T \log T}\right)^{1-\varepsilon}\right)^{[T/a_T]} \leq 2 \exp\left\{-\left(\frac{T}{a_T}\right)^\varepsilon \left(\frac{1}{\log T}\right)^{1-\varepsilon}\right\}.$$

By condition (iii) we have

$$\sum_{j=1}^{\infty} \exp\left\{-\left(\frac{j}{a_j}\right)^\varepsilon \left(\frac{1}{\log j}\right)^{1-\varepsilon}\right\} < \infty,$$

and whence, so far, we have proved

(1.2.19) $\varlimsup_{j \to \infty} C(j) \geq \varlimsup_{j \to \infty} \max_{0 \leq k \leq [j/a_j]-1} \beta_j |W((k+1)a_j) - W(ka_j)| \geq 1$ a.s.

Considering now the case of in-between-times $j \leq T < j+1$, we first observe that $0 \leq a_T - a_j$ and that, by condition (ii), $0 \leq a_T - a_j \leq \frac{a_j}{j} \leq \delta a_j$ for any $\delta > 0$, if $j \leq T < j+1$ and j is big enough. (The latter inequality is immediate, since $\frac{a_T}{T} \leq \frac{a_j}{j}$ by (ii), and so, via $a_T \leq a_j \frac{T}{j}$, we have $a_T - a_j \leq$ $\leq a_j\left(\frac{T}{j} - 1\right) \leq \frac{a_j}{j}$.) Whence, for $j \leq T < j+1$ and j large, we have

(1.2.20) $\quad C(T) \geq \max_{0 \leq k \leq [j/a_j]-1} \beta_{j+1} |W((k+1)a_j) - W(ka_j)|$

$$- \sup_{0 \leq t \leq T - \delta a_T} \sup_{0 \leq s \leq \delta a_T} \beta_T |W(t+s) - W(t)|.$$

On the other hand, by Step 1 we have

$$\varlimsup_{T \to \infty} \sup_{0 \leq t \leq T - \delta a_T} \sup_{0 \leq s \leq \delta a_T} \beta_T |W(t+s) - W(s)|$$

$$\leq \varlimsup_{T \to \infty} \frac{\left(2\delta a_T \left(\log \frac{T}{\delta a_T} + \log \log T\right)\right)^{1/2}}{\left(2 a_T \left(\log \frac{T}{a_T} + \log \log T\right)\right)^{1/2}} = \delta^{1/2}.$$

This, by (1.2.20) and (1.2.19), also completes the proof of (1.2.18) upon observing also that $\beta_{j+1}/\beta_j \to 1$ as $j \to \infty$.

Remark 1.2.2. It is possible to prove that in (1.2.2) and (1.2.4) the $\overline{\lim}$ cannot be changed to a lim, if condition (iii) fails, that is to say in the latter case (1.2.5) and (1.2.6) cannot be true. In fact Deo (1977) has shown that

(1.2.21) $$\lim_{T\to\infty} \sup_{0\leq t\leq T-a_T} \sup_{0\leq s\leq a_T} \beta_T |W(t+s)-W(t)| < 1 \quad \text{a.s.}$$

as well as

(1.2.22) $$\lim_{T\to\infty} \sup_{0\leq t\leq T-a_T} \beta_T |W(t+a_T)-W(t)| < 1 \quad \text{a.s.,}$$

provided

$$\overline{\lim_{T\to\infty}} (\log T/a_T)(\log\log T)^{-1} < \infty.$$

This result suggests the following problem: find the normalizing factor $\delta_T = \delta_T(a_T)$ such that the left hand side r.v. of (1.2.21), resp. that of (1.2.22), should be equal to one almost surely, with δ_T replacing β_T in them. A partial answer concerning (1.2.22) was given by Book and Shore (1978), who showed that

$$\lim_{T\to\infty} \sup_{0\leq t\leq T-a_T} \beta_T |W(t+a_T)-W(t)| \stackrel{\text{a.s.}}{=} \left(\frac{r}{r+1}\right)^{1/2},$$

provided $\lim_{T\to\infty} \log(Ta_T^{-1})/\log\log T = r$, $0 \leq r \leq \infty$.

The similar question in connection with (1.2.21) was studied by Csáki and Révész (1979), who proved that

$$18^{-1} \leq \lim_{T\to\infty} \sup_{0\leq t\leq T-a_T} \sup_{0\leq s\leq a_T} \delta_T |W(t+s)-W(t)| \leq 46 \quad \text{a.s.,}$$

where $\delta_T = (2a_T \log(1+\frac{\pi^2}{16}[Ta_T^{-1}]/\log\log T))^{-1/2}$. The general question of finding the exact value of the above $\lim_{T\to\infty}$ statement appears to be a difficult one. However, if one also has $\lim_{T\to\infty} \log(Ta_T^{-1})/\log\log\log T = \infty$, then the just mentioned $\lim_{T\to\infty}$ is equal to one.

The special case of $a_T = T$ of these questions was studied by Chung (1948) (cf. also Section 1.7) and Hirsch (1965) who evaluated the normalizing factor μ_T resp. ν_T for which

$$\lim_{T\to\infty} \sup_{0\leq t\leq T} \mu_T W(t) \stackrel{\text{a.s.}}{=} \lim_{T\to\infty} \sup_{0\leq t\leq T} \nu_T |W(t)| \stackrel{\text{a.s.}}{=} 1.$$

It should be emphasized that μ_T and ν_T are very different, which is not the case when studying the $\overline{\lim}$ instead of the $\underline{\lim}$ of these functionals.

1.3. The law of iterated logarithm for the Wiener process

Taking $a_T = T$ in (1.2.2) or, equivalently, $c = 1$ in (1.2.8) we get P. Lévy's famous law of iterated logarithm:

Theorem 1.3.1 (P. Lévy 1937, 1948).

(1.3.1) $$\varlimsup_{T \to \infty} \frac{|W(T)|}{\sqrt{2T \log \log T}} \stackrel{a.s.}{=} 1.$$

In fact (1.2.4) with $a_T = T$, also gives

Theorem 1.3.1*.

(1.3.2) $$\varlimsup_{T \to \infty} \sup_{0 \leq t \leq T} \frac{|W(t)|}{\sqrt{2T \log \log T}}$$
$$= \varlimsup_{T \to \infty} \sup_{0 \leq x \leq 1} \frac{|W(xT)|}{\sqrt{2T \log \log T}} \stackrel{a.s.}{=} 1.$$

Remark 1.3.1. Our Remark 1.2.1 is applicable here and says that (1.3.1) and (1.3.2) hold true if T and/or t run over the integers. It also says that if we replace $\varlimsup$ by $\varliminf$, sup by inf and omit the absolute value signs, the right-hand side of (1.3.1) and (1.3.2) will be -1.

A more complete description of the behaviour of $W(xT)$, $0 \leq x \leq 1$, $T \to \infty$, was first given by Strassen. This section is devoted to proving his fundamental theorem.

Let $\mathscr{S}$ be the set of absolutely continuous functions (with respect to Lebesgue measure) such that

$$f(0) = 0 \quad \text{and} \quad \int_0^1 (f'(x))^2 \, dx \leq 1.$$

The set $\mathscr{S}$ is compact and this follows from the following (cf. also Supplementary Remarks, Section 1.3):

Lemma 1.3.1 (Riesz, Sz.-Nagy 1955, p. 75). *Let f be a real valued function on $[0, 1]$. The following two conditions are equivalent:*

(i) f *is absolutely continuous and* $\int_0^1 (f')^2 dx \leq 1$,

(ii) $$\sum_{i=1}^r \frac{\left(f\left(\frac{i}{r}\right) - f\left(\frac{i-1}{r}\right)\right)^2}{1/r} \leq 1 \quad \text{for any} \quad r = 1, 2, \ldots$$

and f is continuous on $[0, 1]$.

Define

$$\eta_n(x) = \frac{W(nx)}{\sqrt{2n \log \log n}}$$

for $x \in [0, 1]$. Then $\{\eta_n(x)\}$ is a sequence of stochastic processes with sample paths almost surely in $C(0, 1)$. In this setup Strassen's theorem is

Theorem 1.3.2 (Strassen 1964). *The sequence $\{\eta_n(x)\}$ is relatively compact in $C(0, 1)$ with probability one, and the set of its limit points is $\mathscr{S}$.*

The meaning of this statement is that there exists an event $\Omega_0 \subset \Omega$ of probability zero with the following two properties:

(i) for any $\omega \notin \Omega_0$ and any sequence of integers $n_1 < n_2 < \ldots$ there exist a subsequence $n_{k_j} = n_{k_j}(\omega)$ and a function $f \in \mathscr{S}$ such that

$$\eta_{n_{k_j}}(x; \omega) \to f(x) \quad \text{uniformly in} \quad x \in [0, 1],$$

(ii) for any $f \in \mathscr{S}$ and $\omega \notin \Omega_0$ there exists a sequence $n_k = n_k(\omega, f)$ such that

$$\eta_{n_k}(x, \omega) \to f(x) \quad \text{uniformly in} \quad x \in [0, 1].$$

Remark 1.3.2. Since $|f(1)| \leq 1$ for any function $f \in \mathscr{S}$, and $f(x) = x \in \mathscr{S}$, Theorem 1.3.2 implies Theorems 1.3.1 and 1.3.1*.

The proof of Theorem 1.3.2 will be based on the following simple

Lemma 1.3.2. *Let d be a positive integer and $\alpha_1, \alpha_2, \ldots, \alpha_d$ be a sequence of real numbers for which*

$$\sum_{i=1}^{d} \alpha_i^2 = 1.$$

Further let

$$S_n = \alpha_1 W(n) + \alpha_2 (W(2n) - W(n)) + \ldots + \alpha_d (W(dn) - W((d-1)n)).$$

Then

(1.3.3)
$$\varlimsup_{n \to \infty} \frac{S_n}{\sqrt{2n \log \log n}} \overset{a.s.}{=} 1$$

and

(1.3.3*)
$$\varliminf_{n \to \infty} \frac{S_n}{\sqrt{2n \log \log n}} \overset{a.s.}{=} -1.$$

Proof. In the first step we prove that

(1.3.4)
$$\varlimsup_{n \to \infty} \frac{|S_n|}{\sqrt{2n \log \log n}} \leq 1.$$

In order to do this, by (1.1.1) we observe that for any $\varepsilon>0$

(1.3.5) $$P\left\{\frac{|S_n|}{\sqrt{2n\log\log n}}\geq 1+\varepsilon\right\}\leq \frac{1}{(\log n)^{(1+\varepsilon)^2}},$$

since $S_n\in\mathcal{N}(0,n)$. Set $N_k=[\theta^k]$ ($\theta>1$). Then (1.3.5) implies

(1.3.6) $$\sum_{k=1}^{\infty}P\left\{\frac{|S_{N_k}|}{\sqrt{2N_k\log\log N_k}}\geq 1+\varepsilon\right\}<\infty.$$

Applying Theorem 1.2.1 one gets

$$\varlimsup_{k\to\infty}\max_{1\leq j\leq d}\max_{N_k\leq n<N_{k+1}}\frac{|W(jn)-W(jN_k)|}{\sqrt{2N_k\log\log N_k}}$$

$$\leq \varlimsup_{k\to\infty}\sup_{0\leq t\leq dN_k}\sup_{0\leq s\leq dN_k(\theta-1)}\frac{|W(t+s)-W(t)|}{\sqrt{2N_k\log\log N_k}}\leq\sqrt{d(\theta-1)},$$

and hence

(1.3.7) $$\varlimsup_{k\to\infty}\sup_{N_k\leq n<N_{k+1}}\frac{|S_n-S_{N_k}|}{\sqrt{2N_k\log\log N_k}}\leq 2d^{3/2}\sqrt{\theta-1}.$$

Choosing θ near enough to 1 (1.3.6) and (1.3.7) imply (1.3.4). Now we turn to the second step and prove

(1.3.8) $$\varlimsup_{n\to\infty}\frac{S_n}{\sqrt{2n\log\log n}}\geq 1 \quad \text{a.s.}$$

Put $N_k=\theta^k$ where, given $0<\varepsilon<1$, we assume that θ is an integer and $\theta>d/\varepsilon$. Put also

$$S_k^*=\alpha_1\big(W(N_k)-W(dN_{k-1})\big)+\alpha_2\big(W(2N_k)-W(N_k)\big)+\ldots$$
$$\ldots+\alpha_d\big(W(dN_k)-W((d-1)N_k)\big).$$

Then
$$E(S_k^*)^2=(\alpha_2^2+\ldots+\alpha_d^2)N_k+\alpha_1^2(N_k-dN_{k-1})=N_k-\alpha_1^2 dN_{k-1}$$

$$\geq N_k\left(1-\frac{d}{\theta}\right)\geq N_k(1-\varepsilon).$$

Whence, by (1.1.1),

$$P\left\{\frac{S_k^*}{\sqrt{2N_k(1-\varepsilon)\log\log N_k}}\geq 1-\varepsilon\right\}\geq\frac{e^{-(1-\varepsilon)^2\log\log N_k}}{2\sqrt{2\pi\log\log N_k}},$$

which, in turn, implies by the Borel–Cantelli lemma that

(1.3.9) $$\varlimsup_{k\to\infty} \frac{S_k^*}{\sqrt{2N_k \log\log N_k}} \geq (1-\varepsilon)^{3/2}.$$

Since ε is arbitrarily small, (1.3.8) now follows from (1.3.1) and (1.3.9). In this way we have (1.3.3) and symmetry of W combined with the latter statement implies (1.3.3*).

Now we introduce some notations. For any real valued function $f \in C(0,1)$ and positive integer d, let $f^{(d)}$ be the linear interpolation of f over the points i/d, that is

$$f^{(d)}(x) = f\left(\frac{i}{d}\right) + d\left[f\left(\frac{i+1}{d}\right) - f\left(\frac{i}{d}\right)\right]\left(x - \frac{i}{d}\right)$$

$$\left(\frac{i}{d} \leq x \leq \frac{i+1}{d},\ i = 0, 1, \ldots, d-1\right).$$

Let
$$C_d = \{f^{(d)} : f \in C(0,1)\} \subset C(0,1),$$
$$\mathscr{S}_d = \{f^{(d)} : f \in \mathscr{S}\},$$

where $\mathscr{S}_d \subset \mathscr{S}$ by Lemma 1.3.1.

The statement of Lemma 1.3.2 easily implies:

Proposition 1.3.1. *The sequence $\{\eta_n^{(d)}(x)\}$ is relatively compact in C_d with probability one, and the set of its limit points is $\mathscr{S}_d$.*

Proof. By Theorem 1.3.1 and continuity of the Wiener process this Proposition holds when $d=1$. We prove it for $d=2$. For larger d the proof is similar and immediate. Let $Z_n = (W(n), W(2n) - W(n))$ $(n=1,2,\ldots)$ and α, β be real numbers such that $\sqrt{\alpha^2 + \beta^2} = 1$. Then by Lemma 1.3.2 and continuity of W the set of limit points of the sequence

$$\left\{\frac{\binom{\alpha}{\beta} Z_n}{\sqrt{2n\log\log n}}\right\}_{n=1}^{\infty} = \left\{\frac{\alpha W(n) + \beta(W(2n) - W(n))}{\sqrt{2n\log\log n}}\right\}_{n=1}^{\infty}$$

is the interval $[-1, +1]$. This implies that the set of limit points of the sequence $\left\{\dfrac{Z_n}{\sqrt{2n\log\log n}}\right\}$ is a subset of the unit circle and the boundary of the unit circle belongs to this limit set.

Now let $Z_n^* = (W(n), W(2n) - W(n), W(3n) - W(2n))$. In the same way as above one can prove that the set of limit points of $\left\{\dfrac{Z_n^*}{\sqrt{2n\log\log n}}\right\}$

is a subset of the unit sphere of R^3 which contains the boundary of this sphere. This fact in itself already implies that the set of limit points of $\left\{\dfrac{Z_n}{\sqrt{2n \log \log n}}\right\}$ is the unit circle of R^2 and this, in turn, is equivalent to our statement.

Proof of Theorem 1.3.2. For each $\omega \in \Omega$ we have
$$\sup_{0 \leq x \leq 1} |\eta_n(x) - \eta_n^{(d)}(x)| \leq \sup_{0 \leq x \leq 1} \sup_{0 \leq s \leq 1/d} |\eta_n(x+s) - \eta_n(x)|,$$
hence, by Corollary 1.2.2, we get
$$\varlimsup_{n \to \infty} \sup_{0 \leq x \leq 1} |\eta_n(x) - \eta_n^{(d)}(x)| \leq d^{-1/2}, \quad \text{a.s.}$$
Consequently we have the theorem by Lemma 1.3.1 and Proposition 1.3.1, where we also use the fact that Lemma 1.3.1 guarantees that $\mathscr{S}$ is closed.

The discreteness of n is inessential in Theorem 1.3.2 and, if we define
$$\eta_t(x) = \frac{W(tx)}{\sqrt{2t \log \log t}}, \quad x \in [0, 1],$$
we have

Theorem 1.3.2* (Strassen 1964). *The net $\eta_t(x)$ is relatively compact in $C(0, 1)$ with probability one, and the set of its limit points is $\mathscr{S}$.*

We should also mention another version of the law of iterated logarithm

Theorem 1.3.3 (P. Lévy 1937, 1948).
$$(1.3.10) \quad \varlimsup_{t \to 0} \frac{|W(t)|}{\sqrt{2t \log \log 1/t}} \stackrel{\text{a.s.}}{=} \lim_{t \to 0} \sup_{0 < s \leq t} \frac{|W(s)|}{\sqrt{2s \log \log 1/s}} \stackrel{\text{a.s.}}{=} 1.$$

This theorem is an immediate consequence of Theorem 1.3.1 and the following

Lemma 1.3.3. *Define*
$$(1.3.11) \quad \tilde{W}(t) = \begin{cases} tW(1/t) & \text{if } t > 0, \\ 0 & \text{if } t = 0 \end{cases}$$
where $\{W(t); 0 \leq t < \infty\}$ is a Wiener process. Then $\tilde{W}(t)$ is also a Wiener process.

Proof. The three properties (i), (ii) and (iii) of Section 1 of a Wiener process are easily verified for the above defined stochastic process.

Clearly Theorem 1.3.3 can be formulated in the following form too:

Theorem 1.3.3*. *For any* $t_0 > 0$ *we have*

$$(1.3.10^*) \qquad \overline{\lim_{h \to 0}} \frac{|W(t_0+h) - W(t_0)|}{\sqrt{2h \log \log 1/h}} \stackrel{a.s.}{=} 1.$$

It is interesting to compare Theorems 1.1.1 and 1.3.3*. The latter one states that the continuity modulus of $W(t)$ for any fixed t_0 is not more than $(2h \log \log 1/h)^{1/2}$ (local continuity modulus). On the other hand, (1.1.5) of Theorem 1.1.1 tells us that at some random points the continuity modulus can be much larger, namely $(2h \log 1/h)^{1/2}$ (global continuity modulus). This means that the sample paths of a Wiener process violate the law of iterated logarithm at some random points. A paper of Orey and Taylor (1974) investigates "How often on a Brownian path does the law of iterated logarithm fail?".

1.4. Brownian bridges

A stochastic process $\{B(t); 0 \le t \le 1\}$ is called a Brownian bridge if
(i) the joint distribution of $B(t_1), B(t_2), \ldots, B(t_n)$ ($0 \le t_1 < t_2 < \ldots < t_n \le 1$; $n = 1, 2, \ldots$) is Gaussian, with $EB(t) \equiv 0$,
(ii) the covariance function of $B(t)$ is

$$R(s, t) = EB(s)B(t) = s \wedge t - st,$$

(iii) the sample path function of $B(t; \omega)$ is continuous in t with probability one.

We note that (ii) above implies $B(0) = B(1) = 0$ a.s.

The existence of such a Gaussian process is a simple consequence of the following:

Lemma 1.4.1. *Let* $\{W(t); 0 \le t < \infty\}$ *be a Wiener process. Then*

$$(1.4.1) \qquad B(t) = W(t) - tW(1) \quad (0 \le t \le 1)$$

is a Brownian bridge.

Proof. The above three conditions are easily verified for the representation (1.4.1).

Moreover, the continuity modulus of $B(t)$ can also be obtained from (1.4.1). Namely we have:

Theorem 1.4.1.

$$(1.4.2) \quad \lim_{h\to 0} \sup_{0\le s\le 1-h} \sup_{0<t\le h} \frac{|B(s+t)-B(s)|}{\sqrt{2h\log 1/h}} \stackrel{a.s.}{=} \lim_{h\to 0} \sup_{0\le s\le 1-h} \frac{|B(s+h)-B(s)|}{\sqrt{2h\log 1/h}}$$

$$\stackrel{a.s.}{=} \overline{\lim_{h\to 0}} \sup_{0<t\le h} \frac{|B(s_0+t)-B(s_0)|}{\sqrt{2h\log\log 1/h}} \stackrel{a.s.}{=} \overline{\lim_{h\to 0}} \frac{|B(s_0+h)-B(s_0)|}{\sqrt{2h\log\log 1/h}} \stackrel{a.s.}{=} 1,$$

where $s_0 \in [0, 1-h]$ is fixed.

Proof. (1.4.2) follows immediately from (1.4.1) and from our earlier, similar, results for a Wiener process.

(1.4.1) exhibits a useful relationship between the Wiener process and the Brownian bridge. Now we display two further connections for the sake of later reference. The first one is

Proposition 1.4.1. *Let* $\{B_i(t)\}_{i=0}^\infty$ *be a sequence of independent Brownian bridges and let* $X_0 \equiv 0$ *and* $\{X_i\}_{i=1}^\infty$ *be a sequence of independent* $\mathcal{N}(0,1)$ *r.v. which is also independent of* $\{B_i(t)\}$. *For any fixed* $0=t_0<t_1<t_2<\ldots$ *define the stochastic process*

$$(1.4.3) \quad W(t) = \sum_{i=0}^{j-1} \sqrt{t_{i+1}-t_i}\, X_{i+1} + \frac{t-t_j}{\sqrt{t_{j+1}-t_j}} X_{j+1} + \sqrt{t_{j+1}-t_j}\, B_j\!\left(\frac{t-t_j}{t_{j+1}-t_j}\right)$$

if $t_j \le t \le t_{j+1}$ ($j=0,1,2,\ldots$). *Then* $\{W(t); 0\le t<\infty\}$ *is a Wiener process.*

Proof. The three properties (i), (ii) and (iii) of a Wiener process in Section 1 (or, equivalently, the covariance function of W) are easily verified for the above defined stochastic process $\{W(t); 0\le t<\infty\}$.

The second connection between a Wiener process and a Brownian bridge is a special form of Doob's transformation (1949).

Proposition 1.4.2. *Let* $B(t)$ *be a Brownian bridge and define*

$$(1.4.4) \quad W(t) = (t+1)B\!\left(\frac{t}{t+1}\right) \quad (t\ge 0).$$

Then $W(t)$ *is a Wiener process.*

Proof. Check again the three properties or the covariance function of a Wiener process.

Remark 1.4.1. (1.4.4) clearly implies

$$(1.4.5) \quad B(t) = (1-t)W\!\left(\frac{t}{1-t}\right) \quad (0\le t<1).$$

1.5. The distributions of some functionals of the Wiener and Brownian bridge processes

For the sake of further reference we summarize here some classical distribution results.

Theorem 1.5.1. Let $\{W(T); 0 \leq t < \infty\}$ be a Wiener process and $\{B(t); 0 \leq t \leq 1\}$ a Brownian bridge. Then, for $u > 0$, we have

(1.5.1) $\quad P\{\sup_{0 \leq t \leq T} W(t) > u\} = 2P(W(T) \geq u) = 2\left(1 - \Phi\left(\frac{u}{\sqrt{T}}\right)\right).$

(1.5.2) $\quad P\{\sup_{0 \leq t \leq 1} |W(t)| \leq u\} = \frac{1}{\sqrt{2\pi}} \int_{-u}^{+u} \sum_{k=-\infty}^{\infty} (-1)^k \exp\left(-\frac{(x - 2ku)^2}{2}\right) dx$

$$= \frac{4}{\pi} \sum_{k=0}^{\infty} \frac{(-1)^k}{2k+1} \exp(-\pi^2 (2k+1)^2 / 8u^2),$$

(1.5.3) $\quad P\{\sup_{0 \leq t \leq 1} B(t) \geq u\} = e^{-2u^2},$

(1.5.4) $\quad P\{\sup_{0 \leq t \leq 1} |B(t)| > u\} = \sum_{k \neq 0} (-1)^{k+1} e^{-2k^2 u^2}.$

These statements are usually proved by the so-called reflection principle (cf. Doob 1949, Billingsley 1968) and their proof will not be repeated here.

Some further distribution results for functionals of a Wiener process and a Brownian bridge follow without proof. The first one gives the distribution of the square integral $\omega^2 = \int_0^1 B^2(x) dx$ of a Brownian bridge. The second one characterizes the maximal deviation $\varkappa = \sup_x B(x) - \inf_x B(x)$ of a Brownian bridge. The third one is the celebrated arc sine law of P. Lévy.

Theorem 1.5.2 (Smirnov 1937, Anderson–Darling 1952).

$$P(\omega^2 \leq u) = 1 - \frac{2}{\pi} \sum_{k=1}^{\infty} (-1)^{k+1} \int_{(2k-1)\pi}^{2k\pi} \frac{e^{-\frac{t^2 u}{2}}}{\sqrt{-t \sin t}} dt$$

$$= \frac{1}{\pi \sqrt{u}} \sum_{k=0}^{\infty} (-1)^k \binom{-1/2}{k} \sqrt{4k+1} \, e^{-\frac{(4k+1)^2}{16u}} b_{1/4}\left(\frac{(4k+1)^2}{16u}\right)$$

where $b_{1/4}(\cdot)$ is the Bessel function of parameter $1/4$.

Theorem 1.5.3 (Kuiper 1960).

$$P(\varkappa \leq u) = 1 - \sum_{j=1}^{\infty} 2(4(ju)^2 - 1) e^{-2j^2 u^2}.$$

Theorem 1.5.4 (P. Lévy 1937, 1948). *Let* $U = \lambda\{t: W(t) \geq 0, 0 \leq t \leq 1\}$ *and* $V = \sup\{t: W(t) = 0, 0 \leq t \leq 1\}$. *Then*

$$P\{U \leq x\} = P\{V \leq x\} = \frac{1}{\pi} \int_0^x \frac{ds}{\sqrt{s(1-s)}} = \frac{2}{\pi} \arcsin \sqrt{x}, \quad 0 < x < 1.$$

Theorem 1.5.5 (Qualls, Watanabe 1972). *For* $-\infty < y < +\infty$ *we have*

$$\lim_{T \to \infty} P\{\sup_{0 \leq t \leq T} (W(t+1) - W(t)) \leq a(y, T)\} = \exp(-e^{-y}),$$

$$\lim_{T \to \infty} P\{\sup_{0 \leq t \leq T} |W(t+1) - W(t)| \leq a(y, T)\} = \exp(-2e^{-y}),$$

where

$$a(y, T) = (y + 2 \log T + \tfrac{1}{2} \log \log T - \tfrac{1}{2} \log \pi)(2 \log T)^{-1/2}.$$

1.6. The modulus of non-differentiability of the Wiener process

In this section we intend to prove the following analogue of Theorem 1.1.1.

Theorem 1.6.1 (Csörgő, Révész 1979a).

$$\lim_{h \to 0} \inf_{0 \leq s \leq 1-h} \sup_{0 \leq t \leq h} \sqrt{\frac{8 \log h^{-1}}{\pi^2 h}} |W(s+t) - W(s)| \stackrel{\text{a.s.}}{=} 1.$$

This theorem implies the well-known

Theorem 1.6.2. *Almost all sample functions of a Wiener process are nowhere differentiable.*

Theorem 1.6.1 actually gives the exact "modulus of non-differentiability" of a Wiener process.

The proof of Theorem 1.6.1 is based on the following lemma which, in turn, is a simple consequence of (1.5.2).

Lemma 1.6.1.

$$\frac{2}{\pi} e^{-\pi^2/8x^2} \leq \frac{4}{\pi}\left(e^{-\pi^2/8x^2} - \frac{1}{3}e^{-9\pi^2/8x^2}\right)$$

$$\leq P\left\{\sup_{0\leq t\leq T} T^{-1/2}|W(t)| \leq x\right\} \leq \frac{4}{\pi} e^{-\pi^2/8x^2}.$$

The proof of Theorem 1.6.1 will be presented in two steps.

Step 1. For any $\varepsilon > 0$ we have

(1.6.1) $\quad \lim_{h\to 0} \inf_{0\leq s\leq 1-h} \sup_{0<t\leq h} \sqrt{\frac{8\log h^{-1}}{\pi^2 h}} |W(s+t)-W(s)| \geq 1-\varepsilon \quad$ a.s.

Proof. Put
$$s_i = ih(\log h^{-1})^{-3} \quad (i = 0, 1, 2, \ldots, \varrho_h),$$
where $\varrho_h = [h^{-1}(\log h^{-1})^3]$. Then, by Lemma 1.6.1, we have

$$P\left\{\min_{0\leq i\leq \varrho_h} \sup_{0<t\leq h} \left(\frac{8\log h^{-1}}{\pi^2 h}\right)^{1/2} |W(s_i+t)-W(s_i)| < 1-\varepsilon\right\}$$

$$\leq (\varrho_h+1)\frac{4}{\pi}\exp\left\{-\frac{1}{(1-\varepsilon)^2}\log h^{-1}\right\} = O(h^\delta(\log h^{-1})^3),$$

where $\delta = (1-\varepsilon)^{-2} - 1 > 0$.

Now let $h_n = n^{-T}$ where $T > \delta^{-1}$. Then the above inequality implies:

(1.6.2) $\quad \lim_{n\to\infty} \min_{0\leq i\leq \varrho_{h_n}} \sup_{0<t\leq h_n} \left(\frac{8\log h_n^{-1}}{\pi^2 h_n}\right)^{1/2} |W(s_i+t)-W(s_i)| \geq 1-\varepsilon \quad$ a.s.

where
$$s_i = ih_n(\log h_n^{-1})^{-3}.$$

Consider the interval $s_i \leq s \leq s_{i+1}$. Then applying Theorem 1.1.1 with $h_n/\left(\log \frac{1}{h_n}\right)^3$ instead of h, we get

$$\overline{\lim_{n\to\infty}} \max_{0\leq i\leq \varrho_{h_n}} \sup_{s_i\leq s<s_{i+1}} \left(\frac{(\log h_n^{-1})^2}{2h_n}\right)^{1/2} |W(s)-W(s_i)| \leq 1 \quad \text{a.s.}$$

which, together with (1.6.2), implies

(1.6.3) $\quad \lim_{n\to\infty} \inf_{0\leq s\leq 1-h_n} \sup_{0<t\leq h_n} \left(\frac{8\log h_n^{-1}}{\pi^2 h_n}\right)^{1/2} |W(s+t)-W(s)| \geq 1-\varepsilon \quad$ a.s.

Finally, choosing $h_{n+1} \leq h < h_n$ and taking into account that $h_n/h_{n+1} \to 1$ ($n \to \infty$) and that

$$\inf_{0 \leq s \leq 1-h_{n+1}} \sup_{0 \leq t \leq h_{n+1}} \left(\frac{8 \log h^{-1}}{\pi^2 h}\right)^{1/2} |W(s+t)-W(s)|$$

$$\leq \inf_{0 \leq s \leq 1-h} \sup_{0 \leq t \leq h} \left(\frac{8 \log h^{-1}}{\pi^2 h}\right)^{1/2} |W(s+t)-W(s)|,$$

we get (1.6.1).

Step 2. *For any* $\varepsilon > 0$ *we have*

(1.6.4) $\quad \varlimsup_{h \to 0} \inf_{0 \leq s \leq 1-h} \sup_{0 < t \leq h} \left(\frac{8 \log h^{-1}}{\pi^2 h}\right)^{1/2} |W(s+t)-W(s)| \leq 1+\varepsilon \quad a.s.$

Proof. Put
$$s_i = ih \quad (i = 0, 1, 2, \ldots, [h^{-1}]).$$

Then, by Lemma 1.6.1, we have

$$P\left\{\min_{0 \leq i \leq [h^{-1}]} \sup_{0 < t \leq h} \left(\frac{8 \log h^{-1}}{\pi^2 h}\right)^{1/2} |W(s_i+t)-W(s_i)| > 1+\varepsilon\right\}$$

$$\leq \left[P\left\{\sup_{0 < t \leq h} \left(\frac{8 \log h^{-1}}{\pi^2 h}\right)^{1/2} |W(t)| > 1+\varepsilon\right\}\right]^{[1/h]+1}$$

$$\leq \left[1 - \frac{2}{\pi} \exp\left(-\frac{1}{(1+\varepsilon)^2} \log \frac{1}{h}\right)\right]^{[1/h]} = \left(1 - \frac{2}{\pi} h^\delta\right)^{[h^{-1}]}$$

$$\leq \exp\left\{-\frac{2}{\pi} h^\delta \left[\frac{1}{h}\right]\right\},$$

where $\delta = (1+\varepsilon)^{-2} < 1$. Now let $h_n = n^{-1}$. Then the above inequality implies

(1.6.5) $\quad \varlimsup_{n \to \infty} \inf_{0 \leq s \leq 1-h_n} \sup_{0 < t \leq h_n} \left(\frac{8 \log h_n^{-1}}{\pi^2 h_n}\right)^{1/2} |W(s+t)-W(s)|$

$$\leq \varlimsup_{n \to \infty} \min_{0 \leq i \leq [h_n^{-1}]} \sup_{0 < t \leq h_n} \left(\frac{8 \log h_n^{-1}}{\pi^2 h_n}\right)^{1/2} |W(s_i+t)-W(s_i)|$$

$$\leq 1+\varepsilon.$$

Finally, choosing $h_{n+1} \leq h < h_n$ and taking into account that $h_n/h_{n+1} \to 1$ $(n \to \infty)$ and

$$\inf_{0 \leq s \leq 1-h} \sup_{0 < t \leq h} \left(\frac{8 \log h^{-1}}{\pi^2 h}\right)^{1/2} |W(s+t) - W(s)|$$

$$\leq \inf_{0 \leq s \leq 1-h} \sup_{0 < t \leq h_n} \left(\frac{8 \log h^{-1}}{\pi^2 h}\right)^{1/2} |W(s+t) - W(s)|,$$

we get (1.6.4).

1.7. How small are the increments of a Wiener process?

The connection between the results of this Section and those of Section 2 is similar to that between Theorem 1.6.1 and Theorem 1.1.1.

Let

$$\mathscr{I}_1 = \mathscr{I}_1(t) = |W(t+a_T) - W(t)|$$

and

$$\mathscr{I}_2 = \mathscr{I}_2(t) = \sup_{0 < s \leq a_T} |W(t+s) - W(t)|.$$

Now the increment $\mathscr{I}_1$ can be much smaller than the increment $\mathscr{I}_2$. In this section we investigate only the question "How small are the increments $\mathscr{I}_2(t)$ $(0 \leq t \leq T - a_T)$?" and, as an answer to it, we prove

Theorem 1.7.1 (Csörgő, Révész 1979a). *Let a_T be a non-decreasing function of T for which*

(i) $0 < a_T \leq T$ $(T \geq 0)$,
(ii) a_T/T *is non-increasing.*

Then

(1.7.1) $$\lim_{T \to \infty} \gamma_T I(T) \stackrel{a.s.}{=} 1,$$

where

$$I(T) = \inf_{0 \leq t \leq T - a_T} \mathscr{I}_2(t)$$

and

$$\gamma_T = \left(\frac{8(\log T a_T^{-1} + \log \log T)}{\pi^2 a_T}\right)^{1/2}.$$

If we also have

(iii) $$\frac{\log T/a_T}{\log \log T} \nearrow +\infty$$

then

(1.7.2) $$\lim_{T\to\infty} \gamma_T I(T) \stackrel{a.s.}{=} 1.$$

The following examples illustrate what this theorem is all about.

Example 1. $a_T = \frac{8}{\pi^2} \log T$. Then $\gamma_T \to 1$ and our Theorem 1.7.1 says that for all T big enough, for any $\varepsilon > 0$ and for almost all ω there exists a $0 \leq t = t(T, \varepsilon, \omega) \leq T - a_T$ such that

$$\sup_{0 \leq s \leq \frac{8}{\pi^2}\log T} |W(t+s) - W(t)| \leq 1 + \varepsilon,$$

but, for all $t \in [0, T - a_T]$, with probability 1,

$$\sup_{0 \leq s \leq \frac{8}{\pi^2}\log T} |W(t+s) - W(t)| \geq 1 - \varepsilon.$$

At the same time our Theorem 1.2.1 stated the existence of a $t \in [0, T - a_T]$ such that, with probability 1,

$$\left| W\left(t + \frac{8}{\pi^2}\log T\right) - W(t) \right| \geq \left(\frac{4}{\pi} - \varepsilon\right) \log T,$$

and hence

$$\sup_{0 \leq s \leq \frac{8}{\pi^2}\log T} |W(t+s) - W(t)| \geq \left(\frac{4}{\pi} - \varepsilon\right) \log T$$

but, for all $t \in [0, T - a_T]$,

$$\sup_{0 < s \leq \frac{8}{\pi^2}\log T} |W(t+s) - W(t)| \leq \left(\frac{4}{\pi} + \varepsilon\right) \log T.$$

Example 2. Let $a_T = T$. Then our Theorem 1.7.1 says

$$\lim_{T\to\infty} \left(\frac{8 \log \log T}{\pi^2 T}\right)^{1/2} \sup_{0 \leq t \leq T} |W(t)| = 1 \quad \text{a.s.},$$

which is the law of iterated logarithm of Chung (1948) when it is applied to the Wiener process.

Example 3. Let $a_T = (\log T)^{1/2}$. Then $\gamma_T \approx \left(\frac{8}{\pi^2}\sqrt{\log T}\right)^{1/2}$, and our Theorem 1.7.1 says that for all T big enough, for any $\varepsilon > 0$ and for almost all ω there exists a $t=t(T,\varepsilon,\omega)\in[0, T-a_T]$ such that

$$\sup_{0\le s\le(\log T)^{1/2}} |W(t+s)-W(t)| \le (1+\varepsilon)\frac{\pi}{\sqrt{8}}(\log T)^{-1/4}.$$

That is to say the interval $[0, T-a_T]$ has a subinterval of length $(\log T)^{1/2}$ where the sample function of the Wiener process is nearly constant; more precisely, the fluctuation from a constant is so small as $(1+\varepsilon)\pi 8^{-1/2}(\log T)^{-1/4}$.

This result is sharp in the sense that for all T big enough, and all $t\in[0, T-a_T]$, we have with probability 1

$$\sup_{0\le s\le(\log T)^{1/2}} |W(t+s)-W(t)| \ge (1-\varepsilon)\frac{\pi}{\sqrt{8}}(\log T)^{-1/4}.$$

Just like that of Theorem 1.6.1, the proof of Theorem 1.7.1 is also based on Lemma 1.6.1 and will be presented in three steps.

Step 1. For any $\varepsilon>0$ we have

(1.7.3) $$\lim_{T\to\infty} \gamma_T I(T) \ge 1-\varepsilon \quad a.s.$$

Proof. Let $T_n=\theta^n$ $(1<\theta<(1-\varepsilon)^{-2})$, $\varphi(T)=\log T a_T^{-1}+\log\log T$ and

$$t_i = t_i^{(n)} = i a_{T_n}(\varphi(T_n))^{-3} \quad (i=0,1,2,\ldots,\varrho_{T_n}),$$

where

$$\varrho_{T_n} = [T_{n+1}a_{T_n}^{-1}(\varphi(T_n))^3].$$

Then by Lemma 1.6.1 we have

$$P\left\{\min_{0\le i\le\varrho_{T_n}} \sup_{0\le s\le a_{T_n}} \left(\frac{8\varphi(T_n)}{\pi^2 a_{T_{n+1}}}\right)^{1/2} |W(t_i+s)-W(t_i)| \le 1-\varepsilon\right\}$$

$$\le (\varrho_{T_n}+1)\frac{4}{\pi}\exp\left\{-(1-\varepsilon)^{-2}\frac{a_{T_n}}{a_{T_{n+1}}}\varphi(T_n)\right\}$$

$$\le \frac{8}{\pi}\theta\frac{T_n}{a_{T_n}}(\varphi(T_n))^3 \exp\left\{-\frac{1}{\theta(1-\varepsilon)^2}\varphi(T_n)\right\}$$

$$\le O(1)(\log n)^3 n^{-\frac{1}{\theta(1-\varepsilon)^2}},$$

where the inequality $\dfrac{a_{T_n}}{a_{T_{n+1}}} = \dfrac{a_{T_n}}{T_n}\dfrac{T_{n+1}}{a_{T_{n+1}}}\dfrac{1}{\theta} \ge \dfrac{1}{\theta}$ was applied.

Hence we get

(1.7.4) $$\lim_{n\to\infty} \min_{0\le i\le\varrho_{T_n}} \sup_{0\le s\le a_{T_n}} \left(\frac{8\varphi(T_n)}{\pi^2 a_{T_{n+1}}}\right)^{1/2} |W(t_i+s)-W(t_i)| \ge 1-\varepsilon, \quad a.s.$$

Consider now the interval $t_i \leq t < t_{i+1}$. Then by Theorem 1.2.1 we get almost surely that

$$\varlimsup_{n\to\infty} \max_{0\leq i\leq \varrho T_n} \sup_{t_i\leq t\leq t_{i+1}} \left\{\frac{(\varphi(T_n))^3}{2a_{T_n}\left(\log\left(\frac{T_{n+1}}{a_{T_n}}(\varphi(T_n))^3\right)+\log\log T_{n+1}\right)}\right\}^{1/2} \times |W(t)-W(t_i)| \leq 1,$$

which, together with (1.7.4), implies

(1.7.5) $\varliminf_{n\to\infty} \inf_{0\leq t\leq T_{n+1}-a_{T_{n+1}}} \sup_{0\leq s\leq a_{T_n}} \left(\frac{8\varphi(T_n)}{\pi^2 a_{T_n}}\right)^{1/2} |W(t+s)-W(t)| \geq 1-\varepsilon$ a.s.

Finally, choosing $T_n \leq T \leq T_{n+1}$ and taking into account that

$$\varphi(T)a_T^{-1} \geq \varphi(T_n)a_{T_{n+1}}^{-1}, \quad T-a_T \leq T_{n+1}-a_{T_{n+1}}$$

and

$$\sup_{0<s\leq a_T} |W(t+s)-W(t)| \geq \sup_{0\leq s\leq a_{T_n}} |W(t+s)-W(t)|,$$

we get (1.7.3) by (1.7.5).

Step 2. Let

$$B(T) = \gamma_T \sup_{0\leq s\leq a_T} |W(T-a_T+s)-W(T-a_T)|.$$

Then for any $\varepsilon>0$ we have

(1.7.6) $\varlimsup_{T\to\infty} B(T) \leq 1+\varepsilon$ a.s.

Proof. By Lemma 1.6.1 we have

$$P(B(T)\leq 1+\varepsilon) \geq \frac{2}{\pi}\exp\left\{-\frac{1}{(1+\varepsilon)^2}\varphi(T)\right\} = \frac{2}{\pi}\left(\frac{a_T}{T\log T}\right)^{\frac{1}{(1+\varepsilon)^2}}.$$

Since conditions (i) and (ii) imply that $T-a_T$ is a continuous nondecreasing function of T, we can define the sequence $\{T_k\}$ as follows: Let $T_1 = 1$ and define T_{k+1} by

$$T_{k+1}-a_{T_{k+1}} = T_k \quad \text{if} \quad \frac{a_T}{T}\to \varrho < 1,$$

$$T_{k+1} = e^{(k+1)\log(k+1)} \quad \text{if} \quad \frac{a_T}{T}\to \varrho = 1.$$

In case of $\varrho<1$, (1.7.6) follows from the simple fact (cf. Step 2 of Theorem 1.2.1) that

$$\sum_{k=2}^{\infty}\left(\frac{a_{T_k}}{T_k \log T_k}\right)^{\frac{1}{(1+\varepsilon)^2}} = \infty$$

and that the r.v.'s $B(T_k)$ ($k=1,2,\ldots$) are independent.

In case of $\varrho=1$, $a_{T_{k+1}} \geq T_{k+1}-T_k$. Hence
$$B(T_{k+1}) \leq \gamma_{T_{k+1}} \sup_{T_k \leq s \leq T_{k+1}} |W(s)-W(T_k)|$$
$$+ \gamma_{T_{k+1}} \sup_{0 \leq s \leq T_k} |W(T_k)-W(s)|.$$

By the law of iterated logarithm
$$\varlimsup_{k \to \infty} \gamma_{T_{k+1}} \sup_{0 \leq s \leq T_k} |W(T_k)-W(s)| = 0 \quad \text{a.s.}$$

and by Lemma 1.6.1
$$P\left\{\gamma_{T_{k+1}} \sup_{T_k \leq s \leq T_{k+1}} |W(s)-W(T_k)| \leq 1+\varepsilon\right\}$$
$$\geq \frac{2}{\pi} \exp\left\{-\frac{1}{(1+\varepsilon)^2} \log\log T_{k+1} \frac{T_{k+1}-T_k}{T_{k+1}}\right\} \geq \text{Const.} \left(\frac{1}{k \log k}\right)^{\frac{1}{(1+\varepsilon)^2}}.$$

The latter combined with our preceding two statements imply (1.7.6) if $\varrho=1$.

Step 3. For any $\varepsilon > 0$ we have

(1.7.7) $$\varlimsup_{T \to \infty} \gamma_T I(T) \leq 1+\varepsilon \quad \text{a.s.,}$$

provided (iii) *holds true*.

Proof. Choose the sequence T_n ($n=1, 2, \ldots$) such that
$$T_n a_{T_n}^{-1} = n$$
and put
$$t_i = i a_{T_{n+1}} \quad (i = 0, 1, 2, \ldots, \varrho_{T_n}),$$
where $\varrho_{T_n} = [T_n a_{T_{n+1}}^{-1}]$.

Then by Lemma 1.6.1 we have
$$P\left\{\min_{0 \leq i \leq \varrho_{T_n}} \sup_{0 < s \leq a_{T_{n+1}}} \left(\frac{8 \log T_{n+1} a_{T_{n+1}}^{-1}}{\pi^2 a_{T_n}}\right)^{1/2} |W(t_i+s)-W(t_i)| > 1+\varepsilon\right\}$$
$$\leq \left(P\left\{\sup_{0 < s \leq a_{T_{n+1}}} \left(\frac{8 \log T_{n+1} a_{T_{n+1}}^{-1}}{\pi^2 a_{T_n}}\right)^{1/2} |W(s)| > 1+\varepsilon\right\}\right)^{\varrho_{T_n}+1}$$
$$\leq \left(1 - \frac{2}{\pi} \exp\left\{-(1+\varepsilon)^{-2} \frac{a_{T_{n+1}}}{a_{T_n}} \log(T_{n+1} a_{T_{n+1}}^{-1})\right\}\right)^{\varrho_{T_n}}$$
$$= \left\{1 - \frac{2}{\pi} \left(\frac{a_{T_{n+1}}}{T_{n+1}}\right)^{\frac{a_{T_{n+1}}}{a_{T_n}}(1+\varepsilon)^{-2}}\right\}^{\varrho_{T_n}} = \exp(-O(1) n^{1-(1+\varepsilon)^{-2}+o(1)}),$$

where the last line follows from observing that $a_{T_{n+1}}/a_{T_n} \to 1$, which, in turn, follows from $T_{n+1}/T_n \to 1$ by the definition of T_n in this proof. In order to see that $T_{n+1}/T_n \to 1$, let

$$\frac{\log T/a_T}{\log \log T} = b(T),$$

and recall that $b(T) \nearrow \infty$ by condition (iii). Then $T_n = \exp\left(n^{\frac{1}{b(T_n)}}\right)$. Whence

$$\frac{T_{n+1}}{T_n} = \exp\left((n+1)^{\frac{1}{b(T_{n+1})}} - n^{\frac{1}{b(T_n)}}\right) \leq \exp\left((n+1)^{\frac{1}{b(T_n)}} - n^{\frac{1}{b(T_n)}}\right)$$

$$\leq \exp\left[n^{\frac{1}{b(T_n)}}\left(\left(1+\frac{1}{n}\right)^{\frac{1}{b(T_n)}} - 1\right)\right] \leq \exp\left(n^{\frac{1}{b(T_n)}}/n\right) \to 1 \quad \text{as} \quad n \to \infty.$$

Hence we get

$$\varlimsup_{n \to \infty} \min_{0 \leq i \leq \varrho_{T_n}} \sup_{0 < s \leq a_{T_{n+1}}} \left(\frac{8 \log T_{n+1} a_{T_{n+1}}^{-1}}{\pi^2 a_{T_n}}\right)^{1/2} |W(t_i+s) - W(t_i)| \leq 1 + \varepsilon, \quad \text{a.s.}$$

This implies (1.7.7) immediately.

In Section 1.2 and in the present Section we studied the properties of some increments of a Wiener process. In order to present some further problems, let

$$\mathcal{I}_1^{(1)}(t) = |W(t+a_T) - W(t)|,$$
$$\mathcal{I}_2^{(1)}(t) = \sup_{0 < s \leq a_T} |W(t+s) - W(t)|,$$
$$\mathcal{I}_1^{(2)}(t) = W(t+a_T) - W(t),$$
$$\mathcal{I}_2^{(2)}(t) = \sup_{0 < s \leq a_T} (W(t+s) - W(t)),$$
$$I_i^{(j,1)}(T) = \sup_{0 < t \leq T-a_T} \mathcal{I}_i^{(j)}(t) \quad (i=1,2; j=1,2),$$
$$I_i^{(j,2)}(T) = \inf_{0 < t \leq T-a_T} \mathcal{I}_i^{(j)}(t) \quad (i=1,2; j=1,2).$$

Now, our question is to find the normalizing factors $\mu_T(i,j,k)$ and $\nu_T(i,j,k)$ ($i=1,2; j=1,2; k=1,2$) for which

$$\varlimsup_{T \to \infty} \mu_T(i,j,k) I_i^{(j,k)}(T) = 1 \quad \text{a.s.}$$

and

$$\lim_{T \to \infty} \nu_T(i,j,k) I_i^{(j,k)}(T) = 1 \quad \text{a.s.}$$

Of the here mentioned eight lim sup problems four were solved in Section 1.2, namely the cases: $k=1$, $i=1, 2$, $j=1, 2$. One of the eight mentioned lim inf problems is solved in the present section, namely the case of $(k=2, i=2, j=1)$. For a partial solution of the case $(i=2, j=1, k=1)$ we refer to Remark 1.2.2. Also, for small a_T i.e. when condition (iii) of Theorem 1.7.1 holds, the lim sup $=$ the lim inf in the just mentioned completely solved five cases. Thus, for a_T satisfying (i) and (ii), five and, for a_T satisfying also (iii), ten of the above problems are completely solved.

1.8. Infinite series representations of the Wiener process and Brownian bridge

Our construction of a Wiener process in Section 1.1 can be slightly modified so that it also gives an infinite series representation of W. In order to see this, we restrict our procedure to the unit interval $[0, 1]$. In this setup then, the construction of Section 1.1 uses a sequence of independent r.v. $X_{r_n} \in \mathcal{N}(0, 1)$, where r_n runs over the dyadic rational numbers of the form $k/2^n$ ($k=1, 3, 5, \ldots, 2^n-1$; $n=0, 1, 2, \ldots$). In the first step we defined $W(t)$ at $t=1$ by $W(1)=X_1$. Now, instead of the latter, we say that the first approximation of $W(t)$ should be

$$W_0(t) = tX_1 \quad (0 \leq t \leq 1).$$

In the second step, there we defined $W(\tfrac{1}{2})=\tfrac{1}{2}X_1+\tfrac{1}{2}X_{1/2}$. Now we say that this step should be replaced by the approximation

$$W_1(t) = tX_1 + h_1(t)X_{1/2} = h_0(t)X_1 + h_1(t)X_{1/2},$$

where

$$h_0(t) = \int_0^t w_0(x)\, dx = t, \quad 0 \leq t \leq 1,$$

and

$$h_1(t) = \int_0^t w_1(x)\, dx = \begin{cases} t, & 0 \leq t \leq \tfrac{1}{2} \\ 1-t, & \tfrac{1}{2} \leq t \leq 1, \end{cases}$$

with w_j standing for the jth Walsh function.

Our third step in Section 1.1 resulted in the definition of $W(\tfrac{1}{4})$ as

$$W(\tfrac{1}{4}) = \tfrac{1}{2}W(\tfrac{1}{2}) + \tfrac{1}{\sqrt{8}}X_{1/4} = \tfrac{1}{2}W(\tfrac{1}{2}) + \frac{X_{1/4}-X_{3/4}}{2\sqrt{8}} + \frac{X_{1/4}+X_{3/4}}{2\sqrt{8}}.$$

Now we say that this third approximation of W should be replaced by

$$W_3(t) = h_0(t)X_1 + h_1(t)X_{1/2} + h_2(t)\frac{X_{1/4}+X_{3/4}}{\sqrt{2}} + h_3(t)\frac{X_{1/4}-X_{3/4}}{\sqrt{2}} =$$
$$= h_0(t)Y_0 + h_1(t)Y_1 + h_2(t)Y_2 + h_3(t)Y_3,$$

where

$$h_2(t) = \int_0^t w_2(x)\,dx = \begin{cases} t, & 0 \le t \le \tfrac{1}{4} \\ \tfrac{1}{2}-t, & \tfrac{1}{4} \le t \le \tfrac{1}{2} \\ t-\tfrac{1}{2}, & \tfrac{1}{2} \le t \le \tfrac{3}{4} \\ 1-t, & \tfrac{3}{4} \le t \le 1, \end{cases}$$

$$h_3(t) = \int_0^t w_3(x)\,dx = \begin{cases} t, & 0 \le t \le \tfrac{1}{4} \\ \tfrac{1}{2}-t, & \tfrac{1}{4} \le t \le \tfrac{3}{4} \\ t-1, & \tfrac{3}{4} \le t \le 1 \end{cases}$$

and

$$Y_0 = X_1, \quad Y_1 = X_{1/2}, \quad Y_2 = \frac{X_{1/4}+X_{3/4}}{\sqrt{2}}, \quad Y_3 = \frac{X_{1/4}-X_{3/4}}{\sqrt{2}}.$$

We observe also that Y_0, Y_1, Y_2, Y_3 are independent $\mathcal{N}(0,1)$ r.v.

Reformulating each step of our construction in Section 1.1 as indicated above, the nth approximation of $W(t)$ is

(1.8.1) $$W_n(t) = \sum_{k=0}^n h_k(t)Y_k = \sum_{k=0}^n Y_k \int_0^t w_k(x)\,dx,$$

where $Y_0, Y_1, \ldots, Y_n$ are independent $\mathcal{N}(0,1)$ r.v. Moreover,

$$W\left(\frac{k}{2^n}\right) = W_N\left(\frac{k}{2^n}\right) \quad \text{whenever} \quad N \ge 2^n - 1,$$

where $W(\cdot)$ is the Wiener process as constructed in Section 1.1. Whence, applying the fact that our construction of $W(\cdot)$ in Section 1.1 is a uniformly (in t) convergent one (with probability one), it follows that (1.8.1) is also uniformly (in t) convergent with probability one. This, in turn, implies the following infinite series representation of W

(1.8.2) $$W(t) = \sum_{k=0}^\infty Y_k \int_0^t w_k(x)\,dx,$$

where $Y_0, Y_1, \ldots$ is a sequence of independent $\mathcal{N}(0,1)$ r.v.

Since the Walsh functions $\{w_k\}$ form a complete orthonormal system, it is only natural to ask whether $\{w_k\}$ in (1.8.2) could be replaced by any other complete orthonormal system $\{\varphi_k\}$. Indeed, it is clear that, for any such system of functions $\{\varphi_k\}$, the series $\sum_{k=0}^{\infty} Y_k \int_0^t \varphi_k(x)\,dx$ converges with probability one for each fixed $t \in [0, 1]$. It is also clear (direct calculations) that the covariance function of the latter series is that of a Wiener process. On the other hand, it is not clear at all that the latter convergence should hold uniformly in t. However, Ito and Nisio (1968) showed that it is so for any complete orthonormal system $\{\varphi_k\}$ and also that the thus defined limit is a Wiener process, i.e., we have with probability one and uniformly in $t \in [0, 1]$ that

$$(1.8.3) \qquad W(t) = \sum_{k=0}^{\infty} Y_k \int_0^t \varphi_k(x)\,dx,$$

for any sequence $\{Y_i\}$ of independent $\mathcal{N}(0, 1)$ r.v.

As an important special case of (1.8.3), we take $\{\varphi_0(x)=1,\ \varphi_k(x) = \sqrt{2}\cos \pi k x;\ 0 \leq x \leq 1,\ k=1, 2, \ldots\}$ as our complete orthonormal system on $[0, 1]$, and get

$$(1.8.4) \qquad W(t) = Y_0 t + \sqrt{2} \sum_{k=1}^{\infty} Y_k \int_0^t \cos k\pi x\,dx$$

$$= Y_0 t + \sqrt{2} \sum_{k=1}^{\infty} Y_k \frac{\sin k\pi t}{k\pi},$$

the classical representation of W by Paley and Wiener (1934).

The latter representation immediately implies a similar representation for a Brownian bridge. Since $W(1) = Y_0$ by (1.8.4), and $B(t) = W(t) - tW(1)$ by (1.4.1), we get

$$(1.8.5) \qquad B(t) = \sqrt{2} \sum_{k=1}^{\infty} Y_k \frac{\sin k\pi t}{k\pi}, \quad 0 \leq t \leq 1.$$

1.9. The Ornstein–Uhlenbeck process

Consider the Gaussian process $\{V(t) = W(t)/\sqrt{t};\ 0 < t < \infty\}$. Then $EV(t) = 0$, $EV^2(t) = 1$ and $EV(t)V(s) = \sqrt{s/t}$, $s < t$. The form of this covariance function immediately suggests that, in order to get a stationary Gaussian process out of $V(t)$, we should consider

$$(1.9.1) \qquad U_\alpha(t) = V(e^{\alpha t}), \quad -\infty < t < +\infty \quad (\alpha \text{ fixed } > 0).$$

This latter process is a stationary Gaussian process, for $EU_\alpha(t)U_\alpha(s) = e^{-\alpha|t-s|/2}$, and it is called the Ornstein–Uhlenbeck process. We will use the notation $U(t)=U_2(t)$, and mention, without proof, the following

Theorem 1.9.1 (Darling, Erdős 1956).

(1.9.2) $$\lim_{T\to\infty} P\left\{\sup_{0\leq t\leq T} U(t) \leq a(y,T)\right\} = \exp(-e^{-y}),$$

(1.9.3) $$\lim_{T\to\infty} P\left\{\sup_{0\leq t\leq T} |U(t)| \leq a(y,T)\right\} = \exp(-2e^{-y}),$$

where

$$a(y,T) = (y + 2\log T + \tfrac{1}{2}\log\log T - \tfrac{1}{2}\log\pi)(2\log T)^{-1/2}, \quad -\infty < y < \infty.$$

It follows from definition (1.9.1) that several properties of the Wiener process are inherited by $U_\alpha(t)$. For example, the latter process is also continuous, non-differentiable and Markovian.

Remark 1.9.1. Darling and Erdős (1956) evaluated the limit distribution of $\max_{1\leq k\leq n} S_k k^{-1/2}$ and that of $\max_{1\leq k\leq n} |S_k| k^{-1/2}$, where S_k is the kth partial sum of i.i.d.r.v. with mean zero, variance one and finite third moment. Doing this, they have actually proved Theorem 1.9.1, without stating it explicitly. It is not difficult to see that Theorem 1.9.1 can be proved easily from their main results:

(1.9.4) $$\lim_{n\to\infty} P\left\{\max_{1\leq k\leq n} k^{-1/2} S_k \leq a(y,\log n)\right\} = \exp(-e^{-y}),$$

(1.9.5) $$\lim_{n\to\infty} P\left\{\max_{1\leq k\leq n} k^{-1/2} |S_k| \leq a(y,\log n)\right\} = \exp(-2e^{-y}),$$
$$-\infty < y < +\infty.$$

The above introduction of the Ornstein–Uhlenbeck process via a Wiener process suggests a similar investigation of the standardized Brownian bridge $\{B(y)/\sqrt{y(1-y)};\ 0<y<1\}$. First we observe

(1.9.6) $$\left\{\sqrt{\frac{1-y}{y}}\, W\!\left(\frac{y}{1-y}\right);\ 0<y<1\right\} \stackrel{\mathscr{D}}{=} \left\{\frac{B(y)}{\sqrt{y(1-y)}};\ 0<y<1\right\},$$

via checking the respective covariance functions. Letting now $e^t = y(1-y)^{-1}$, we get

(1.9.7) $$\{U(t);\ -\infty < t < +\infty\} \stackrel{\mathscr{D}}{=} \left\{(1+e^t)e^{-t/2} B\!\left(\frac{e^t}{1+e^t}\right);\ -\infty < t < +\infty\right\}.$$

Consequently, we have also

Corollary 1.9.1. Let ε_n be a decreasing sequence of numbers such that $\varepsilon_n \to 0$. Then, with $-\infty < y < +\infty$, we have

$$(1.9.8) \quad \lim_{n\to\infty} P\left\{ \sup_{\varepsilon_n < x < 1-\varepsilon_n} \frac{B(x)}{\sqrt{x(1-x)}} \leq a\left(y, 2\log\frac{1-\varepsilon_n}{\varepsilon_n}\right)\right\} = \exp(-e^{-y}),$$

$$(1.9.9) \quad \lim_{n\to\infty} P\left\{ \sup_{\varepsilon_n < x < 1-\varepsilon_n} \frac{|B(x)|}{\sqrt{x(1-x)}} \leq a\left(y, 2\log\frac{1-\varepsilon_n}{\varepsilon_n}\right)\right\} = \exp(-2e^{-y}).$$

Proof. By (1.9.6), (1.9.7) and stationarity of the Ornstein–Uhlenbeck process $U(t)$ we have

$$\lim_{n\to\infty} P\left\{ \sup_{\varepsilon_n < x < 1-\varepsilon_n} \frac{B(x)}{\sqrt{x(1-x)}} \leq a\left(y, 2\log\frac{1-\varepsilon_n}{\varepsilon_n}\right)\right\}$$

$$= \lim_{n\to\infty} P\left\{ \sup_{\log\frac{\varepsilon_n}{1-\varepsilon_n} < t < \log\frac{1-\varepsilon_n}{\varepsilon_n}} U(t) \leq a\left(y, 2\log\frac{1-\varepsilon_n}{\varepsilon_n}\right)\right\}$$

$$= \lim_{n\to\infty} P\left\{ \sup_{0 < t < 2\log\frac{1-\varepsilon_n}{\varepsilon_n}} U(t) \leq a\left(y, 2\log\frac{1-\varepsilon_n}{\varepsilon_n}\right)\right\}.$$

Hence (1.9.8) follows from (1.9.2) and a similar argument yields (1.9.9).

1.10. On the notion of a two-parameter Wiener process

Consider the lattice points $\underline{n} = (n_1, n_2)$ ($n_i = 0, 1, 2, \ldots$; $i = 1, 2$) of $R_+^2 = [0, \infty) \times [0, \infty)$. For each $\underline{n}$ define a r.v. $X_{\underline{n}}$ such that the r.v. $X_{\underline{n}}$ are independent with

$$P(X_{\underline{n}} = +1) = P(X_{\underline{n}} = -1) = 1/2.$$

Further let

$$S_{\underline{n}} = \sum_{\underline{i} \leq \underline{n}} X_{\underline{i}} = \sum_{i_1=1}^{n_1} \sum_{i_2=1}^{n_2} X_{(i_1, i_2)}, \quad (\underline{n} = (n_1, n_2)).$$

This model is a natural two-parameter analogue of the random walk model of Section 1.0 and a continuous version of it will serve as a model of the two-parameter Wiener process.

1.11. Definition and existence of a two-parameter Wiener process

Let $X(\underline{z})$ ($\underline{z}=(x,y)\in R_+^2$) be a two-parameter stochastic process and consider the rectangle $R=[x_1,x_2)\times[y_1,y_2)\subset R_+^2$ ($0\leq x_1<x_2<\infty$, $0\leq y_1<y_2<\infty$). Define the "X-measure" $X(R)$ of R by

$$X(R) = X(x_2,y_2) - X(x_1,y_2) - X(x_2,y_1) + X(x_1,y_1).$$

A stochastic process $\{W(\underline{z}), \underline{z}\in R_+^2\}$ is called a (two-parameter) Wiener process if

(i) $W(R)\in \mathcal{N}(0,\lambda(R))$ for all $R=[x_1,x_2)\times[y_1,y_2)$ where $\lambda(R)=(x_2-x_1)(y_2-y_1)$,

(ii) $W(0,y)\equiv W(x,0)\equiv 0$ ($0\leq x,y<\infty$),

(iii) $W(\underline{z})$ is an independent increment process, that is $W(R_1), W(R_2), \ldots, W(R_n)$ ($n=2,3,\ldots$) are independent r.v. if $R_1, R_2, \ldots, R_n$ are disjoint rectangles,

(iv) the sample path function $W(\underline{z};\omega)$ is continuous in $\underline{z}$ with probability 1.

We note that (i)–(iii) imply that the covariance function of a Wiener process $W(\underline{z})$ is

$$R(\underline{z}_1,\underline{z}_2) = EW(\underline{z}_1)W(\underline{z}_2) = (x_1\wedge x_2)(y_1\wedge y_2)$$

where $\underline{z}_1=(x_1,y_1)$, $\underline{z}_2=(x_2,y_2)$.

We also note that for any fixed $0<x_0<\infty$ the process $\{x_0^{-1/2}W(x_0,y), 0\leq y<\infty\}$ is a (one-parameter) Wiener process and the same can be said about $\{y_0^{-1/2}W(x,y_0), 0\leq x<\infty\}$.

The aim of this Section is to give a constructive proof for the existence of the two-parameter Wiener process. The idea of construction will be mainly that of Section 1.1. The existence of the one-parameter Wiener process will be also used.

Let $\{r_n\}$ be the sequence of positive dyadic rational numbers and let $\{W_{r_n}(x)\}$ be independent one-parameter Wiener processes. For any positive integer k, let

$$W(x,k) = W_1(x) + W_2(x) + \ldots + W_k(x)$$

and

$$W(x,k+1/2) = \frac{W(x,k)+W(x,k+1)}{2} + \frac{W_{k+1/2}(x)}{\sqrt{4}}.$$

Now we wish to define $W(x,k/2^n)$ for $k=1,2,\ldots$ and $n=1,2,\ldots$. Assume that it is already defined for $k=1,2,\ldots$ and $n=1,2,\ldots,n_0$.

Then for $k=1, 2, \ldots$ and $n=n_0+1$ we let

$$W\left(x, \frac{2k+1}{2^n}\right) = \frac{W\left(x, \frac{k}{2^{n_0}}\right)+W\left(x, \frac{k+1}{2^{n_0}}\right)}{2} + \frac{W_{\frac{2k+1}{2^n}}(x)}{\sqrt{2^{n+1}}}.$$

Whence, by induction, we have defined our Wiener process at every (x, r), $0 \leq x < \infty$, where r is a non-negative dyadic rational number. For an arbitrary $y>0$ we define

$$W(x, y) = \lim_{n \to \infty} W\left(x, \frac{[2^n y]}{2^n}\right),$$

where the existence of the limit on the right hand side immediately follows from Kolmogorov's Three Series Theorem exactly the same way as in Section 1.1. Also, the problem of uniform convergence can be posed and settled exactly the same way here, namely via showing that

$$\sum_{r=0}^{\infty} \sup_{(x,y) \in I^2} |W(x, y) - W(x, y_r)| < \infty \quad \text{a.s.} \quad \left(y_r = \frac{[2^r y]}{2^r}\right).$$

The thus defined process obviously satisfies conditions (i)–(iii). In the rest of this Section we intend to prove that our process also satisfies (iv). At first we prove the following analogue of Lemma 1.1.1.

Lemma 1.11.1. *For any $\varepsilon > 0$ there exists a constant $C = C(\varepsilon) > 0$ such that the inequality*

(1.11.1) $\quad P\left\{ \sup_{0 \leq s \leq h} \sup_{(x,y) \in I^2} |W(x, y+s) - W(x, y)| \geq v h^{1/2} \right\} \leq C h^{-1} e^{-\frac{v^2}{2+\varepsilon}}$

holds for every positive v and $0 < h < 1$.

Proof. Using the notations introduced in the proof of Lemma 1.1.1 we have

$$|W(x, y+s) - W(x, y)| \leq |W(x, y+s) - W(x, (y+s)_r)|$$
$$+ |W(x, (y+s)_r) - W(x, y_r)| + |W(x, y_r) - W(x, y)|$$

$$\leq |W(x, (y+s)_r) - W(x, y_r)| + \sum_{j=0}^{\infty} |W(x, (y+s)_{r+j+1}) - W(x, (y+s)_{r+j})|$$

$$+ \sum_{j=0}^{\infty} |W(x, y_{r+j+1}) - W(x, y_{r+j})|.$$

Now, by (1.5.1) and (1.1.1) for any positive h, u, z_j and integers r, j, we have

$$P\left\{\sup_{0 \leq s \leq h} \sup_{(x,y) \in I^2} |W(x, (y+s)_r) - W(x, y_r)| \geq u(h+R^{-1})^{1/2}\right\}$$

$$\leq R(Rh+1) P\left\{\sup_{0 < y < 1} |W(x, (y+h)_r) - W(x, y_r)| \geq u(h+R^{-1})^{1/2}\right\}$$

$$\leq 4e^{-u^2/2} R(Rh+1),$$

and

$$P\left\{\sup_{0 \leq s \leq h} \sup_{(x,y) \in I^2} |W(x, (y+s)_{r+j+1}) - W(x, (y+s)_{r+j})| \geq z_j 2^{-(r+j+1)/2}\right\}$$

$$\leq 2^{r+j+3} e^{-z_j^2/2}.$$

From here on the proof is pretty much the same as that of Lemma 1.1.1. Hence the details will be omitted.

Lemma 1.11.1 easily implies

Lemma 1.11.2. *We have*

$$\limsup_{h \to 0} \sup_{0 \leq s \leq h} \sup_{(x,y) \in I^2} \frac{|W(x, y+s) - W(x, y)|}{\sqrt{2h \log 1/h}} \overset{a.s.}{=} 1.$$

Proof is the same as that of Theorem 1.1.1, and will be omitted.

This lemma clearly means that, for almost all ω, $W(x, y)$ is continuous in y ($0 \leq y \leq 1$), with the usual modulus of continuity $(2h \log 1/h)^{1/2}$, for every $x \in [0, 1]$. The next lemma will say that, for almost all ω, $W(x, y)$ is continuous in x and y, where $x \in [0, 1]$ and y is running over the dyadic rationals of $[0, 1]$.

Lemma 1.11.3. *We have*

$$\limsup_{h \to 0} \sup_{\substack{0 \leq s \leq h \\ y \in I_r}} \sup_{x \in [0,1]} |W(x+s, y) - W(x, y)| \overset{a.s.}{=} 0$$

where I_r is the set of the dyadic rationals of $[0, 1]$.

Proof. This lemma is a straight consequence of the above construction of $W(x, y)$ and the following elementary

Lemma 1.11.4. *Let $W_1(x), W_2(x), \ldots$ be a sequence of independent Wiener processes. Then*

$$\limsup_{n \to \infty} \sup_{0 \leq x \leq 1} \frac{|W_n(x)|}{\log n} \overset{a.s.}{=} 0.$$

Lemmas 1.11.2 and 1.11.3 together prove that our above constructed process $W(x, y)$ is continuous with probability 1. Thus, it also satisfies condition (iv). Hence this $W(x, y)$ is indeed a Wiener process. While Lemmas 1.11.2 and 1.11.3 prove the continuity of W, they do not say anything about its modulus of continuity. A kind of modulus of continuity will be evaluated in Section 1.13, where we give an analogue of Theorem 1.1.1. An analogue of Theorem 1.2.1 is presented in the next section.

1.12. How big are the increments of a two-parameter Wiener process?

In order to formulate a possible two-time parameter analogue of Theorem 1.2.1 we introduce the following notations:

Let $\mathbf{R}_T = \mathbf{R}(a_T)$ be the set of rectangles

$$R = [x_1, x_2] \times [y_1, y_2] \quad (0 \leq x_1 < x_2 \leq T^{1/2}, 0 \leq y_1 < y_2 \leq T^{1/2})$$

for which $\lambda(R) = (x_2 - x_1)(y_2 - y_1) \leq a_T$. Let $\mathbf{R}_T^* = \mathbf{R}^*(a_T) \subset \mathbf{R}_T$ be the set of those elements R of $\mathbf{R}_T$ for which $\lambda(R) = a_T$.

Theorem 1.12.1. *Let $W(x, y)$ $(0 \leq x, y < \infty)$ be a Wiener process and let a_T be a non-decreasing function of T satisfying conditions* (i)–(ii) *of Theorem 1.2.1. Then*

(1.12.1) $\quad \overline{\lim}_{T \to \infty} \sup_{R \in \mathbf{R}_T} \beta_T |W(R)| \stackrel{a.s.}{=} \overline{\lim}_{T \to \infty} \sup_{R \in \mathbf{R}_T^*} \beta_T |W(R)| \stackrel{a.s.}{=} 1$

where $\beta_T = (2a_T(\log T a_T^{-1} + \log \log T))^{-1/2}$.

If a_T also satisfies condition (iii) *of Theorem 1.2.1, then*

(1.12.2) $\quad \lim_{T \to \infty} \sup_{R \in \mathbf{R}_T} \beta_T |W(R)| \stackrel{a.s.}{=} \lim_{T \to \infty} \sup_{R \in \mathbf{R}_T^*} \beta_T |W(R)| \stackrel{a.s.}{=} 1.$

It is clear that this theorem can be considered as an analogue of Theorem 1.2.1 in the 2-parameter case. However it does not imply the law of iterated logarithm for the multi-parameter Wiener process in its full richness. Especially the following result does not follow from our Theorem 1.12.1.

Theorem 1.12.2 (Paranjape–Park 1973, Park 1974, Pruitt–Orey 1973, Wichura 1973, Zimmermann 1972). *We have*

$$\overline{\lim}_{\substack{x \to \infty \\ y \to \infty}} \frac{|W(x, y)|}{\sqrt{4xy \log \log xy}} \stackrel{a.s.}{=} 1,$$

that is to say
$$\limsup_{\substack{T\to\infty \\ x\geq T \\ y\geq T}} \frac{|W(x,y)|}{\sqrt{4xy\log\log xy}} \overset{\text{a.s.}}{=} 1.$$

It is somewhat surprising that in this theorem the usual constant 2 of the denominator is replaced by 4. Some explanation of this phenomenon is given in Park (1974) and our Theorems 1.12.3 and 1.12.4 will provide further explanation. We also emphasize that in Theorem 1.12.2 it is assumed that both x and y go to infinity simultaneously. It is natural to ask what happens if this is not the case. Our next theorem is somewhat stronger than Theorem 1.12.2 and gives an answer to the latter question.

Theorem 1.12.3. *For any $\alpha > \tfrac{1}{2}$ we have*

$$(1.12.3) \quad \varlimsup_{T\to\infty} \sup_{(x,y)\in D_T} \frac{|W(x,y)|}{\sqrt{4T\log\log T}} \overset{\text{a.s.}}{=} \varlimsup_{T\to\infty} \sup_{(x,y)\in D_T^*} \frac{|W(x,y)|}{\sqrt{4T\log\log T}} \overset{\text{a.s.}}{=} 1,$$

where
$$D_T = D_T(T^\alpha) = \{(x,y): xy \leq T,\ 0 \leq x \leq T^\alpha,\ 0 \leq y \leq T^\alpha\},$$
$$D_T^* = D_T^*(T^\alpha) = \{(x,y): xy = T,\ 0 \leq x \leq T^\alpha,\ 0 \leq y \leq T^\alpha\}.$$

Applying this theorem for $\alpha = 1$, it can be seen that it is not necessary to assume in Theorem 1.12.2 that both variables go to infinity (cf. also Consequence 1.12.2). In order to see that Theorem 1.12.3 implies Theorem 1.12.2, we note that for $\alpha = 1$ in the former we get $[T, \infty) \times [T, \infty) \subset \subset \bigcup_{U\geq T^2} D_U^*$, and for $\alpha = 0.6$, say, $\bigcup_{U\geq T^3} D_U^* \subset [T, \infty) \times [T, \infty)$. In our next theorem we investigate the question of how the function T^α of Theorem 1.12.3 can be replaced by an arbitrary increasing function b_T. We have

Theorem 1.12.4. *Let $b_T \geq T^{1/2}$ be a non-decreasing function of T and define*

$$\gamma_T = (2T[\log(\log b_T T^{-1/2}+1) + \log\log T])^{-1/2},$$
$$D_T = D_T(b_T) = \{(x,y): xy \leq T,\ 0 \leq x \leq b_T,\ 0 \leq y \leq b_T\},$$
$$D_T^* = D_T^*(b_T) = \{(x,y): xy = T,\ 0 \leq x \leq b_T,\ 0 \leq y \leq b_T\}.$$

Suppose that
(i) *γ_T is a non-increasing function of T,*
(ii) *for any $\varepsilon > 0$ there exists a $\theta_0 = \theta_0(\varepsilon) > 1$ such that*

$$\varlimsup_{k\to\infty} \frac{\gamma_{\theta^{k+1}}}{\gamma_{\theta^k}} \leq 1 + \varepsilon$$

if $1 < \theta \leq \theta_0$.

Then

(1.12.4) $\overline{\lim\limits_{T\to\infty}}\sup\limits_{(x,y)\in D_T}\gamma_T|W(x,y)|\stackrel{a.s.}{=}\overline{\lim\limits_{T\to\infty}}\sup\limits_{(x,y)\in D_T^*}\gamma_T|W(x,y)|\stackrel{a.s.}{=}1.$

If we also have

(iii) $$\lim_{T\to\infty}\frac{\log(\log b_T T^{-1/2}+1)}{\log\log T}=\infty,$$

then

(1.12.5) $\lim\limits_{T\to\infty}\sup\limits_{(x,y)\in D_T}\gamma_T|W(x,y)|\stackrel{a.s.}{=}\lim\limits_{T\to\infty}\sup\limits_{(x,y)\in D_T^*}\gamma_T|W(x,y)|\stackrel{a.s.}{=}1.$

We mention some special cases of Theorem 1.12.4:

1° if $b_T=T^{1/2}$, we get the simplest form of the law of iterated logarithm (the constant in the denominator is the usual 2);

2° if $b_T = T^{1/2}e^{(\log T)^\gamma}$ $(\gamma \geq 0)$,

then $\gamma_T \approx (2(\gamma+1)T\log\log T)^{-1/2}$, that is to say for $\gamma=0$ we get again the law of iterated logarithm with the constant 2 and the constant is increasing as γ is increasing; we get the constant 4 of Theorem 1.12.2 (or Theorem 1.12.3) when $\gamma=1$;

3° if $b_T=e^T$ then $\gamma_T \approx (2T\log T)^{-1/2}$; that is even the order of magnitude of γ_T has changed now. In this case (iii) of Theorem 1.12.4 holds, that is (1.12.5) holds true;

4° if $b_T = e^{e^T}$ then $\gamma_T \approx 2^{-1/2}T^{-1}$.

Clearly, Theorem 1.12.4 is a generalization of Theorem 1.12.3 (and, a fortiori, that of Theorem 1.12.2). However, it is not a generalization of Theorem 1.12.1. Now we formulate our main result, which is a generalization of both Theorems 1.12.4 and 1.12.1.

Theorem 1.12.5 (Csörgő, Révész 1978). *Let $0<a_T\leq T$, $b_T\geq T^{1/2}$ be non-decreasing functions of T and define*

$$\delta_T = (2a_T(\log Ta_T^{-1}+\log(\log b_T a_T^{-1/2}+1)+\log\log T))^{-1/2}.$$

Further let $L_T=L_T(a_T,b_T)$ (resp. $L_T^*=L_T^*(a_T,b_T)$) be the set of rectangles $R=[x_1,x_2]\times[y_1,y_2]\subset D_T(b_T)$ for which $\lambda(R)\leq a_T$ (resp. $\lambda(R)=a_T$).
Suppose that
  (i) *δ_T is a non-increasing function of T,*
  (ii) *Ta_T^{-1} is a non-decreasing function of T,*

(iii) *for any* $\varepsilon > 0$ *there exists a* $\theta_0 = \theta_0(\varepsilon) > 1$ *such that*

$$\varlimsup_{k \to \infty} \frac{\delta_{\theta^k}}{\delta_{\theta^{k+1}}} \leq 1 + \varepsilon$$

if $1 < \theta \leq \theta_0$.

Then

(1.12.6) $\qquad \varlimsup_{T \to \infty} \sup_{R \in L_T} \delta_T |W(R)| \overset{a.s.}{=} \varlimsup_{T \to \infty} \sup_{R \in L_T^*} \delta_T |W(R)| \overset{a.s.}{=} 1.$

If we also have

(iv) $\qquad \lim_{T \to \infty} \frac{\log T a_T^{-1} + \log(\log b_T a_T^{-1/2} + 1)}{\log \log T} = \infty$

then

(1.12.7) $\qquad \lim_{T \to \infty} \sup_{R \in L_T} \delta_T |W(R)| \overset{a.s.}{=} \lim_{T \to \infty} \sup_{R \in L_T^*} \delta_T |W(R)| \overset{a.s.}{=} 1.$

The proof of this Theorem is based on an inequality, which is an analogue of (1.1.2) and is formulated in the following way.

Theorem 1.12.6. *For any* $\varepsilon > 0$ *there exists a* $C = C(\varepsilon) > 0$ *such that*

(1.12.8) $\qquad P\left\{ \sup_{R \in L_T} |W(R)| \geq u a_T^{1/2} \right\}$

$\leq C \dfrac{T}{a_T} (1 + \log T a_T^{-1})(1 + \log b_T a_T^{-1/2}) e^{-u^2/(2+\varepsilon)} \quad (u > 0),$

where $L_T = L_T(a_T, b_T)$ *is the class of rectangles defined in Theorem 1.12.5 and* a_T *and* b_T *also satisfy the conditions of the latter.*

At first we introduce some notations and prove a lemma.
Let $\mu = \mu(T)$ be the smallest integer for which

$$\mu \geq \log b_T a_T^{-1/2}$$

and, for any integer q, let $Q = Q(q) = 2^q$. Define the following sequences of real numbers

$$z_i = z_i(q) = z_i(q, T) = a_T^{1/2} e^{i/Q} \quad (i = 0, \pm 1, \pm 2, \ldots, \pm Q\mu),$$
$$x_j(i) = x_j(i, T) = j z_i Q^{-1} \quad (j = 0, 1, 2, \ldots),$$
$$y_j(i) = y_j(i, T) = j a_T z_i^{-1} Q^{-1} \quad (j = 0, 1, 2, \ldots),$$

and the following rectangles

$$R_i = R_i(q) = R_i(q, 0, 0) = [0, z_i] \times [0, a_T z_i^{-1}],$$
$$R_i(j, l) = R_i(q, j, l) = R_i + (x_j(i), y_l(i)) = \{(x, y): (x - x_j(i), y - y_l(i)) \in R_i\}.$$

Let $L_T^* = L_T^*(q)$ be the set of rectangles $R_i(q, j, l)$ contained in the domain $D_T(b_T)$. For any $R = [x_1, x_2] \times [y_1, y_2] \in L_T$ define the rectangle $R(q) \in L_T^*(q)$ as follows: let $i_0 = i_0(R)$ denote the smallest integer for which: $z_{i_0} \geq x_2 - x_1$ and let $j_0 = j_0(R)$, $l_0 = l_0(R)$ denote the largest integers for which $x_{j_0}(i_0) \leq x_1$, $y_{l_0}(i_0) \leq y_1$ and now let

$$R(q) = R_{i_0}(q, j_0, l_0) = (x_{j_0}(i_0), y_{l_0}(i_0)) + [0, z_{i_0}] \times [0, a_T z_{i_0}^{-1}].$$

Lemma 1.12.1.

(1.12.9) $\quad \text{card } L_T^*(q) \leq 8Q^3 \, T a_T^{-1}(1 + \log T a_T^{-1})(1 + \log b_T a_T^{-1/2})$,

(1.12.10) *for each $R \in L_T^*$ we have $\lambda(R \circ R(q)) \leq 6 a_T Q^{-1}$, where λ is the Lebesgue measure and the operation $\circ$ stands for symmetric difference,*

(1.12.11) $\quad \lambda(R) = a_T \quad \text{for each} \quad R \in L_T^*(q)$.

Proof. At first we evaluate the number of rectangles $R_i(q, j, l)$ belonging to $L_T^*(q)$ for a fixed i. Clearly if $R_i(q, j, l)$ belongs to the set $L_T^*(q)$ then its right-upper vertex belongs to the domain

$$A = \{(x, y): z_i \leq x \leq z_i T a_T^{-1}, \; a_T z_i^{-1} \leq y \leq Tx^{-1}\}.$$

Let $M_T(i)$ be the number of elements of the double array $(x_j(i), y_l(i))$, $(j = 0, 1, 2, \ldots; \; l = 0, 1, 2, \ldots)$, contained in A and $N_T(i)$ be the number of those pairs (j, l) for which we have

$$[x_j(i), x_{j+1}(i)] \times [y_l(i), y_{l+1}(i)] \subset A.$$

Then

$$N_T(i)(x_{j+1}(i) - x_j(i))(y_{l+1}(i) - y_l(i)) = N_T(i) \frac{z_i}{Q} \frac{a_T}{z_i Q}$$

$$\leq \lambda(A) \leq T \log \frac{T}{a_T},$$

that is

$$N_T(i) \leq Q^2 \frac{T}{a_T} \log \frac{T}{a_T}.$$

We also have

$$M_T(i) - N_T(i) \leq \left(z_i \frac{T}{a_T} - z_i\right) \frac{Q}{z_i} + 1 = \frac{T - a_T}{a_T} Q + 1,$$

and hence

$$M_T(i) \leq 2 Q^2 \frac{T}{a_T} \left(\log \frac{T}{a_T} + 1\right).$$

That is for a fixed i the number of rectangles $R_i(q, j, l)$ belonging to $L_T^*(q)$ is not more than $M_T(i)$. Since the number of possible values of i is not more than $2Q\mu + 1 \leq 1 + 2Q(\log b_T a_T^{-1/2} + 1) \leq 4Q(\log b_T a_T^{-1/2} + 1)$, we get (1.12.9).

Now (1.12.10) resp. (1.12.11) simply follow from the definition of $R(q)$ resp. that of $L_T^*(q)$.

In the proof of Theorem 1.12.6 the following result will also be used:

Lemma 1.12.2 (Pruitt, Orey 1973). *Let $R = [x_1, x_2] \times [y_1, y_2]$ be any rectangle and let $S = [s_1, s_2] \times [t_1, t_2]$ ($0 \leq x_1 \leq s_1 < s_2 \leq x_2$, $0 \leq y_1 \leq t_1 < t_2 \leq y_2$). Then we have for any $u > 0$,*

$$P\{\sup_{S \subset R} |W(S)| \geq u\} \leq 4P\{|W(R)| \geq u\}.$$

Now we turn to the

Proof of Theorem 1.12.6. For any $R \in L_T$, the symmetric difference $R(q) \circ R(q+1)$ is the sum of at most 4 rectangles say $R(q) \circ R(q+1) = R^{(1)}(q) + R^{(2)}(q) + R^{(3)}(q) + R^{(4)}(q)$. Denote this class of rectangles $R^{(i)}(q)$ ($i = 1, 2, 3, 4$) by $\tilde{L}_T^*(q)$. Since $R(q) \to R$ as $q \to \infty$ for any R in L_T^*, we have

$$(1.12.12) \quad \sup_{R \in L_T} |W(R)| \leq \sup_{R \in L_T^*} \sup_{S \subset R} |W(S)| + 4 \sum_{i=0}^{\infty} \sup_{R \in \tilde{L}_T^*(q+i)} \sup_{S \subset R} |W(S)|,$$

where S is a rectangle with edges parallel to the coordinate axes.

Then by Lemmas 1.12.1 and 1.12.2 we have

$$(1.12.13) \quad P\{\sup_{R \in L_T^*(q)} \sup_{S \subset R} |W(S)| \geq x a_T^{1/2}\} \leq 4 \operatorname{card} L_T^*(q) e^{-x^2/2}$$

and

$$(1.12.14)$$

$$P\{\sup_{R \in \tilde{L}_T^*(q+i)} \sup_{S \subset R} |W(S)| \geq y_i (6 a_T Q^{-1} 2^{-i})^{1/2}\} \leq 4 \operatorname{card} \tilde{L}_T^*(q+i) e^{-y_i^2/2}.$$

Since $\operatorname{card} \tilde{L}_T^*(q+i) \leq 4 \operatorname{card} L_T^*(q+i)$, by (1.12.12), (1.12.13) and (1.12.14) we get

$$(1.12.15) \quad P\{\sup_{R \in L_T} |W(R)| \geq x a_T^{1/2} + 4 \sum_{i=0}^{\infty} y_i (6 a_T Q^{-1} 2^{-i})^{1/2}\}$$

$$\leq 4 \operatorname{card} L_T^*(q) e^{-x^2/2} + 16 \sum_{i=0}^{\infty} \operatorname{card} L_T^*(q+i) e^{-y_i^2/2}.$$

Choosing $y_i = (6i + x^2)^{1/2}$, we have

$$(1.12.16) \quad xa_T^{1/2} + 4 \sum_{i=0}^{\infty} y_i (6a_T Q^{-1} 2^{-i})^{1/2}$$

$$\leq xa_T^{1/2} \left\{ 1 + 4(6Q^{-1})^{1/2} \sum_{i=0}^{\infty} 2^{-i/2} \right\} + 24 a_T^{1/2} Q^{-1/2} \sum_{i=0}^{\infty} (i 2^{-i})^{1/2}$$

$$\leq xa_T^{1/2}(1 + Q^{-1/2} A) + a_T^{1/2} Q^{-1/2} B \leq (1+\varepsilon) x a_T^{1/2},$$

provided that Q is big enough and $x \geq 1$, where $A = 4\sqrt{6} \sum_{i=0}^{\infty} 2^{-i/2}$ and $B = 24 \sum_{i=0}^{\infty} (i 2^{-i})^{1/2}$; further, by (1.12.9),

$$(1.12.17) \quad 4 \operatorname{card} L_T^*(q) e^{-x^2/2} + 16 \sum_{i=0}^{\infty} \operatorname{card} L_T^*(q+i) e^{-y_i^2/2}$$

$$\leq C T a_T^{-1} (1 + \log T a_T^{-1})(1 + \log b_T a_T^{-1/2}) e^{-x^2/2}.$$

Now given $u > 1$ we let $(1+\varepsilon) x = u$, and (1.12.8) follows from (1.12.15), (1.12.16) and (1.12.17); otherwise, i.e., when $u \leq 1$, (1.12.8) is trivially true.

Proof of Theorem 1.12.5. This will be given in three steps.

Step 1. Let
$$A(T) = \sup_{R \in L_T} \delta_T |W(R)|.$$

Suppose that conditions (i), (ii), (iii) of Theorem 1.12.5 are fulfilled. Then

$$(1.12.18) \quad \varlimsup_{T \to \infty} A(T) \leq 1 \quad \text{a.s.}$$

Proof. By Theorem 1.12.6 we have

$$P\{A(T) \geq 1 + \varepsilon\} \leq C \left(\frac{a_T}{T} \right)^{\varepsilon} (1 + \log T a_T^{-1})(1 + \log b_T a_T^{-1/2})^{-\varepsilon} (\log T)^{-1-\varepsilon}.$$

Let $T_k = \theta^k$ ($\theta > 1$). Then

$$\sum_{k=1}^{\infty} P\{A(T_k) \geq 1 + \varepsilon\} < \infty$$

for every $\varepsilon > 0$, $\theta > 1$, hence, by the Borel–Cantelli lemma,

$$(1.12.19) \quad \varlimsup_{k \to \infty} A(T_k) \leq 1 \quad \text{a.s.}$$

Since $\sup_{R \in L_T} |W(R)|$ is non-decreasing in T and

$$(1.12.20) \quad 1 \leq \delta_{T_{k+1}}^{-1} \sup_{T_k \leq T \leq T_{k+1}} \delta_T = \delta_{T_k}/\delta_{T_{k+1}} \leq 1+\varepsilon$$

for any $\varepsilon>0$ if θ is near enough to 1, (1.12.18) follows from (1.12.19) and (1.12.20).

Step 2. *Suppose that conditions* (i), (ii), (iii) *of Theorem 1.12.5 are satisfied. Then for any* $\varepsilon>0$

(1.12.21) $$\varlimsup_{T\to\infty} \sup_{R\in L_T^*} \delta_T |W(R)| \geq 1-\varepsilon.$$

Proof. First we assume that $\lim \dfrac{a_T}{T} = \varrho < 1$. Given any $0<\varepsilon<1$, define the sequence $\{T_k\}$ by $T_1=1$ and $T_{k-1} = \varepsilon(T_k - a_{T_k})$, $k=2, 3, \ldots$. (The latter definition of T_k is feasible, since, just like in the case of Theorem 1.2.1, $T - a_T$ is a continuous non-decreasing function of T.) Define also $L=L(k)$ to be the largest integer for which we have

$$\frac{T_k^{L+1}}{(T_k - a_{T_k})^L b_{T_k}} \leq b_{T_k} \quad \text{for any given } k,$$

and the rectangles
$$S_i(k) = [x_1(i), x_2(i)] \times [y_1(i), y_2(i)]$$
$$= \left[\left(\frac{T_k - a_{T_k}}{T_k}\right)^{i+1} b_{T_k}, \left(\frac{T_k - a_{T_k}}{T_k}\right)^i b_{T_k}\right] \times \left[\frac{T_{k-1} T_k^{i+1}}{(T_k - a_{T_k})^{i+1} b_{T_k}}, \frac{T_k^{i+1}}{(T_k - a_{T_k})^i b_{T_k}}\right],$$
where $i=0, 1, \ldots, L=L(k)$.

We observe that
$$T_{k-1} = x_1(i) y_1(i) < x_2(i) y_2(i) = T_k,$$
$$0 < x_1(i) < x_2(i) \leq b_{T_k}, \quad 0 < y_1(i) < y_2(i) \leq b_{T_k}, \quad i=0, 1, \ldots, L=L(k).$$
Hence
$$S_i(k) \subset D_{T_k} - D_{T_{k-1}}$$
and, for each k, the $S_i(k)$ are disjoint rectangles. From the definition of T_k ($k=1, 2, \ldots$) it also follows that

(1.12.22) $\quad (1-\varepsilon) a_{T_k} \leq (x_2(i) - x_1(i))(y_2(i) - y_1(i)) \leq a_{T_k} \quad$ for all i.

Whence we have for each k

(1.12.23) $\quad P\{\max_{0\leq i\leq L} \delta_{T_k} |W(S_i(k))| \geq 1-\varepsilon\}$

$\geq 1 - 2\{1 - \Phi(\sqrt{1-\varepsilon}(2(\log T_k a_{T_k}^{-1} + \log(\log b_{T_k} a_{T_k}^{-1/2} + 1) + \log\log T_k))^{1/2})\}^{L+1}$

$\geq 1 - \left\{1 - \left(\dfrac{a_{T_k}}{T_k} \dfrac{1}{\log b_{T_k} a_{T_k}^{-1/2} + 1} \dfrac{1}{\log T_k}\right)^{1-\varepsilon}\right\}^{L+1}$

$\geq 1 - \exp\left\{-\dfrac{a_{T_k}}{T_k} \dfrac{1}{\log b_{T_k} a_{T_k}^{-1/2} + 1} \dfrac{1}{\log T_k}\right)^{1-\varepsilon} (L+1)\right\}.$

It follows from the definition of L that
$$L+1 = L(k)+1 \geqq \text{Const.} \frac{T_k}{a_{T_k}} \log \frac{b_{T_k}^2}{T_k},$$
and a simple calculation shows that the exponential lower bound of (1.12.23) has its minimum at $b_{T_k} = T_k^{1/2}$. Whence, by (1.12.23), we have

(1.12.24)
$$P\{\max_{0 \leqq i \leqq L} \delta_{T_k} |W(S_i(k))| \geqq 1-\varepsilon\} \geqq \text{Const.} \left(\frac{a_{T_k}}{T_k \log T_k}\right)^{1-\varepsilon} \frac{1}{(\log T_k/a_{T_k})^{1-\varepsilon}}.$$

Since $\sum \left(\frac{a_{T_k}}{T_k \log T_k}\right)^{1-\varepsilon} (\log T_k/a_{T_k})^{-(1-\varepsilon)}$ diverges (which can be shown exactly the same way as the divergence of $\sum \left(\frac{a_{T_k}}{T_k \log T_k}\right)^{1-\varepsilon}$ in Step 2 of Theorem 1.2.1), we get (1.12.21) when $\varrho < 1$ by the Borel–Cantelli lemma, (1.12.22) and Step 1 of this proof.

Considering now the case of $\lim_{T \to \infty} a_T/T = \varrho = 1$ (or, equivalently, the case of $a_T = T$), define $T_k = \theta^k$ ($\theta > 1$) and $L = L(k)$ as the largest integer for which we have
$$T_k^{1/2} M^{L+1} = a_{T_k}^{1/2} M^{L+1} \leqq b_{T_k},$$
where $1 < M < \theta$ is a given fixed number. Define also the rectangles
$$S_i(k) = [x_1(i), x_2(i)] \times [y_1(i), y_2(i)]$$
$$= [T_k^{1/2} M^i, T_k^{1/2} M^{i+1}] \times [T_{k-1} T_k^{-1/2} M^{-i}, T_k^{1/2} M^{-i-1}], \quad i = 0, 1, \ldots, L(k).$$
We observe that
$$T_{k-1} = x_1(i) y_1(i) < x_2(i) y_2(i) = T_k,$$
$$0 < x_1(i) < x_2(i) \leqq b_{T_k}, \quad 0 < y_1(i) < y_2(i) \leqq b_{T_k}, \quad i = 0, 1, \ldots, L(k).$$
Hence, $S_i(k) \subset D_{T_k} - D_{T_{k-1}}$, and the $S_i(k)$ are disjoint rectangles for each k. Choosing now M and θ big enough so that M/θ is small enough, it follows that (1.12.22) and (1.12.23) hold again. Our present definition of L gives that
$$L+2 = L(k)+2 \geqq \frac{1}{\log M} \log b_{T_k} a_{T_k}^{-1/2}.$$

From here on this proof continues along the lines of that of $\varrho < 1$ above.

Step 3. Suppose that conditions (i)–(iv) of Theorem 1.12.5 are satisfied. Then for any $\varepsilon > 0$ we have

(1.12.25)
$$\varlimsup_{T \to \infty} \sup_{R \in L_T^*} \delta_T |W(R)| \geqq 1-\varepsilon \quad \text{a.s.}$$

Proof. Let ϱ be as in Step 2 and let $L=L(T)$ be the largest integer for which we have

$$\frac{T^{L+1}}{(T-a_T)^L b_T} < b_T \quad \text{if} \quad \varrho < 1,$$

$$a_T^{1/2} M^{L+1} = T^{1/2} M^{L+1} < b_T \quad \text{if} \quad \varrho = 1.$$

Define the rectangles

$$S_i = S_i(T) = [x_1(i), y_1(i)] \times [x_2(i), y_2(i)]$$

$$= \begin{cases} \left[\left(\frac{T-a_T}{T}\right)^{i+1} b_T, \left(\frac{T-a_T}{T}\right)^i b_T\right] \times \left[0, \frac{T^{i+1}}{(T-a_T)^i b_T}\right] & \text{if} \quad \varrho < 1, \\ [T^{1/2} M^i, T^{1/2} M^{i+1}] \times [0, T^{1/2} M^{-i-1}] & \text{if} \quad \varrho = 1, \end{cases}$$

$$i = 0, 1, \ldots, L = L(T).$$

We observe that

$$0 = x_1(i) y_1(i) < x_2(i) y_2(i) = T$$

$$0 < x_1(i) < x_2(i) \leq b_T$$

$$0 = y_1(i) < y_2(i) \leq b_T, \quad i = 0, 1, \ldots, L = L(T),$$

and (1.12.22) also holds.

Since the sets $S_i(T)$ ($i=0, 1, 2, \ldots, L$) are disjoint, we have

$$P\{\max_{0 \leq i \leq L} \delta_T |W(S_i(T))| \leq 1-\varepsilon\}$$

$$\leq (1 - \Phi(\sqrt{1-\varepsilon}[2(\log T a_T^{-1} + \log(\log b_T a_T^{-1/2} + 1) + \log \log T)]^{1/2}))^{L+1},$$

and from here on the proof is completed along the lines of Step 3 of Theorem 1.2.1.

In the sequel we are going to need such a version of Theorem 1.12.5 where the symmetric domain $D_T = D_T(b_T)$ is replaced by the non-symmetric one:

$$D_{2,T} = D_{2,T}(b_T^{(1)}, b_T^{(2)}) = \{(x, y): xy \leq T, \ 0 < x \leq b_T^{(1)}, 0 < y \leq b_T^{(2)}\}.$$

Hence, we formulate

Theorem 1.12.7. *Let* $0 < a_T \leq T$, $b_T^{(1)} b_T^{(2)} \geq T$ *be non-decreasing functions of* T *and define*

$$\delta_{2,T} = \left(2 a_T \left(\log \frac{T}{a_T} + \log\left(\log \sqrt{\frac{b_T^{(1)} b_T^{(2)}}{a_T}} + 1\right) + \log \log T\right)\right)^{-1/2}.$$

Further let $L_{2,T}=L_{2,T}(a_T, b_T^{(1)}, b_T^{(2)})$ (resp. $L_{2,T}^*=L_{2,T}^*(a_T, b_T^{(1)}, b_T^{(2)})$) *be the set of rectangles* $R=[x_1, x_2]\times[y_1, y_2]\subset D_{2,T}$ *for which* $\lambda(R)\leq a_T$ (resp. $\lambda(R)=a_T$). *Suppose that conditions* (i), (ii) *and* (iii) *of Theorem 1.12.5 hold when* δ_T *is replaced by* $\delta_{2,T}$ *in them. Then*

(1.12.26) $\quad \varlimsup_{T\to\infty} \sup_{R\in L_{2,T}} \delta_{2,T}|W(R)| \overset{a.s.}{=} \varlimsup_{T\to\infty} \sup_{R\in L_{2,T}^*} \delta_{2,T}|W(R)| \overset{a.s.}{=} 1.$

If we also have

(iv) $\quad \lim_{T\to\infty} \dfrac{\log T a_T^{-1}+\log\left(\log\sqrt{\dfrac{b_T^{(1)} b_T^{(2)}}{a_T}}+1\right)}{\log\log T}=\infty,$

then

(1.12.27) $\quad \lim_{T\to\infty} \sup_{R\in L_{2,T}} \delta_{2,T}|W(R)| \overset{a.s.}{=} \lim_{T\to\infty} \sup_{R\in L_{2,T}^*} \delta_{2,T}|W(R)| \overset{a.s.}{=} 1.$

The proof of this theorem is similar to that of Theorem 1.12.5 and will not be repeated here.

In our previous theorem we considered the class $L_{2,T}$ (resp. $L_{2,T}^*$) of *all* rectangles belonging to the set $D_{2,T}$ having an area less than (resp. equal to) a_T. Now, we replace the above class $L_{2,T}$ (resp. $L_{2,T}^*$) by the class $L_{3,T}$ (resp. $L_{3,T}^*$) containing rectangles of some special shape *only*. We let

$L_{3,T} = L_{3,T}(a_T, b_T^{(1)}, b_T^{(2)}, c_T^{(1)}, c_T^{(2)})$, (resp. $L_{3,T}^* = L_{3,T}^*(a_T, b_T^{(1)}, b_T^{(2)}, c_T^{(1)}, c_T^{(2)})$)

be the set of rectangles $R=[x_1, x_2]\times[y_1, y_2]\subset D_{2,T}$ for which $x_2-x_1\leq c_T^{(1)}$, $y_2-y_1\leq c_T^{(2)}$ and $\lambda(R)\leq a_T$ (resp. $\lambda(R)=a_T$).

Theorem 1.12.8. *Let* $0<a_T\leq T$, $b_T^{(1)}b_T^{(2)}\geq T$, $c_T^{(1)}\leq b_T^{(1)}$, $c_T^{(2)}\leq b_T^{(2)}$, $c_T^{(1)}c_T^{(2)}\leq a_T$ *be non-decreasing functions of* T *and define*

$$\delta_{3,T} = \left(2a_T\left(\log\frac{T}{a_T}+\log\left(\log\sqrt{\frac{c_T^{(1)} c_T^{(2)}}{a_T}}+1\right)+\log\log T\right)\right)^{-1/2}.$$

Suppose that conditions (i), (ii) *and* (iii) *of Theorem 1.12.5 hold when* δ_T *is replaced by* $\delta_{3,T}$ *in them. Then*

(1.12.28) $\quad \varlimsup_{T\to\infty} \sup_{R\in L_{3,T}} \delta_{3,T}|W(R)| \overset{a.s.}{=} \varlimsup_{T\to\infty} \sup_{R\in L_{3,T}^*} \delta_{3,T}|W(R)| \overset{a.s.}{=} 1.$

If we also have

(iv) $$\lim_{T\to\infty} \frac{\log T a_T^{-1} + \log\left(\log\sqrt{\frac{c_T^{(1)} c_T^{(2)}}{a_T}+1}\right)}{\log\log T} = \infty,$$

then

(1.12.29) $$\limsup_{T\to\infty}\sup_{R\in L_{3,T}} \delta_{3,T}|W(R)| \overset{a.s.}{=} \limsup_{T\to\infty}\sup_{R\in L^*_{3,T}} \delta_{3,T}|W(R)| \overset{a.s.}{=} 1.$$

The proof of this theorem is again similar to that of Theorem 1.12.5 and will be omitted.

We note that in case of $c_T^{(1)}=b_T^{(1)}$ and $c_T^{(2)}=b_T^{(2)}$, Theorem 1.12.8 implies Theorem 1.12.7. Also, if $b_T^{(1)}=b_T^{(2)}=b_T$, then Theorem 1.12.7 reduces to Theorem 1.12.5.

Choosing specific forms for the parameters of the above theorems of this Section, we list a few consequences of them.

Corollary 1.12.1. *Let b_T be a function of T satisfying the conditions of Theorem 1.12.4 and define*

$$D_T = \bar{D}_T(b_T) = \bigcup_{S\geq T} D_S^*$$

with D_S^* as in Theorem 1.12.3. Then

$$\limsup_{T\to\infty}\sup_{(x,y)\in \bar{D}_T} \gamma_{xy}|W(x,y)| \overset{a.s.}{=} 1.$$

The domain $\bar{D}_T$ seems to be a rather artificial one. However using this corollary one can get similar results for many concrete domains. As an example we give:

Corollary 1.12.2. *Let*

$$E_U = \{(x,y): x\geq 1,\ y\geq U\}.$$

Then

(1.12.30) $$\limsup_{U\to\infty}\sup_{(x,y)\in E_U} \frac{|W(x,y)|}{\sqrt{4xy\log\log xy}} \overset{a.s.}{=} 1,$$

or equivalently

(1.12.30*) $$\varlimsup_{y\to\infty}\sup_{x\geq 1} \frac{|W(x,y)|}{\sqrt{4xy\log\log xy}} \overset{a.s.}{=} 1.$$

Proof. (1.12.30) follows from Corollary 1.12.1 and from the trivial relationship

$$\bar{D}_T(T^{3/4}) \subset E_U \subset \bar{D}_U(U)$$

if T is big enough (for example if $T\geq U^4$).

Corollary 1.12.3. Let $0 < a_T \leq T$ and T/a_T be non-decreasing functions. Then

$$(1.12.31) \quad \varlimsup_{T \to \infty} \sup_{0 \leq y \leq T - a_T} \sup_{0 \leq s \leq a_T} \sup_{0 \leq x_1 < x_2 \leq 1} \frac{|W(x_2, y+s) - W(x_1, y+s) - W(x_2, y) + W(x_1, y)|}{\sqrt{2 a_T \left(\log \frac{T}{a_T} + \log \log T \right)}} \stackrel{a.s.}{=} 1.$$

If, in addition, we also have $\log(T/a_T)/\log \log T \to \infty$ as $T \to \infty$, then the $\varlimsup_{T \to \infty}$ in (1.12.31) can be replaced by $\lim_{T \to \infty}$.

Proof. Let $b_T^{(2)} = T$, $b_T^{(1)} = 1$, $c_T^{(1)} = 1$ and $c_T^{(2)} = a_T$ in Theorem 1.12.8.

Corollary 1.12.4. Let $0 < a_T \leq T$ and T/a_T be non-decreasing functions. Then

$$(1.12.32) \quad \varlimsup_{T \to \infty} \sup_{0 \leq y \leq T - a_T} \sup_{0 \leq s \leq a_T} \sup_{0 \leq x \leq 1} \frac{|W(x, y+s) - W(x, y)|}{\sqrt{2 a_T \left(\log \frac{T}{a_T} + \log \log T \right)}} \stackrel{a.s.}{=} 1.$$

If, again, $\log(T/a_T)/\log \log T \to \infty$ as $T \to \infty$, then the $\varlimsup_{T \to \infty}$ in (1.12.32) can be replaced by $\lim_{T \to \infty}$.

Proof. Apply Corollary 1.12.3 with $x_1 = 0$, $x_2 = x$, and Theorem 1.2.1 to the process $W(1, y)$.

Corollary 1.12.5. Let $f(y)$ be a non-decreasing function of y tending to ∞, define $g(x) = x f(x)$ and

$$(1.12.33) \quad \delta(u) = \left(2u \left(\log \left(\log \frac{u}{\text{inv } g(u)} + 1 \right) + \log \log u \right) \right)^{-1/2}.$$

Assume that $\delta(u)$ is a non-decreasing function of u and for any $\varepsilon > 0$ there exists a $\theta_0 = \theta_0(\varepsilon) > 1$ such that

$$\varlimsup_{k \to \infty} \frac{\delta_{\theta^k}}{\delta_{\theta^{k+1}}} \leq 1 + \varepsilon \quad \text{if} \quad 1 < \theta \leq \theta_0.$$

Then

$$(1.12.34) \quad \varlimsup_{y \to \infty} \sup_{1 \leq x \leq f(y)} \delta(xy) |W(x, y)| \stackrel{a.s.}{=} 1,$$

or, equivalently,

(1.12.35) $$\lim_{T\to\infty}\sup_{(x,y)\in V_T} \delta(xy)|W(x,y)| \stackrel{\text{a.s.}}{=} 1,$$

where $V_T = \{(x,y): 1 \leq x \leq f(y), y \geq T\}$.

Proof. Let $b_T^{(2)} = T$, $b_T^{(1)} = T/\text{inv } g(T)$, $a_T = T$. Then the appropriate conditions of Theorem 1.12.7 are satisfied and we have also

$$\bigcup_{u \geq Tf(T)} D_{2,u}(b_u^{(1)}, b_u^{(2)}) \subset V_T \subset \bigcup_{u \geq T} D_{2,u}(b_u^{(1)}, b_u^{(2)}).$$

Whence we have (1.12.35) by Theorem 1.12.7.

1.13. A continuity modulus of $W(x,y)$

In this section an analogue of Theorem 1.1.1 is given.

Theorem 1.13.1 (Pruitt, Orey 1973). *Let $\mathbf{R}(h)$ (resp. $\mathbf{R}^*(h)$) be the set of rectangles*

$$R = [x_1, x_2] \times [y_1, y_2] \quad (0 \leq x_1 < x_2 \leq 1;\ 0 \leq y_1 < y_2 \leq 1)$$

with $\lambda(R) \leq h$ (resp. $\lambda(R) = h$). Then

$$\lim_{h \to 0} \sup_{R \in \mathbf{R}(h)} \frac{|W(R)|}{\sqrt{2h \log 1/h}}$$

$$\stackrel{\text{a.s.}}{=} \lim_{h \to 0} \sup_{R \in \mathbf{R}^*(h)} \frac{|W(R)|}{\sqrt{2h \log 1/h}} \stackrel{\text{a.s.}}{=} 1.$$

The proof of this theorem is based on the following analogue of Lemma 1.1.1, which is a more natural and stronger analogue of the latter than Lemma 1.11.1 is.

Lemma 1.13.1. *For any $\varepsilon > 0$ there exists a constant $C = C(\varepsilon) > 0$ such that the inequality*

(1.13.1) $$P\left(\sup_{R \in \mathbf{R}(h)} |W(R)| \geq v h^{1/2}\right) \leq C h^{-1} (\log h^{-1})^2 e^{-\frac{v^2}{2+\varepsilon}}$$

holds for every positive v and $h < 1$.

Proof. Choosing $b_T = T^{1/2}$ and $a_T = Th$, this lemma follows from Theorem 1.12.6 and from the following trivial

Observation. For any fixed $T > 0$ we have

$$\{W(x, y); \ 0 \leq x \leq T^{1/2}, 0 \leq y \leq T^{1/2}\} \stackrel{\mathscr{D}}{=}$$
$$\{T^{1/2} W(xT^{-1/2}, yT^{-1/2}); \ 0 \leq x \leq T^{1/2}, 0 \leq y \leq T^{1/2}\}.$$

The proof of Theorem 1.13.1 follows the lines of that of Theorem 1.1.1 and will not be presented here.

1.14. The limit points of $W(x, y)$ as $y \to \infty$

Several extensions of Theorem 1.3.2 can be formulated in the multivariate case (cf. Wichura 1973). We concentrate on one of these, which will be applied when investigating similar properties of the empirical process (cf. Chapter 5).

Theorem 1.14.1. *Consider the process*

$$\xi_y(x) = \frac{W(x, y)}{\sqrt{2y \log \log y}} \quad (0 \leq x \leq 1, y \geq 3)$$

as a function of y, taking values in $C(0, 1)$. Then $\xi_y(x)$ ($y \to \infty$) is relatively compact in $C(0, 1)$ with probability one and the set of its limit points is $\mathscr{S}$.

Here, $\mathscr{S}$ is again that set of absolutely continuous functions which was defined in Section 1.3.

The proof of Theorem 1.14.1 will be based on a number of lemmas.

Lemma 1.14.1. *Let $\{X_i\}$ and $\{Y_i\}$ be two sequences of r.v. with $EX_i = EY_i = 0$, and assume that X_{n+1} and Y_{n+1} are both independent of $(X_1, \ldots, X_n, Y_1, \ldots, Y_n)$. Put $S_n = X_1 + \ldots + X_n$, $T_n = Y_1 + \ldots + Y_n$ and $M_n = \max(|S_n|, |T_n|)$. Then the sequence $\{M_n\}$ is a sub-martingale sequence, i.e., we have*

$$E(M_{n+1} | M_1, \ldots, M_n) \geq M_n \quad \text{a.s.}$$

Proof. It suffices to show that

$$E(M_{n+1} | X_1, \ldots, X_n, Y_1, \ldots, Y_n) \geq M_n \quad \text{a.s.,}$$

since the σ-algebra generated by $X_1, \ldots, X_n, Y_1, \ldots, Y_n$ is larger than that generated by $M_1, \ldots, M_n$. We have

$$E(M_{n+1}|X_1, \ldots, X_n, Y_1, \ldots, Y_n)$$
$$= E\{\max(|S_n+X_{n+1}|, |T_n+Y_{n+1}|)|X_1, \ldots, X_n, Y_1, \ldots, Y_n)\}$$
$$\geq E(|S_n+X_{n+1}||X_1, \ldots, X_n, Y_1, \ldots, Y_n)$$
$$= E(|S_n+X_{n+1}||S_n) \geq |E(S_n+X_{n+1}|S_n)| = |S_n| \quad \text{a.s.}$$

One shows similarly, that

$$E(M_{n+1}|X_1, \ldots, X_n, Y_1, \ldots, Y_n) \geq |T_n| \quad \text{a.s.},$$

and our lemma is proved.

Lemma 1.14.2. *For every $\varepsilon > 0$ there exists an $A = A(\varepsilon) > 0$ so that for any $u > 0$ we have*

$$(1.14.1) \quad P\left\{\sup_{1 \leq n \leq N} \sup_{0 \leq x \leq 1-h} \sup_{0 \leq t \leq h} \frac{|W(x+t, n) - W(x, n)|}{\sqrt{N}} \geq u\sqrt{h}\right\}$$
$$\leq \frac{A}{h} e^{-\frac{u^2}{2+\varepsilon}}$$

where n and N run over the positive integers.

Proof. First we observe that, by a straightforward generalization of Lemma 1.14.1,

$$M_n = \sup_{0 \leq x \leq 1-h} \sup_{0 \leq t \leq h} |W(x+t; n) - W(x; n)|$$

is a sub-martingale, that is to say

$$E(M_{n+1}|M_n, M_{n-1}, \ldots, M_1) \geq M_n \quad \text{a.s.}$$

Consequently, $\exp\left\{\frac{tM_n^2}{hN}\right\}$ $(t > 0)$ is also a submartingale. Now observe that

$$(1.14.2) \quad P\left\{\sup_{1 \leq n \leq N} M_n \geq u\sqrt{hN}\right\} = P\left\{\sup_{1 \leq n \leq N} \exp\left\{\frac{tM_n^2}{hN}\right\} \geq e^{tu^2}\right\}$$
$$\leq e^{-tu^2} E \exp\left\{\frac{tM_N^2}{hN}\right\}$$

by the sub-martingale inequality (cf. Doob, 1953, p. 314). Also by Lemma

1.1.1, $E \exp\left\{\dfrac{tM_n^2}{hN}\right\}$ is bounded above whenever $t<\tfrac{1}{2}$, and hence we have our statement.

Lemma 1.14.3. *With the notation of Lemma 1.14.2 we have*

(1.14.3) $$\varlimsup_{N\to\infty} \frac{M_N}{\sqrt{2Nh\log\log N}} \stackrel{a.s.}{=} 1,$$

for any $0<h<1$.

Proof. First we prove that

(1.14.4) $$\varlimsup_{N\to\infty} \frac{M_N}{\sqrt{2Nh\log\log N}} \leq 1 \quad \text{a.s.}$$

Let $N_k=[\theta^k]$, $\theta>1$, and $u=\sqrt{2(1+\delta)\log\log N}$. Then, by Lemma 1.14.2 for any $\delta>0$, we have that the series

$$\sum_{k=1}^{\infty} P\{\sup_{1\leq n\leq N_k} M_n \geq \sqrt{2(1+\delta)\log\log N_k}\sqrt{N_k h}\}$$

converges. The desired estimation for $N\in[N_{k-1}, N_k]$ is carried out the usual way, taking θ near to 1. Whence (1.14.4) is proved. The converse inequality follows by applying (1.3.2) to the process $\{W(h,n)/\sqrt{h}; n=1,2,...\}$.

The next lemma is due to Helen Finkelstein (1971).

Lemma 1.14.4 (Finkelstein 1971). *Let $Z_1, Z_2, \ldots$ be independent identically distributed random vectors with values in d-dimensional Euclidean space R^d with $EZ_1=0$ and assume that the components of Z_i ($i=1, 2, \ldots$) are independent $\mathcal{N}(0, 1)$ random variables. Let*

$$U_n = \frac{\sum_{i=1}^{n} Z_i}{\sqrt{2n\log\log n}}.$$

Then with probability 1 *the sequence* $\{U_n\}$ *is relatively compact and the set B_d of its limit points is the d-dimensional unit ball*

$$B_d = \{x\in R^d: \|x\| \leq 1\}$$

where $\|\cdot\|$ *is the Euclidean norm in* R^d.

Proof. This lemma is true if $d=1$ (see Theorem 1.3.1 and Remark 1.3.1). We prove it for $d=2$. For higher dimensions the proof is immediate.

Let $Z_i = (Z_{i1}, Z_{i2})$ $(i=1, 2, \ldots)$ and α, β be real numbers such that $\alpha^2 + \beta^2 = 1$. Then $\alpha Z_{i1} + \beta Z_{i2}$ are independent $\mathcal{N}(0, 1)$ r.v. So (applying Theorem 1.3.1) the set of limit points of the sequence

$$\left\{ \binom{\alpha}{\beta} U_n \right\} = \frac{\sum_{i=1}^n (\alpha Z_{i1} + \beta Z_{i2})}{\sqrt{2n \log \log n}}$$

is the interval $[-1, +1]$. This implies that the set of limit points B_2 of the sequence $\{U_n\}$ is a subset of the unit circle and the boundary of the unit circle belongs to B_2.

Let now $\{Z_{i3}\}$ be a sequence of independent $\mathcal{N}(0, 1)$ r.v. assumed to be independent from the given sequence $\{Z_i\}$. Let $Z_i^* = (Z_{i1}, Z_{i2}, Z_{i3})$ and

$$U_n^* = \frac{\sum_{i=1}^n Z_i^*}{\sqrt{2n \log \log n}}.$$

In the same way as above one can prove that the set of limit points of $\{U_n^*\}$ is a subset of the unit sphere of R^3 which contains the boundary of this sphere. This fact in itself already implies that B_2 is equal to the unit circle of R^2 as stated in the lemma.

The above Lemmas 1.14.3 and 1.14.4 play the same role in the proof of Theorem 1.14.1 as Corollary 1.2.2 and Proposition 1.3.1 do in the proof of Theorem 1.3.2 respectively.

In the same way as we proved Theorem 1.3.2, we can now prove that $\xi_n(x)$ is relatively compact in $C(0, 1)$ with probability one, and the set of its limit points is $\mathscr{S}$. This, in turn, easily implies Theorem 1.14.1 (cf. Theorem 1.3.2* vs. Theorem 1.3.2).

For later use we present a further analogue of Theorem 1.1.1.

Theorem 1.14.2 (Chan 1977). *Let $\{h_n\}$ be a sequence of positive numbers for which*

(i) $$\lim_{n \to \infty} \frac{\log h_n^{-1}}{\log \log n} = \infty.$$

Then

(1.14.5) $$\lim_{n \to \infty} \sup_{0 \leq t \leq 1 - h_n} \gamma_n |W(t + h_n, n) - W(t, n)| \stackrel{a.s.}{=} 1$$

and

(1.14.6) $$\lim_{n\to\infty} \sup_{0\leq t\leq 1-h_n} \sup_{0\leq s\leq h_n} \gamma_n |W(t+s,n)-W(t,n)| \stackrel{a.s.}{=} 1,$$

where $\gamma_n = (2nh_n \log 1/h_n)^{-1/2}$.

Remark 1.2.1, regarding the question of omitting the absolute value sign and changing the sup to inf, holds true in this case too.

The proof of this theorem and that of Theorem 1.2.1 are quite similar. Hence only the main steps will be presented here.

Proof of Theorem 1.14.2.

Step 1. For any $\varepsilon > 0$ and $\theta = \theta(\varepsilon) > 1$ let

$$A(k) = \sup_{[\theta^k]\leq n < [\theta^{k+1}]} \sup_{0\leq t\leq 1-h_{[\theta^k]}} \sup_{0\leq s\leq h_{[\theta^k]}} \gamma_{[\theta^k]} |W(t+s,n)-W(t,n)|.$$

Then

(1.14.7) $$\varlimsup_{k\to\infty} A(k) \leq 1+\varepsilon \quad a.s.$$

provided θ is near enough to one.

Proof. By Lemma 1.14.2 and condition (i) we have

$$\sum_{k=1}^{\infty} P(A(k) \geq \sqrt{1+\varepsilon}) \leq \sum_{k=1}^{\infty} h_{[\theta^k]}^{\varepsilon/2} < \infty$$

and this proves (1.14.7).

Step 2. Let

$$B(n) = \max_{0\leq k\leq [1/h_n]} \gamma_n |W((k+1)h_n, n) - W(kh_n, n)|.$$

Then

(1.14.8) $$\lim_{n\to\infty} B(n) \geq 1 \quad a.s.$$

Proof. Clearly we have

$$\sum_{n=1}^{\infty} P(B(n) \leq \sqrt{1-\varepsilon}) \leq \sum_{n=1}^{\infty} \left(1 - \frac{h_n^{1-\varepsilon}}{6\sqrt{\log 1/h_n}}\right)^{1/h_n} < \infty$$

and this proves (1.14.8).

1.15. The Kiefer process

Let $W(x, y)$ be a two-parameter Wiener process. A Kiefer process
$$\{K(x, y); \ 0 \leq x \leq 1, \ 0 \leq y < \infty\}$$
is defined by
$$K(x, y) = W(x, y) - xW(1, y).$$

This definition immediately implies the following properties of a Kiefer process:

(i) for any $0 < x_0 < 1$
$$W(y) = \frac{K(x_0, y)}{\sqrt{x_0(1-x_0)}} \quad (y \geq 0)$$
is a Wiener process,

(ii) for any $y_0 > 0$
$$B(x) = \frac{K(x, y_0)}{\sqrt{y_0}} \quad (0 \leq x \leq 1)$$
is a Brownian bridge,

(iii) $B_n(x) = K(x, n) - K(x, n-1)$ $(0 \leq x \leq 1; \ n = 1, 2, \ldots)$ is a sequence of independent Brownian bridges,

(iv) $EK(x, y) = 0$ and the covariance function of $K(x, y)$ is
$$EK(x_1, y_1)K(x_2, y_2) = (x_1 \wedge x_2 - x_1 x_2)(y_1 \wedge y_2),$$

(v) the sample path functions of $K(x, y)$ are continuous with probability 1,

(vi) $\quad W(x, y) = (x+1) K\left(\frac{x}{x+1}, y\right), \quad x \geq 0, y \geq 0,$

(vii) $\quad K(x, y) = (1-x) W\left(\frac{x}{1-x}, y\right), \quad 0 \leq x < 1, y \geq 0.$

In establishing (vi) and (vii), we should also use (1.4.4) and (1.4.5). Consider the process
$$\eta_y(x) = \frac{K(x, y)}{\sqrt{2y \log \log y}} \quad (0 \leq x \leq 1, y \geq 3)$$
as a function of y taking values in $C(0, 1)$. Then we have

Theorem 1.15.1. *The process $\{\eta_y(x)\}$ is relatively compact in $C(0, 1)$ with probability 1 and the set of its limit points (as $y \to \infty$) is $\mathscr{F}$ where*

$\mathscr{F} \subset C(0, 1)$ *is the set of absolutely continuous functions* f *for which*

$$f(0) = f(1) = 0 \quad \text{and} \quad \int_0^1 (f'(x))^2 \, dx \leq 1.$$

Proof. This theorem is a straight consequence of Theorem 1.14.1. From the latter theorem we get

Corollary 1.15.1.

$$\varlimsup_{y \to \infty} \frac{\sup_{0 \leq x \leq 1} |K(x, y)|}{\sqrt{y \log \log y}} \stackrel{\text{a.s.}}{=} 1/\sqrt{2}.$$

Theorem 1.15.2. *The statements* (1.14.5) *and* (1.14.6) *of Theorem 1.14.2 remain true if* $W(x, n)$ *is replaced by* $K(x, n)$ *in them.*

Proof. This theorem is a straight consequence of Theorem 1.14.2.

Applying property (vi) (or (vii)) of the Kiefer process and (1.12.30*), we immediately get the following analogue of the latter.

Corollary 1.15.2.

(1.15.1) $$\varlimsup_{y \to \infty} \sup_{0 < x < 1} \frac{|K(x, y)|}{\sqrt{4x(1-x) y \log \log y / x(1-x)}} \stackrel{\text{a.s.}}{=} 1.$$

On the other hand, Theorem 1.15.1 implies

Corollary 1.15.3. *For any* $0 < \varepsilon < \tfrac{1}{2}$ *we have*

(1.15.2) $$\varlimsup_{y \to \infty} \sup_{\varepsilon < x < 1-\varepsilon} \frac{|K(x, y)|}{\sqrt{2x(1-x) y \log \log y}} \stackrel{\text{a.s.}}{=} 1.$$

Comparing (1.15.1) and (1.15.2), it is natural to ask: how does a Kiefer process behave on an interval $\varepsilon_y < x < 1 - \varepsilon_y$, when ε_y is a non-increasing positive function? An answer to this question is our next corollary which follows from Corollary 1.12.5.

Corollary 1.15.4. *Let* $0 < \varepsilon_y < \tfrac{1}{2}$ *be a non-increasing function of* y *and define* $f(y) = \dfrac{1}{\varepsilon_y} - 1$ *and* $g(x) = xf(x)$. *Then*

(1.15.3) $$\varlimsup_{y \to \infty} \sup_{\varepsilon_y \leq x \leq 1-\varepsilon_y} \left\{ 2yx(1-x) \left[\log \left(\log \frac{y}{x(1-x) \operatorname{inv} g\left(\frac{y}{x(1-x)}\right)} + 1 \right) \right. \right.$$

$$\left. \left. + \log \log \frac{y}{x(1-x)} \right] \right\}^{-1/2} |K(x, y)| \stackrel{\text{a.s.}}{=} 1.$$

Especially if $\varepsilon_y = e^{-(\log y)^\gamma}$ $(0 \leq \gamma \leq 1)$, then

(1.15.4) $\varlimsup\limits_{y \to \infty} \sup\limits_{\varepsilon_y \leq x \leq 1-\varepsilon_y} \left(2yx(1-x)(\gamma+1)\log\log\dfrac{y}{x(1-x)}\right)^{-1/2} |K(x,y)| \stackrel{\text{a.s.}}{=} 1,$

and if $\varepsilon_y = dy^{-1}\log\log y$, $d > 0$, then

(1.15.5) $\varlimsup\limits_{y \to \infty} \sup\limits_{\varepsilon_y \leq x \leq 1-\varepsilon_y} \left(4yx(1-x)\log\log\dfrac{y}{x(1-x)}\right)^{-1/2} |K(x,y)| \stackrel{\text{a.s.}}{=} 1.$

Supplementary remarks

Section 1.0. A Wiener process is frequently called a Brownian process or Brownian motion. The latter terminology was used only in a physical sense by us and the idealized version is called the Wiener process throughout.

Section 1.1. In the definition of a Wiener process condition (i) can be replaced by the weaker condition

(i*) $\qquad W(0) = 0, \quad EW(t) = 0, \quad EW^2(t) = t \quad (t \geq 0).$

For a proof of this fact see for example Ito's book (1960).

A Wiener process generates a measure on the Borel sets of $C(0, 1)$ in a natural way. Several authors consider this measure as the Wiener process.

Our inequality (1.1.1) is far from being best possible. Our method is capable of producing also the following stronger inequality

(S.1.1.1) $P\{\sup\limits_{0 \leq t \leq 1-h} \sup\limits_{0 \leq s \leq h} |W(t+s) - W(t)| \geq vh^{1/2}\} \leq Cv^4 h^{-1} e^{-v^2/2}$

where C is a positive constant. If we are interested only in the distribution of $\sup\limits_{0 \leq t \leq 1-h} |W(t+h) - W(t)|$, then a theorem of Qualls and Watanabe (1972) implies the sharper inequality

(S.1.1.2) $P\{\sup\limits_{0 \leq t \leq 1-h} |W(t+h) - W(t)| \geq vh^{1/2}\} \leq Cvh^{-1} e^{-v^2/2}$

with a positive constant C (cf. also Theorem 1.5.5).

In Lévy's books (1937, 1948) a little bit weaker form of Theorem 1.1.1 is formulated. For this form we refer to Orey and Taylor (1974) or Pruitt and Orey (1973). For describing the behaviour of $\sup\limits_{0 \leq t \leq 1-h} (W(t+h) - W(t))$ in a more exact manner than as it is done in Theorem 1.1.1, we refer to the so-called Chung, Erdős, Sirao (1959) test, cf. also Révész (1979a).

Another way of describing the increments of a Wiener process is via the following quadratic variation theorems:

Theorem S. 1.1.1 (Lévy 1940, Baxter 1956).

(S.1.1.3) $$\lim_{n\to\infty} \sum_{k=1}^{2^n} (W(k/2^n) - W((k-1)/2^n))^2 \stackrel{a.s.}{=} 1.$$

Theorem S. 1.1.2 (Dudley 1973). *Let $0 = x_0^{(n)} < x_1^{(n)} < \ldots < x_{k_n}^{(n)} = 1$ be a sequence of partitions of the unit interval with $k_n \nearrow \infty$ and assume that*

$$\max_{0 \leq i \leq k_n - 1} |x_{i+1}^{(n)} - x_i^{(n)}| = o(1/\log n).$$

Then

(S.1.1.4) $$\lim_{n\to\infty} \sum_{k=1}^{k_n - 1} (W(x_{i+1}^{(n)}) - W(x_i^{(n)}))^2 \stackrel{a.s.}{=} 1.$$

Section 1.2. The statements of (1.2.3) were also proved by Lai (1973) under somewhat stronger restrictions on a_T.

Applying the method of proof of Theorem 1.2.1, the following, somewhat more general theorem can be also proved.

Theorem S. 1.2.1. *Let $0 < a_T \leq b_T$ and assume that b_T/a_T is a non-decreasing function of T. Assume also that one of the following three conditions holds true:*

(S.1.2.1) $\quad b_T \nearrow +\infty, \quad a_T$ *is non-decreasing,*

(S.1.2.2) $\quad b_T/a_T \nearrow \infty$ *and* a_T, b_T *are non-increasing,*

(S.1.2.3) $\quad b_T \searrow 0, \quad a_T \searrow 0,$

as $T \to \infty$. Then

(S.1.2.4) $$\overline{\lim_{T\to\infty}} \sup_{0 \leq t \leq b_T - a_T} \sup_{0 \leq s \leq a_T} \gamma_T |W(t+s) - W(t)|$$
$$\stackrel{a.s.}{=} \overline{\lim_{T\to\infty}} \sup_{0 \leq t \leq b_T - a_T} \gamma_T |W(t+a_T) - W(t)| \stackrel{a.s.}{=} 1,$$

where $\gamma_T = \left(2a_T \left(\log \frac{a_T}{b_T} + \log(|\log b_T| + 1)\right)\right)^{-1/2}$. If we also have $\log\left(\frac{b_T}{a_T}\right) \cdot (\log(|\log b_T| + 1))^{-1} \to \infty$ as $T \to \infty$, then (S.1.2.4) holds with $\lim_{T\to\infty}$ instead of $\overline{\lim}_{T\to\infty}$.

We note that this theorem is a generalization of our Theorem 1.2.1 and that it also contains two classical theorems, namely Theorems 1.1.1 and

1.3.3. Indeed, if $b_T=T$ we get Theorem 1.2.1, if $b_T=1$ and $a_T=1/T$ then Theorem 1.1.1 follows and, when $b_T=a_T=1/T$, then Theorem 1.3.3 is obtained.

Section 1.3. In the book of Riesz and Sz.-Nagy (1955) Lemma 1.3.1 is formulated in a slightly different way. Using the same method of proof the quoted form can also be obtained.

In connection with Theorem 1.3.1 we should note that stronger results are also available in terms of the so-called upper and lower classes of functions, introduced by P. Lévy (cf. Supplementary Remarks to Section 3.2).

Exactly the same way as Theorem 1.3.2 is a generalization of Theorem 1.3.1, the following two theorems are Strassen type generalizations of Theorem 1.2.1.

Theorem S. 1.3.1. *Assume conditions* (i), (ii) *and* (iii) *of Theorem 1.2.1. Let $\mathscr{S}$ be as in Theorem 1.3.2. Then, for every $\varepsilon>0$, $f\in\mathscr{S}$, for almost all $\omega\in\Omega$ and for all T large enough there exists a $t=t(\omega, f, \varepsilon, T)$ such that $0\leq t(\omega, f, \varepsilon, T)\leq T-a_T$ and*

$$(S.1.3.1) \qquad \sup_{0\leq x\leq 1}\left|\frac{W(t+xa_T)-W(t)}{\sqrt{2a_T(\log Ta_T^{-1}+\log\log T)}}-f(x)\right|<\varepsilon.$$

Conversely, for every $\varepsilon>0$, every $t\in[0, T-a_T]$ and for almost all $\omega\in\Omega$, there exists an $f\in\mathscr{S}$ such that (S.1.3.1) *holds true whenever T is large enough.*

Theorem S. 1.3.2. *Assume only the conditions* (i) *and* (ii) *of Theorem 1.2.1. Then, for every $\varepsilon>0$, $f\in\mathscr{S}$ and for almost all $\omega\in\Omega$ there exist $T=T(\varepsilon, \omega, f)$ and $t=t(\varepsilon, \omega, f)\in[0, T-a_T]$ such that* (S.1.3.1) *holds true. However, the converse statement of Theorem S.1.3.1 is true as stated.*

We note that the important difference between Theorems S.1.3.1 and S. 1.3.2 is the fact that in the former we state that for *every* T big enough and for *every* $f\in\mathscr{S}$ there exists a $t\in[0, T-a_T]$ such that the function

$$\Gamma_{t,T}(x)=\frac{W(t+xa_T)-W(t)}{\sqrt{2a_T(\log Ta_T^{-1}+\log\log T)}}, \quad x\in[0, 1]$$

approximates the given f, while in Theorem S.1.3.2 we only state that for *every* $f\in\mathscr{S}$ there exists a T (in fact there exists infinitely many T) and a $t\in[0, T-a_T]$ such that $\Gamma_{t,T}(x)$ approximates the given f.

For a preliminary version of Theorem S.1.3.2 we refer to Chan, Csörgő, Révész (1978) and for the here quoted form to Révész (1979b).

Rates of convergence results in the context of Theorems 1.3.1 and 1.3.2 are also available (cf. Supplementary Remarks of Section 3.2).

Section 1.4. A Brownian bridge is frequently called tied-down Brownian motion or tied-down Brownian process.

Section 1.6. Our Theorem 1.6.1 is related to a result of Dvoretzky (1963) and that of Taylor (1974).

Theorem S. 1.6.1 (Dvoretzky 1963). *There exists a universal constant $C>0$ such that*

(S.1.6.1) $$\inf_{0 \leq t \leq 1} \varlimsup_{h \searrow 0} \frac{|W(t+h)-W(t)|}{h^{1/2}} \geq C \quad a.s.$$

Our Theorem 1.6.1 clearly implies

(S.1.6.2) $$\varlimsup_{h \searrow 0} \inf_{0 \leq t \leq 1-h} \left(\frac{8 \log h^{-1}}{\pi^2 h}\right)^{1/2} |W(t+h)-W(t)| \leq 1 \quad a.s.$$

Comparing the respective statements of (S.1.6.1) and (S.1.6.2), the former states that every neighbourhood of every $t \in [0, 1]$ contains a point $t+h(t)$ such that $|W(t+h(t))-W(t)|$ is at least so big as $Ch^{1/2}(t)$. On the other hand, (S.1.6.2) states that for all $h>0$ small enough there exists a $t=t(h)$ such that $|W(t+h)-W(t)|$ is so small as $\left(\dfrac{\pi^2 h}{8 \log h^{-1}}\right)^{1/2}$.

Now the mentioned result of Taylor is

Theorem S. 1.6.2 (Taylor 1974). *There exists a universal constant $C>0$ such that*

$$\lim_{h \searrow 0} \left(\frac{\log h^{-1}}{h}\right)^{1/2} \inf_{0 \leq t \leq 1-h} \sup_{t \leq u < v \leq t+h} |W(v)-W(u)| \stackrel{a.s.}{=} C.$$

It follows from our Theorem 1.6.1 that Taylor's constant $C \leq \pi/\sqrt{2}$.

Section 1.7. It is an interesting question to pose whether Chung's law of iterated logarithm (cf. Example 2 of Section 1.7) had also a functional form, like Theorem 1.3.2 is a functional form of Theorem 1.3.1. A solution of this problem was given by Donsker and Varadhan (1977) in terms of local times of a Wiener process. For another approach and solution to this problem we refer to Csáki (1981).

Section 1.9. Comparing Theorems 1.5.5 and 1.9.1 we observe that the asymptotic behaviour of the processes $\sup_{0 \leq t \leq T} |W(t+1) - W(t)|$ and $\sup_{0 \leq t \leq T} |U(t)|$ and that of $\sup_{0 \leq t \leq T} (W(t+1) - W(t))$ and $\sup_{0 \leq t \leq T} U(t)$ is the same. This surprising fact is due to the following general

Theorem S. 1.9.1 (Qualls, Watanabe 1972). *If $\{X(t); 0 \leq t < \infty\}$ is a stationary Gaussian process with*

$$EX(t) = 0, \quad E(X(t+s) - X(t))^2 > 0$$

and

$$EX(t)X(t+\tau) = \varrho(\tau) = 1 - |\tau| + O(\tau^2) \quad as \quad |\tau| \to 0,$$

and also with

$$\int_{-\infty}^{+\infty} \varrho^2(t) \, dt < \infty \quad or \quad \lim_{t \to \infty} \varrho(t) \log t = 0,$$

then

$$\lim_{T \to \infty} P\{\sup_{0 \leq t \leq T} X(t) \leq a(y, T)\} = \exp(-e^{-y}),$$

$$\lim_{T \to \infty} P\{\sup_{0 \leq t \leq T} |X(t)| \leq a(y, T)\} = \exp(-2e^{-y}),$$

where $a(y, T)$ $(-\infty < y < +\infty)$ *is as in Theorem 1.9.1.*

Obviously, the two processes $W(t+1) - W(t)$ and $U(t)$ satisfy the conditions of Theorem S.1.9.1, and whence Theorems 1.5.5 and 1.9.1 are special cases of the latter.

Section 1.10. The process $W(x, y)$ is sometimes called the Yeh process (cf. Yeh 1960 and Čencov 1956) or Brownian (Wiener) sheet.

Section 1.12. As it stands now, Theorem 1.12.8 is comparable to Theorem 1.2.1. It would be desirable to extend the former the same way as Theorem S.1.2.1 is an extension of the latter.

Section 1.14. Theorems 1.14.1 and 1.15.1 are closely related to Finkelstein's theorem on the empirical process, and both of them can be considered as her results. The idea of using the martingale technique in the proof of Lemma 1.14.2 is borrowed from a paper of Csáki (1968).

A natural generalization of Theorem 1.14.2 is

Theorem S. 1.14.2 (Chan 1977). *Let $0 < h_T \leq 1$, $0 < a_T \leq T$ be functions of T so that T/a_T, a_T are non-decreasing and h_T is non-increasing. Let $R = R(x, y; s, t) = [x, x+s] \times [y, y+t]$. Then*

(S.1.14.1) $\quad \overline{\lim}_{T \to \infty} \sup_{0 \leq y \leq T-a_T} \sup_{0 \leq x \leq 1-h_T} \sup_{0 \leq t \leq a_T} \sup_{0 \leq s \leq h_T} \beta_T |W(R)| \stackrel{a.s.}{=} 1,$

where
$$\beta_T = \left[2 h_T a_T \left(\log \frac{T}{a_T h_T} + \log \log T \right) \right]^{-1/2}.$$

If we also have

$$\lim_{T \to \infty} \frac{\log(T/a_T h_T)}{\log \log T} = \infty,$$

then $\overline{\lim}_{T \to \infty}$ in (S.1.14.1) can be replaced by $\lim_{T \to \infty}$.

Section 1.15. The Kiefer process appears in a paper by Kiefer (1972) proving the first strong embedding theorem for the empirical process. A weak convergence version of Kiefer's theorem was given by Müller (1970).

An analogue of Theorem S.1.14.2 can be stated also for a Kiefer process. Namely we have

Theorem S. 1.15.1 (Chan 1977). *Let $0 < \varepsilon_T < \frac{1}{2}$, $0 < a_T \leq T$ be functions of T such that ε_T and a_T/T are non-increasing and a_T is non-decreasing. Define $K((x_1, x_2], t) = K(x_2, t) - K(x_1, t)$ $(0 \leq x_1 < x_2 \leq 1)$. Then*

(S.1.15.1)

$\overline{\lim}_{T \to \infty} \sup_{0 \leq t \leq T-a_T} \sup_{0 \leq x \leq 1-\varepsilon_T} \sup_{0 \leq s \leq \varepsilon_T} \beta_T |K((x, x+s], t+a_T) - K((x, x+s], t)|$

$= \overline{\lim}_{T \to \infty} \sup_{0 \leq t \leq T-a_T} \sup_{0 \leq x \leq 1-\varepsilon_T} \beta_T |K((x, x+\varepsilon_T], t+a_T) - K((x, x+\varepsilon_T], t)| \stackrel{a.s.}{=} 1,$

where
$$\beta_T = \left[2 a_T \varepsilon_T (1 - \varepsilon_T) \left(\log \frac{T}{\varepsilon_T a_T} + \log \log T \right) \right]^{-1/2}.$$

If we also have

(S.1.15.2) $\quad \displaystyle\lim_{T \to \infty} \frac{\log \frac{1}{\varepsilon_T} \frac{T}{a_T}}{\log \log T} = \infty,$

then $\overline{\lim}_{T \to \infty}$ in (S.1.15.1) can be replaced by $\lim_{T \to \infty}$.

2. Strong Approximations of Partial Sums of I.I.D.R.V. by Wiener Processes

2.0. Notations

Throughout this Chapter $X_1, X_2, \ldots$ will denote a sequence of i.i.d.r.v. with $EX_1=0$, $EX_1^2=1$. $S_n=X_1+X_2+\ldots+X_n$ $(n=1, 2, \ldots)$, $S_0=0$ stand for the partial sums and $S_n(t)$ is defined by (0.3).

2.1. A proof of Donsker's theorem with Skorohod's embedding scheme

In the Introduction we formulated Donsker's invariance principle (Theorem 0.1). In this Section we present a proof of this theorem. This proof is different from the original idea of Donsker and produces a somewhat stronger result (Theorem 2.1.2). The basic tool of the present proof is the so-called Skorohod embedding scheme. The idea of proving Donsker's theorem via Skorohod's embedding scheme is due to Breiman (1968).

Skorohod's theorem (1961) essentially states that for any distribution function F with first moment 0 and finite second moment, one can define a probability space $(\Omega, \mathcal{A}, P)$ with a Wiener process W and a stopping time τ with finite expectation such that the distribution function of $W(\tau)$ is the given F. (The r.v. τ is called a stopping time, if the event $\{\omega: \tau(\omega) \leq t\}$ is an element of the σ-algebra generated by $\{W(s); s \leq t\}$.) We need the following general form of this theorem.

Theorem 2.1.1 (Skorohod 1961). *There exists a probability space $(\Omega, \mathcal{A}, P)$ with a Wiener process $\{W(t); 0 \leq t < \infty\}$ and a sequence $\tau_1, \tau_2, \ldots$ of nonnegative i.i.d.r.v. defined on it such that*

(i) $\{W(\tau_1+\ldots+\tau_k); k=1, 2, \ldots\} \stackrel{\mathcal{D}}{=} \{S_k; k=1, 2, \ldots\}$,
(ii) $\{\tau_1+\ldots+\tau_k; k=1, 2, \ldots\}$ *is a stopping time sequence,*
(iii) $E\tau_1=1$.

Also, if $EX_1^{2\nu} < \infty$ then $E\tau_1^\nu < \infty$ $(\nu=1, 2, \ldots)$.

We do not prove this theorem here. For an elegant proof of it we refer to Breiman's book (1968). Using his proof one can also get the following version of Skorohod's theorem.

Theorem 2.1.1*. *There exists a probability space with a Wiener process $\{W(t); 0 \leq t < \infty\}$ and a triangular sequence $\{\tau_1^{(n)}, \ldots, \tau_n^{(n)}\}$ of non-negative, r.v. defined on it such that*

(i) $\{\tau_1^{(n)}, \ldots, \tau_n^{(n)}\}$ *are i.i.d.r.v. for each* n, *and* $\tau_1^{(n)} \stackrel{\mathscr{D}}{=} \tau_1^{(1)}$, $n=1, 2, \ldots$

(ii) $$\left\{ W\left(\frac{\tau_1^{(n)} + \ldots + \tau_k^{(n)}}{n}\right); \ k = 1, \ldots, n \right\} \stackrel{\mathscr{D}}{=}$$
$$\stackrel{\mathscr{D}}{=} \{S_k/\sqrt{n}; \ k = 1, \ldots, n\} \quad \text{for each} \quad n = 1, 2, \ldots,$$

and the natural analogues of the further statements of Theorem 2.1.1 also hold.

We emphasize that this theorem of Skorohod has opened a completely new chapter in the area of invariance theorems. Already Skorohod (1961, Chapter 7.3) uses it to prove (0.5) with a rate of convergence for a special functional h. It provided also the basic tool for Strassen (1964) to prove his famous strong invariance principle.

In order to prove Donsker's theorem we first prove

Theorem 2.1.2. *There exists a probability space with a Wiener process $\{W(t); 0 \leq t \leq 1\}$ and a sequence of stochastic processes $\{\tilde{S}_n(t); 0 \leq t \leq 1\}$ such that*

(2.1.1) $$\{\tilde{S}_n(t); 0 \leq t \leq 1\} \stackrel{\mathscr{D}}{=} \{S_n(t); 0 \leq t \leq 1\}$$

for each $n = 1, 2, \ldots$, *and*

(2.1.2) $$\sup_{0 \leq t \leq 1} |\tilde{S}_n(t) - W(t)| \stackrel{P}{\longrightarrow} 0.$$

Proof. Using the notation of Theorem 2.1.1*, let

$$\tilde{S}_n(t) = W\left(\frac{\tau_1^{(n)} + \ldots + \tau_{[nt]}^{(n)}}{n}\right) + \left(W\left(\frac{\tau_1^{(n)} + \ldots + \tau_{[nt]+1}^{(n)}}{n}\right)\right.$$
$$\left. - W\left(\frac{\tau_1^{(n)} + \ldots + \tau_{[nt]}^{(n)}}{n}\right)\right)(nt - [nt]).$$

Then the relationship of (2.1.1) holds, because of Theorem 2.1.1* (ii).

Now to verify (2.1.2) it suffices to show

(2.1.3) $$\sup_{0\le t\le 1}\left|W\left(\frac{\tau_1^{(n)}+\ldots+\tau_{[nt]}^{(n)}}{n}\right)-W(t)\right|\xrightarrow{P} 0$$

and

(2.1.4) $$\sup_{0\le t\le 1}\left|\left(W\left(\frac{\tau_1^{(n)}+\ldots+\tau_{[nt]+1}^{(n)}}{n}\right)-W\left(\frac{\tau_1^{(n)}+\ldots+\tau_{[nt]}^{(n)}}{n}\right)\right)(nt-[nt])\right|\xrightarrow{P} 0.$$

To show that (2.1.3) holds true, we note that, via Theorem 2.1.1* (i), we have

$$\sup_{0\le t\le 1}\left|\frac{\tau_1^{(n)}+\ldots+\tau_{[nt]}^{(n)}}{n}-t\right|\xrightarrow{P} 0.$$

This latter statement combined with continuity of the Wiener process $W(\cdot)$ (cf. Theorem 1.1.1) gives (2.1.3). One can verify (2.1.4) in a similar way.

This theorem implies Donsker's invariance principle as follows:

Proof of Theorem 0.1. From (2.1.2) it follows that

(2.1.5) $$h(\tilde{S}_n(t))\xrightarrow{P} h(W(t))$$

and from (2.1.1) we have

(2.1.6) $$h(\tilde{S}_n(t))\stackrel{\mathscr{D}}{=} h(S_n(t))$$

for every continuous functional $h: C(0, 1)\to R^1$. Now (2.1.5) and (2.1.6) together imply (0.5).

Remark 2.1.1. If $t=1$, we get, from (2.1.1) and (2.1.2), that

(2.1.7) $$\tilde{S}_n(1)\xrightarrow{P} W(1)\in \mathcal{N}(0, 1)$$

and

(2.1.8) $$\tilde{S}_n(1)\stackrel{\mathscr{D}}{=} S_n(1) = S_n/\sqrt{n} \quad \text{for each} \quad n.$$

This, however, does not imply that $S_n/\sqrt{n}$ itself converges to a r.v. The reason for the different behaviour of $\tilde{S}_n(1)$ from that of $S_n/\sqrt{n}$ (i.e., (2.1.7) does not necessarily hold for the latter one) is that the relationship

$$\{\tilde{S}_n(1); n=1, 2, \ldots\}\stackrel{\mathscr{D}}{=}\{S_n/\sqrt{n}; n=1, 2, \ldots\}$$

is not true. In fact, $S_n/\sqrt{n}$ cannot converge in probability to any random variable. This follows easily, for example, from the following:

Theorem 2.1.3 (Rényi 1958). *For every event A of positive probability we have*

$$P\{S_n/\sqrt{n} \leq y | A\} \to \Phi(y).$$

In order to see how the latter implies that $S_n/\sqrt{n}$ cannot converge to a r.v., assume that $S_n/\sqrt{n} \xrightarrow{P} X$, and apply Theorem 2.1.3 with $A = \{\omega : X(\omega) \leq 0\}$. This results in a contradiction.

2.2. The strong invariance principle appears

As we have already mentioned in the Introduction, Strassen (1964) was the first one who introduced the notion of strong invariance principle, when proving Theorem 0.2, which we now restate as follows:

Theorem 2.2.1 (Strassen 1964). *A probability space $(\Omega, \mathcal{A}, P)$ with a sequence $\{\hat{S}_n\}$ and a Wiener process $\{W(t); 0 \leq t < \infty\}$ on it can be so constructed that*

(2.2.1) $\qquad \{\hat{S}_n;\ n = 1, 2, \ldots\} \stackrel{\mathscr{D}}{=} \{S_n;\ n = 1, 2, \ldots\}$

and

(2.2.2) $\qquad \dfrac{|\hat{S}_n - W(n)|}{\sqrt{n \log \log n}} \stackrel{a.s.}{=\!=} 0.$

Remark 2.2.1. If we are given a probability space $\{\Omega, \mathcal{A}, P\}$ with a sequence $X_1, X_2, \ldots$ of i.i.d.r.v., then it is not sure at all that a Wiener process $W(t)$ can be defined on the underlying Ω for which the relationship

(2.2.2*) $\qquad \dfrac{|S_n - W(n)|}{\sqrt{n \log \log n}} \xrightarrow{a.s.} 0$

would be true. This is the reason why, as a first step, we define a new probability space and a new sequence of r.v. (on this new space) which is equivalent to the original one in the sense of (2.2.1), and then the statement (2.2.2) can be stated for this new sequence. Having only (2.2.2), we can still prove any result for S_n itself which (2.2.2*) could have directly produced. Indeed, for this very reason, and for the sake of simplicity, we will simply say from now on that S_n (itself) can be approximated by a Wiener process. That is to say, from now on, we will state our results along the lines of (2.2.2*) instead of those of (2.2.2), and we ask the reader to remember that they will be meant á la the latter.

In order to illuminate the relationship of Theorems 2.1.2 and 2.2.1, we restate the latter in the following form:

Theorem 2.2.1*. *There exists a Wiener process* $W(\cdot)$ *such that*

$$(2.2.3) \qquad \sup_{0 \leq t \leq 1} \frac{|S_n(t) - n^{-1/2} W(nt)|}{\sqrt{\log \log n}} \xrightarrow{\text{a.s.}} 0.$$

Caution! Remark 2.2.1 must be applied to understand (2.2.3) correctly.

Remark 2.2.2. Comparing Theorems 2.1.2 and 2.2.1* we note that a disadvantage of the former is that only in-distribution type statements can be obtained from it. On the other hand, a disadvantage of (2.2.3) is that the rate of convergence in it is weaker than that in (2.1.2). Consequently, Donsker's theorem does not follow from Theorem 2.2.1*. (Cf. also Theorem 2.2.3.)

We know (Theorem 2.1.3) that $S_n(t)$ cannot be approximated by a single Wiener process $W(t)$ such that

$$\sup_{0 \leq t \leq 1} |S_n(t) - W(t)| \xrightarrow{P} 0$$

should hold. Applying the law of iterated logarithm to $S_n(1)$, one sees immediately that the sequence $W_n(t) = n^{-1/2} W(nt)$ cannot be replaced by a single Wiener process in (2.2.3) either. However, it is crucial in (2.2.2) or in (2.2.2*) that S_n is approximated by a single Wiener process $\{W(t); 0 \leq t < \infty\}$.

Proof of Theorem 2.2.1. The proof of this theorem again hinges on Skorohod's embedding scheme. Let

$$(2.2.4) \qquad S_n = W(\tau_1 + \ldots + \tau_n), \quad n = 1, 2, \ldots,$$

with the τ_i as in Theorem 2.1.1. Using Kolmogorov's strong law of large numbers we can write: $\tau_1 + \ldots + \tau_n = n + \eta_n$, where $\eta_n \stackrel{\text{a.s.}}{=} o(n)$. With this definition of S_n, (2.2.2) follows from Theorem 1.3.2 ((2.2.2) also follows easily from Theorem 1.2.1).

Proof of Theorem 2.2.1*. Using our definition of S_n in (2.2.4), in order to verify (2.2.3), it suffices to show

(2.2.5) $$\sup_{0\leq t\leq 1} \frac{|S_{[nt]}-W([nt])|}{\sqrt{n\log\log n}} \xrightarrow{\text{a.s.}} 0,$$

(2.2.6) $$\sup_{0\leq t\leq 1} \frac{|W(nt)-W([nt])|}{\sqrt{n\log\log n}} \xrightarrow{\text{a.s.}} 0,$$

(2.2.7) $$\sup_{0\leq t\leq 1} \frac{|(S_{[nt]+1}-S_{[nt]})(nt-[nt])|}{\sqrt{n\log\log n}} \xrightarrow{\text{a.s.}} 0.$$

The statement of (2.2.5) immediately follows from that of (2.2.2), while (2.2.6) and (2.2.7) are trivial.

We have already remarked that the rate of convergence in Theorem 2.2.1 is not strong enough to prove Donsker's theorem from it. Indeed to achieve this, one needs to replace the denominator of (2.2.2) by $n^{1/2}$ at least. This, however, is impossible (cf., however, Supplementary Remarks, Section 2.2, Theorem S.2.2.1) if we assume only the existence of the second moment of X_1, for Breiman proved

Theorem 2.2.2 (Breiman 1967). *There exists a distribution function F with mean 0 and variance 1 such that for any i.i.d. sequence $\{X_i\}$ having this distribution and for any Wiener process $W(t)$ we have*

$$\varlimsup_{n\to\infty} \frac{|S_n-W(n)|}{n^{1/2}} > 0 \quad \text{a.s.}$$

A stronger result of this type was obtained by Major (1976b) who proved that the rate in (2.2.2) cannot be improved if we assume the existence of two moments only. His result says

Theorem 2.2.3 (Major 1976b). *For any sequence $\{a_n\}$ of real numbers with $a_n \nearrow \infty$ there exists a distribution function F with mean 0 and variance 1 such that for any i.i.d. sequence $\{X_i\}$ having this distribution F and for any Wiener process $W(t)$ we have*

$$\varlimsup_{n\to\infty} a_n \frac{|S_n-W(n)|}{(n\log\log n)^{1/2}} \overset{\text{a.s.}}{=} \infty.$$

The above results suggest that one should investigate the possibility of getting better rates for (2.2.3) when higher than second moments are assumed for X_1. Towards this end we prove

Theorem 2.2.4. Given i.i.d.r.v. $X_1, X_2, \ldots$ with $EX_1=0$, $EX_1^2=1$, $E|X_1|^p<\infty$, $2<p\leq 4$, there exists a Wiener process W such that

$$(2.2.8) \qquad \frac{|S_n-W(n)|}{n^{1/p}(\log n)^{1/2}} \xrightarrow{\text{a.s.}} 0 \quad \text{if} \quad 2<p<4,$$

$$(2.2.9) \qquad \frac{|S_n-W(n)|}{(n \log \log n)^{1/4}(\log n)^{1/2}} \xrightarrow{\text{a.s.}} O(1) \quad \text{if} \quad p=4.$$

The statement of (2.2.8) (resp. (2.2.9)) was proved by Breiman (1967) (resp. Strassen (1965b)), using again the Skorohod Embedding Scheme. We do the same here, proving both (2.2.8) and (2.2.9) at the same time. In the proof we need the following

Lemma 2.2.1 (Loève 1963, p. 243). Let $T_1, T_2, \ldots$ be i.i.d.r.v. with $ET_1=0$ and $E|T_1|^q<\infty$, $1\leq q<2$. Then

$$(2.2.10) \qquad \frac{T_1+\ldots+T_n}{n^{1/q}} \xrightarrow{\text{a.s.}} 0.$$

Proof of Theorem 2.2.4. Again, we define S_n as in (2.2.4). Now the role played by Kolmogorov's law of large numbers in the proof of Theorem 2.2.1 is taken over here by the above lemma when proving (2.2.8), while in the case of (2.2.9) it is replaced by the Hartman–Wintner law of the iterated logarithm for i.i.d.r.v. with finite second moment (cf. (3.2.6)). After this, the proof follows the line of thought of that of Theorem 2.2.1, definitely using now Theorem 1.2.1.

We give some details for $2<p<4$. Applying Lemma 2.2.1 with $T_n=\tau_n-1$, we get $\tau_1+\ldots+\tau_n \stackrel{\text{a.s.}}{=} n+\eta_n$, where $\eta_n \stackrel{\text{a.s.}}{=} o(n^{2/p})$. Hence it suffices to show that

$$\frac{|W(n+o(n^{2/p}))-W(n)|}{n^{1/p}\sqrt{\log n}} \xrightarrow{\text{a.s.}} 0.$$

Towards this end we note that for every $\varepsilon>0$ there exist $\Omega_\varepsilon \subset \Omega$ and a sequence of positive numbers $a_n=o(n^{2/p})$, satisfying the conditions of Theorem 1.2.1, such that

$$|\eta_n| \leq a_n \quad \text{if} \quad \omega \in \Omega_\varepsilon \quad \text{and} \quad P(\Omega_\varepsilon) \geq 1-\varepsilon.$$

Thus Theorem 1.2.1 implies (2.2.8).

A Theorem 2.2.1* type version of Theorem 2.2.4 is immediate and from it the statement of Donsker's theorem follows when assuming the indicated higher than second moments.

Since Lemma 2.2.1 is not true for $q \geq 2$, it is clear that the method of proof of Theorem 2.2.4 will not give a better rate than that of (2.2.9) no matter how many moments are assumed to exist. In fact Strassen conjectured that the rate of (2.2.9) could not be improved with any further conditions and posed the

Question (Strassen 1965b). Let $X_1, X_2, \ldots$ be i.i.d.r.v. with mean zero and variance one, and let $\{W(t); 0 \leq t < \infty\}$ be a Wiener process such that

$$(2.2.11) \qquad |S_n - W(n)| \stackrel{a.s.}{=} o\bigl((n \log \log n)^{1/4} (\log n)^{1/2}\bigr).$$

Is then the distribution of the X_1 standard normal?

Similar questions were asked by Breiman (1967) and also by Borovkov (1973). As to the question of Strassen, Kiefer (1969a) proved that the order of (2.2.9) is indeed the best, provided the S_n are constructed via a Skorohod type stopping time procedure.

In the light of Strassen's problem, one can also ask the

Question. Should there exist any $f(n) \nearrow \infty$ such that

$$(2.2.12) \qquad |S_n - W(n)| \stackrel{a.s.}{=} o\bigl(f(n)\bigr)$$

should imply normality of the X_1?

2.3. The stochastic Geyser problem as a lower limit to the strong invariance principle

Rényi (1962) posed a problem which has nothing to do with our question of (2.2.12). A solution of a more general form of it, however, turned out to be also an answer to our problem. The original question of Rényi went like this: Let $X_1, X_2, \ldots$ be i.i.d. positive and bounded r.v., let $\{S_n\}$ be their partial sum sequence; can one then determine the distribution function of the X_i with probability one, observing only the sequence $\{[S_n]\}$? This problem was motivated by the following story: Robinson Crusoe had a geyser on his island, which kept on erupting at random time points. After he had observed the number of eruptions per day for a long time, it occurred to him that he should now be able to predict the geyser's behaviour, i.e., he should be able to estimate the distribution function of the time length between two eruptions.

We now formulate a more general form of the geyser problem. Let $X_1, X_2, \ldots$ be i.i.d.r.v. and let $F(\cdot)$ be their distribution function. Put

$$V_n = S_n + R_n,$$

where $\{R_n\}$ is also a random variable sequence, not necessarily independent of S_n. Then we can ask whether it is possible to determine the distribution function $F(\cdot)$ with probability one via some Borel function of $\{V_n;\ n=1, 2, \ldots\}$. In statistical terminology $\{R_n;\ n=1, 2, \ldots\}$ can be viewed as a random error sequence when trying to observe S_n in order to estimate $F(\cdot)$. Answering this question Bártfai proved

Theorem 2.3.1 (Bártfai 1966). *Assume that the moment generating function $R(t) = \int e^{tx} dF(x)$ of X_1 exists in a neighborhood of $t=0$ and $R_n \stackrel{a.s.}{=} o(\log n)$. Then, given the values of $\{V_n;\ n=1, 2, \ldots\}$, the distribution function $F(\cdot)$ is determined with probability one, i.e., there exists a r.v. $L(x) = L(V_1, V_2, \ldots; x)$ measurable with respect to the σ-algebra generated by $V_1, V_2, \ldots$ such that for any given real x, $L(x) \stackrel{a.s.}{=} F(x)$.*

The proof of this theorem is given in the next section. Here we show why Bártfai's result is also an answer to the question of (2.2.12), via stating the following

Theorem 2.3.2. *Let $X_1, X_2, \ldots$ be i.i.d.r.v. with mean zero and variance one. Denote their distribution function by $F(\cdot)$ and let $\{W(t);\ 0 \leq t < \infty\}$ be a Wiener process such that*

(2.3.1) $$S_n - W(n) \stackrel{a.s.}{=} o(\log n).$$

Then $F(\cdot) = \Phi(\cdot)$.

Proof. Let $R_n = W(n) - S_n$. Then $W(n) = V_n = S_n + R_n$ and, assuming the existence of the moment generating function of the X_1 in a neighborhood of zero, we should be able to determine $F(\cdot)$ by Theorem 2.3.1 with probability one, having observed $\{W(n);\ n=1, 2, \ldots\}$. This, however, is impossible unless $F(\cdot) = \Phi(\cdot)$.

In order to complete our proof we show that (2.3.1) is also impossible if the moment generating function of X_1 does not exist. Assume then that for any $t > 0$, $Ee^{tX_1} = \infty$. Then

$$\sum_{n=1}^{\infty} P\left\{e^{\frac{1}{c}X_n} > n\right\} = \sum_{n=1}^{\infty} P\{X_n > c \log n\} = \infty, \quad \text{for any} \quad c > 0,$$

and the Borel–Cantelli Lemma implies that the inequality $S_{n+1} - S_n > c \log n$ occurs infinitely often with probability 1. On the other hand $W(n+1) - W(n) \leq 2\sqrt{\log n}$ with probability one for all but finitely many n (cf. (1.2.3)). Consequently $S_n - W(n) \geq c \log n$ infinitely often, for any $c > 0$, with probability one.

The second part of our proof followed the lines of Breiman (1967).

We note also that the last lines of our proof above show that, for whatever $W(t)$, one has

(2.3.2) $$\varlimsup_n \frac{S_n - W(n)}{\log n} \stackrel{\text{a.s.}}{=} \infty,$$

provided $R(t)$ does not exist for any $t > 0$.

2.4. The longest runs of pure heads and the stochastic Geyser problem

In connection with a teaching experiment in mathematics, T. Varga posed a problem which has nothing to do with the Stochastic Geyser Problem. A solution of a more general form of it, however, turned out to be also an answer to the latter. The experiment goes like this: his class of secondary school children is divided into two sections. In one of the sections each child is given a coin which they then throw two hundred times, recording the resulting head and tail sequence on a piece of paper. In the other section the children do not receive coins but are told instead that they should try to write down a "random" head and tail sequence of length two hundred. Collecting these slips of paper, he then tries to subdivide them into their original groups. Most of the time he succeeds quite well. His secret is that he had observed that in a randomly produced sequence of length two hundred, there are, say, head-runs of length seven. On the other hand, he had also observed that most of those children who were to write down an imaginary random sequence are usually afraid of putting down runs of longer than four. Hence, in order to find the slips coming from the coin tossing group, he simply selects the ones which contain runs longer than five.

This experiment led T. Varga to ask: What is the length of the longest run of pure heads in n Bernoulli trials?

An answer to this question was given by Erdős and Rényi who proved

Theorem 2.4.1 (Erdős, Rényi 1970). *Let $X_1, X_2, \ldots$ be i.i.d.r.v., each taking the values ± 1 with probability $\frac{1}{2}$. Put $S_0 = 0$, $S_n = X_1 + \ldots + X_n$.*

Then for any $c \in (0, 1)$ and for almost all $\omega \in \Omega$ there exists an $n_0 = n_0(c, \omega)$ such that

$$(2.4.1) \qquad \max_{0 \leq k \leq n - [c \log_2 n]} (S_{k + [c \log_2 n]} - S_k) = [c \log_2 n],$$

if $n > n_0$.

That is, this theorem guarantees the existence of a run of length $[c \log_2 n]$ for every $c \in (0, 1)$ with probability one if n is large enough.

On the other hand, they also showed for $c > 1$ that the equality of (2.4.1) can only hold for a finite number of values of n with probability one. They proved

Theorem 2.4.2 (Erdős, Rényi 1970). *With the above notation one has*

$$(2.4.2) \qquad \max_{0 \leq k \leq n - [c \log_2 n]} \frac{S_{k + [c \log_2 n]} - S_k}{[c \log_2 n]} \xrightarrow{a.s.} \alpha(c),$$

where $\alpha(c) = 1$ for $c \leq 1$, and, if $c > 1$, then $\alpha(c)$ is the only solution of the equation

$$(2.4.3) \qquad \frac{1}{c} = 1 - h\left(\frac{1 + \alpha}{2}\right),$$

with $h(x) = -x \log_2 x - (1 - x) \log_2 (1 - x)$, $0 < x < 1$; the herewith defined $\alpha(\cdot)$ is a strictly decreasing continuous function for $c > 1$ with $\lim_{c \searrow 1} \alpha(c) = 1$ and $\lim_{c \to \infty} \alpha(c) = 0$.

As to the problem of the longest runs of pure heads, these two theorems do not give a complete answer (for example, they do not say anything about the existence of pure head-runs of length $[\log_2 n]$; cf. the Supplementary Remarks of Section 2.4). A generalization of Theorem 2.4.2 also produced a new proof of the stochastic geyser problem.

Theorem 2.4.3 (Erdős, Rényi 1970). *Let $X_1, X_2, \ldots$ be i.i.d.r.v. with mean zero and a moment generating function $R(t) = Ee^{tX_1}$, finite in a neighbourhood of $t = 0$. Let*

$$\varrho(x) = \inf_t e^{-tx} R(t),$$

the so-called Chernoff function of X_1. Then for any $c > 0$ we have

$$(2.4.4) \qquad \max_{0 \leq k \leq n - [c \log n]} \frac{S_{k + [c \log n]} - S_k}{[c \log n]} \xrightarrow{a.s.} \alpha(c),$$

where

(2.4.5) $$\alpha(c) = \sup\{x: \varrho(x) \geq e^{-1/c}\}.$$

Remark 2.4.1. Since $\varrho(0)=1$ and $\varrho(x)$ is a strictly decreasing function in the interval where $\varrho(x)>0$, $\alpha(c)$ is well defined for every $c>0$ by (2.4.5), and it is a continuous decreasing function with $\lim_{c\to\infty}\alpha(c)=0$.

The proof of Theorem 2.4.1 is elementary and will not be given here. The statement of Theorem 2.4.2 is a special case of that of Theorem 2.4.3. This can be seen immediately, for in this case $R(t)=\frac{1}{2}(e^t+e^{-t})$ and

$$\varrho(x) = \begin{cases} (1+x)^{-(1+x)/2}(1-x)^{-(1-x)/2} & \text{if } 0 \leq x < 1 \\ 0 & \text{if } x \geq 1. \end{cases}$$

Remark 2.4.2. In case of $X_1 \in \mathcal{N}(0,1)$ one gets easily that $\alpha(c)$ of (2.4.5) is equal to $\sqrt{2/c}$ for all $c>0$. This, naturally, agrees with Theorem 1.2.1 when taking $a_N = c \log N$, $c>0$.

The proof of Theorem 2.4.3 is based on

Theorem 2.4.4 (Chernoff 1952). *Under the assumptions and notations of Theorem 2.4.3 we have*

(2.4.6) $$P(S_n \geq nx) \leq \varrho^n(x)$$

and

(2.4.7) $$(P(S_n \geq nx))^{1/n} \to \varrho(x)$$

for any $x>0$.

Proof of Theorem 2.4.3. As a *first step* we prove

(2.4.8) $$\varlimsup_{n\to\infty} \max_{0 \leq k \leq n-l_n} \frac{S_{k+l_n}-S_k}{l_n} \leq \alpha \quad \text{a.s.}$$

where $l_n = [c \log n]$ and $\alpha = \alpha(c)$. Put

$$A_n = A_n(c,\varepsilon) = \left\{\max_{0 \leq k \leq n-l_n} \frac{S_{k+l_n}-S_k}{l_n} \geq \alpha+\varepsilon\right\}$$

and

$$A_n^* = A_n^*(c,\varepsilon) = \left\{\max_{0 \leq k \leq n-l_n} \frac{S_{k+l_n}-S_k}{l_n+1} \geq \alpha+\varepsilon\right\}.$$

Then by (2.4.6) we have

$$P(A_n) \leq nP\left(\frac{S_{l_n}}{l_n} \geq \alpha+\varepsilon\right) \leq n\varrho^{l_n}(\alpha+\varepsilon).$$

Applying the definition (2.4.5) it follows that there exists a $\delta>0$ such that $\varrho(\alpha+\varepsilon) \leq \exp\left(-\frac{1+\delta}{c}\right)$. Hence

$$P(A_n) \leq n \exp\left(-\frac{1+\delta}{c}[c\log n]\right),$$

and if T is the smallest integer for which $T\delta>1$, then

(2.4.9) $$\sum_{n=1}^{\infty} P(A_{nT}) < \infty.$$

This means that only finitely many of the events A_{nT} can occur with probability 1. Similarly one can see that

(2.4.9)* $$\sum_{n=1}^{\infty} P(A_{nT}^*) < \infty.$$

Taking into account the trivial inequality $l_{(n+1)T}-l_{nT} \leq 1$ for n big, we get (2.4.8).

Our *second step* is to prove

(2.4.10) $$\lim_{n\to\infty} \max_{0\leq k\leq n-l_n} \frac{S_{k+l_n}-S_k}{l_n} \geq \alpha \quad \text{a.s.}$$

Put

$$B_n = B_n(c,\varepsilon) = \left\{\max_{0\leq k\leq n-l_n} \frac{S_{k+l_n}-S_k}{l_n} < \alpha-\varepsilon\right\}.$$

Since $\varrho(\alpha-\varepsilon)>e^{-1/c}$, for any $\varepsilon>0$ and $c>0$ there exists a $\delta=\delta(c,\varepsilon)>0$ such that $\varrho(\alpha-\varepsilon)-\delta \geq \exp\left(-\frac{1-\delta}{c}\right)$. Then by (2.4.7) we obtain

$$P(B_n) \leq \prod_{d=1}^{[n/l_n]} P\left(\frac{S_{dl_n}-S_{(d-1)l_n}}{l_n} < \alpha-\varepsilon\right)$$

$$= \left(1-P\left(\frac{S_{l_n}}{l_n} \geq \alpha-\varepsilon\right)\right)^{[n/l_n]}$$

$$\leq (1-(\varrho(\alpha-\varepsilon)-\delta)^{l_n})^{[n/l_n]} \leq \left(1-\exp\left(-\frac{1-\delta}{c}l_n\right)\right)^{[n/l_n]}$$

$$= O\left(\exp\left(-\frac{n^{\delta/2}}{\log n}\right)\right)$$

if n is big enough. This proves the convergence of the series $\sum_{n=1}^{\infty} P(B_n)$ which, in turn, implies (2.4.10).

(2.4.8) and (2.4.10) together prove the theorem.

It is clear that $\alpha(c)$ of Theorem 2.4.3 is uniquely determined by the moment generating function $R(t)$ of X_1. The converse of this statement is also true and we have

Theorem 2.4.5 (Erdős, Rényi 1970). *The function $\alpha(c)$ of Theorem 2.4.3 uniquely determines the distribution function of X_1.*

Proof. Definition (2.4.5) shows that the function $\alpha(c)$ uniquely determines the Chernoff function $\varrho(x)$. Further, the Chernoff function $\varrho(x)$ uniquely determines the moment generating function $R(t)$ (cf., e.g., Bártfai (1977)). Finally, it is well known that the moment generating function $R(t)$ uniquely determines the distribution function F of X_1.

Now we are in the position to give a

Proof of Theorem 2.3.1. For any $c > 0$ we have

$$\lim_{n \to \infty} \max_{0 \leq k \leq n-[c \log n]} \frac{V_{k+[c \log n]} - V_k}{[c \log n]}$$

$$\stackrel{a.s.}{=} \lim_{n \to \infty} \max_{0 \leq k \leq n-[c \log n]} \frac{S_{k+[c \log n]} - S_k}{[c \log n]} \stackrel{a.s.}{=} \alpha(c).$$

Hence the sequence V_k determines $\alpha(c)$ with probability one which, via Theorem 2.4.5, also terminates our proof.

In passing we note that the way we have proved Theorem 2.3.1 does not provide an immediate handle for the construction of the there mentioned r.v. L. Thus we can say that, while Theorem 2.3.1 theoretically solves the so-called stochastic geyser problem, it does not provide a sequence of estimators for F.

2.5. Improving the upper limit

When talking about Strassen's question of (2.2.11), we have already mentioned that Kiefer (1969a) proved that the rate of convergence of (2.2.9) cannot be improved if one were to use the Skorohod Embedding Scheme, no matter what further restrictions we might put on the distribution function $F(\cdot)$ of the i.i.d.r.v. Hence an improvement could only have come from a different method of construction. Such a method was developed by us (Csörgő, Révész (1975), cf. also Supplementary Remarks to Section 2.5 of this Chapter) when, via improving the rate of convergence of (2.2.9), we gave the first negative answer to Strassen's question, proving the following counterexample.

Theorem 2.5.1 (Csörgő, Révész 1975). *Given i.i.d.r.v.* $X_1, X_2, \ldots$ *with a continuous distribution function* $F(\cdot)$,

$$EX_1 = EX_1^3 = 0, \quad EX_1^2 = 1, \quad EX_1^{10} < \infty \quad \text{and} \quad \varlimsup_{|t| \to \infty} |Ee^{itX_1}| < 1,$$

then there exists a Wiener process $\{W(t); 0 \le t < \infty\}$ *such that*

(2.5.1) $$\frac{|S_n - W(n)|}{n^{1/6} (\log n)^{13/2}} \xrightarrow{a.s.} 0.$$

As a preliminary step to the proof, we wish to approximate $S_n/\sqrt{n}$, for each given n, by a r.v. $N_n \in \mathcal{N}(0, 1)$. Define N_n by

(2.5.2) $$N_n = \text{inv } \Phi \left(F_n(S_n/\sqrt{n}) \right),$$

where $F_n(\cdot)$ is the distribution function of $S_n/\sqrt{n}$. Then $N_n \in \mathcal{N}(0, 1)$, and knowing, by the central limit theorem, that $F_n(\cdot)$ is near to $\Phi(\cdot)$ for n large, one expects that our N_n should be also close to $S_n/\sqrt{n}$. Indeed one would hope that a good rate of convergence of $F_n(\cdot)$ to $\Phi(\cdot)$ should also result in a "good nearness" of N_n to $S_n/\sqrt{n}$. As to a good rate of convergence of $F_n(\cdot)$ to $\Phi(\cdot)$ we have

Theorem 2.5.2 (Cramér 1962, p. 220). *Assuming the conditions of Theorem 2.5.1 we have*

$$F_n(x) - \Phi(x) = \frac{e^{-x^2/2}}{\sqrt{2\pi}} \sum_{i=1}^{8} \frac{Q_i(x)}{n^{i/2}} + o\left(\frac{1}{n^4}\right)$$

uniformly in x, *where* $Q_i(x)$ ($i=1, 2, \ldots, 8$) *is a polynomial of degree* $i+3$ *with coefficients depending only on the first ten moments of* $F(\cdot)$. (*Here* $Q_1(x) = 0$, *as a result of our assumption* $EX_1^3 = 0$.)

Applying this theorem of Cramér the following nearness of N_n to $S_n/\sqrt{n}$ is attained:

Lemma 2.5.1. *With the conditions of Theorem 2.5.1 and* N_n *as in* (2.5.2) *we have*

(2.5.3) $$|N_n - S_n/\sqrt{n}| = O\left(\frac{(\log n)^{11/2}}{n}\right),$$

provided that $|S_n/\sqrt{n}| \le c\sqrt{\log n}$, $0 < c < \sqrt{6}$.

The next two lemmas are needed in the proof of the latter lemma.

Lemma 2.5.2. *With the conditions of Theorem 2.5.1 we have*

$$\Phi(x_n) \approx F_n(x_n) \quad \text{and} \quad 1 - \Phi(x_n) \approx 1 - F_n(x_n),$$

provided that $\{x_n\}$ *is a sequence of real numbers with* $|x_n| \leq c\sqrt{\log n}$, $0 < c < \sqrt{8}$.

Proof. This is a simple consequence of Theorem 2.5.2 (cf. also Rubin, Sethuraman 1965 and Michel 1974).

Lemma 2.5.3. *Let* $\{a_n\}$ *and* $\{b_n\}$ *be two sequences of real numbers for which* $0 < a_n < 1$, $0 < b_n < 1$, $a_n \approx b_n$ *and* $1 - a_n \approx 1 - b_n$. *Then, as* $n \to \infty$, *we have*

$$(\text{inv } \Phi(a_n))^2 - (\text{inv } \Phi(b_n))^2 \to 0.$$

The proof is via elementary calculations.

Proof of Lemma 2.5.1. By Lagrange's mean value theorem we have

$$\left| \text{inv } \Phi(F_n(t)) - t \right| = \left| \text{inv } \Phi(F_n(t)) - \text{inv } \Phi(\Phi(t)) \right|$$

$$= |F_n(t) - \Phi(t)| \left. \frac{d \text{ inv } \Phi(y)}{dy} \right|_{y = \xi_t}$$

$$= |F_n(t) - \Phi(t)| \frac{1}{\Phi'(\text{inv } \Phi(\xi_t))},$$

where $\min(F_n(t), \Phi(t)) \leq \xi_t \leq \max(F_n(t), \Phi(t))$. Suppose that $|t| \leq c(\log n)^{1/2}$, $0 < c < \sqrt{8}$. Hence, by Lemma 2.5.2 $\xi_t \approx \Phi(t)$ and $1 - \xi_t \approx 1 - \Phi(t)$. By Lemma 2.5.3 we have $(\text{inv } \Phi(\xi_t))^2 - t^2 \to 0$, and this combined with Theorem 2.5.2 implies

$$\left| \text{inv } \Phi(F_n(t)) - t \right| \leq \left[\frac{e^{-t^2/2}}{\sqrt{2\pi}} O\left(\frac{(\log n)^{11/2}}{n} \right) + o\left(\frac{1}{n^4} \right) \right] \sqrt{2\pi} e^{t^2/2}$$

$$= O\left(\frac{(\log n)^{11/2}}{n} \right).$$

Thus, so far, we have approximated S_n by $\sqrt{n} N_n$ well enough for each n. Our aim, however, was to construct a Wiener process $\{W(t); 0 \leq t < \infty\}$ such that $\{W(n); n = 1, 2, \ldots\}$ should be near to $\{S_n; n = 1, 2, \ldots\}$. As to our $\sqrt{n} N_n$ we only have

$$\{\sqrt{n} N_n\} \stackrel{\mathscr{D}}{=} \{W(n)\}, \quad n = 1, 2, \ldots$$

instead of the desired equality

(2.5.4) $\qquad \{\sqrt{n}\,N_n;\ n=1,2,\ldots\} \stackrel{\mathscr{D}}{=} \{W(n);\ n=1,2,\ldots\},$

which we have no reason to believe in at all. If it were to be true by any chance, however, we would have also achieved our aim already. Though (2.5.4) is not disproved, it seems that it cannot be true in general. Thus, our desired Wiener process has yet to be constructed.

Towards this end we now construct a sequence U_k such that it will be a good approximation of S_{n_k} for a subsequence of integers $n_k = [k^\alpha]$ $(\alpha > 1)$ and, also, we will have

(2.5.5) $\qquad \{U_k;\ k=1,2,\ldots\} \stackrel{\mathscr{D}}{=} \{W(n_k);\ k=1,2,\ldots\}.$

Define
$$U_k = \sum_{j=1}^{k} \sqrt{m_j}\, R_j,$$

where $m_j = n_j - n_{j-1} \approx a j^{\alpha-1}$ $(j=1,2,\ldots)$, $R_j = \operatorname{inv} \Phi(F_j(Z_j))$,

$$Z_j = \frac{X_{n_{j-1}+1} + X_{n_{j-1}+2} + \ldots + X_{n_j}}{\sqrt{m_j}} \quad \text{and} \quad F_j(t) = P\{Z_j \leq t\}.$$

Since the $R_j \in \mathcal{N}(0,1)$ are independent r.v. we immediately have our desired relation (2.5.5). As to the nearness of the thus constructed U_k to S_{n_k}, we prove:

Lemma 2.5.4.

(2.5.6) $\qquad \dfrac{|U_k - S_{n_k}|}{n_k^{\frac{2-\alpha}{2\alpha}} (\log n_k)^{13/2}} \xrightarrow{a.s.} 0,\quad 1 < \alpha < 2.$

Proof. Let $e_j = R_j - Z_j$, $0 < c < \sqrt{6}$

$$I_j = \begin{cases} 1 & \text{if } |Z_j| > c\sqrt{\log m_j} \\ 0 & \text{otherwise.} \end{cases}$$

First we evaluate the variance of e_j. Consider

$$Ee_j^2 = Ee_j^2 I_j + Ee_j^2(1-I_j).$$

Then, by Lemma 2.5.1

$$Ee_j^2(1-I_j) = O\left(\frac{(\log m_j)^{11}}{m_j^2}\right).$$

Since we are assuming $EX_1^{10} < \infty$, $Ee_j^{10} = O(1)$ and Hölder's inequality implies

$$Ee_j^2 I_j \leq (Ee_j^{10})^{1/5}(EI_j^{5/4})^{4/5} = O(1)(P\{|Z_j| \geq c\sqrt{\log m_j}\})^{4/5}$$

$$= O(m_j^{-\frac{c^2}{2} \cdot \frac{4}{5}}).$$

Hence

$$Ee_j^2 = O\left(\frac{(\log m_j)^{11}}{m_j^2}\right), \quad \text{provided} \quad \sqrt{5} \leq c \leq \sqrt{6}.$$

Now let

$$\varrho_j = \frac{e_j m_j}{(\log m_j)^{11/2}}$$

and

$$\lambda_j = \frac{(\log m_j)^{11/2} j^{1/2} \log j}{\sqrt{m_j}} = O((\log j)^{13/2} j^{\frac{2-\alpha}{2}}).$$

Then $E\varrho_j = 0$, $E\varrho_j^2 = O(1)$, and Kolmogorov's Three Series Theorem implies that the series

$$\sum_{j=1}^{\infty} \frac{\varrho_j}{j^{1/2} \log j} = \sum_{j=1}^{\infty} \frac{\sqrt{m_j} e_j}{\lambda_j}$$

converges with probability one. Since $\alpha < 2$, λ_j is strictly increasing and Kronecker's lemma implies that

$$\frac{1}{\lambda_k} \sum_{j=1}^{k} \sqrt{m_j} e_j \xrightarrow{\text{a.s.}} 0.$$

This also completes the proof of our lemma, because

$$O\left(\frac{1}{\lambda_k}\right) \sum_{j=1}^{k} \sqrt{m_j} e_j = \frac{U_k - S_{n_k}}{k^{\frac{2-\alpha}{2}} (\log k)^{13/2}}.$$

Proof of Theorem 2.5.1. Let $B_1(x), B_2(x), \ldots$ ($0 \leq x \leq 1$) be a sequence of independent Brownian bridges which is also independent from the given sequence $\{X_i\}$. Clearly it is not sure that such a sequence $\{B_i(x)\}$ can be constructed on the probability space where the r.v. $\{X_i\}$ live. However, it is easy to redefine the sequence $\{X_i\}$ on a new probability space where the desired construction is available (cf. Remark 2.2.1). Now we construct a Wiener process $W(t)$ via joining the points (n_{i-1}, U_{i-1}) and (n_i, U_i) by the independent Brownian bridges $B_i(x)$ ($i = 1, 2, \ldots$) (cf. Proposition 1.4.1).

Then $U_k = W(n_k)$, and Lemma 2.5.4 gives

(2.5.7) $$\frac{|W(n_k) - S_{n_k}|}{n_k^{\frac{2-\alpha}{2\alpha}} (\log n_k)^{13/2}} \xrightarrow{\text{a.s.}} 0.$$

In order to complete the proof of our theorem we have to study $W(n) - S_n$ when $n_k \leq n < n_{k+1}$. Clearly,

$$W(n) - S_n = W(n) - W(n_k) + W(n_k) - S_{n_k} + S_{n_k} - S_n$$

and, by Theorem 1.2.1,

(2.5.8) $$\sup_{n_k \leq n < n_{k+1}} \frac{|W(n) - W(n_k)|}{\sqrt{m_k} (\log m_k)^{13/2}} \xrightarrow{\text{a.s.}} 0.$$

Also,

(2.5.9) $$\sup_{n_k \leq n < n_{k+1}} \frac{|S_n - S_{n_k}|}{\sqrt{m_k} (\log m_k)^{13/2}} \xrightarrow{\text{a.s.}} 0.$$

This latter statement can be checked easily via elementary calculations, or apply Theorem 2 of Michel (1974).

Choosing now $\alpha = \frac{3}{2}$, (2.5.1) follows from (2.5.7), (2.5.8) and (2.5.9).

Remark 2.5.1. The above proof hinges on the choice of the sequence $\{n_j\}$. The larger we choose $\alpha > 1$ in $n_j = [j^\alpha]$, the better is the normal approximation for our blocks $X_{n_j+1} + \ldots + X_{n_{j+1}}$. The random fluctuation of the Brownian bridges $B_j(x)$ connecting the points (n_{j-1}, U_{j-1}) and (n_j, U_j) however, tends to destroy the gains produced by α too large. The choice $\alpha = \frac{3}{2}$ is obtained as a compromise.

2.6. The best rates emerge

Re-examining the method of the proof of Theorem 2.5.1 it becomes clear that, for any given $\varepsilon > 0$, there exist further moment conditions which, when assumed, enable one to prove

(2.6.1) $$\frac{|S_n - W(n)|}{n^\varepsilon} \xrightarrow{\text{a.s.}} 0$$

instead of (2.5.1) of Theorem 2.5.1.

Recalling also Theorem 2.3.2 it is reasonable to ask whether there should exist any distribution function $F(\cdot) \neq \Phi(\cdot)$ such that

(2.6.2) $$|S_n - W(n)| \stackrel{\text{a.s.}}{=} O(\log n),$$

for a suitably constructed $W(\cdot)$. It is clear from (2.3.2) that, if such a distribution function $F(\cdot)$ were to exist, then it must have a moment generating function. Indeed, the surprising fact that (2.6.2) holds for all those distributions which do have a moment generating function turned out to be true as a result of

Theorem 2.6.1 (Komlós, Major, Tusnády 1975, 1976). *Assume that $R(t) = Ee^{tX_1}$ exists in a neighborhood of $t=0$. Then there exists a Wiener process $\{W(t); 0 \leq t < \infty\}$ such that (2.6.2) holds.*

This theorem is a consequence of the following sharp result:

Theorem 2.6.2 (Komlós, Major, Tusnády 1975, 1976). *Given the conditions of Theorem 2.6.1, there exists a Wiener process W such that for all real x and every n we have*

$$(2.6.3) \qquad P\{\max_{1 \leq k \leq n} |S_k - W(k)| > C \log n + x\} < Ke^{-\lambda x},$$

where C, K, λ are positive constants, depending only on the distribution function of X_1.

As we have already seen, Theorem 2.6.1 is the best possible in the sense that no further restrictions on the distribution function of X_1 can improve the rate of (2.6.2), unless X_1 is already a $\mathcal{N}(0, 1)$ r.v. It is also the best possible in the sense that the assumption that the moment generating function of X_1 should exist, cannot be dropped (cf. (2.3.2)). Returning, however, to (2.6.1) we can still ask what minimal set of moment-assumptions should guarantee the there indicated rate for a given $\varepsilon > 0$. The answer to this question is given by

Theorem 2.6.3 (Komlós, Major, Tusnády 1975, 1976 and Major 1976a). *Replacing the assumption of Theorem 2.6.1 that the moment generating function exists by $E|X_1|^p < \infty$, $p > 2$, we have*

$$(2.6.4) \qquad \frac{|S_n - W(n)|}{n^{1/p}} \xrightarrow{a.s.} 0.$$

Notice that the latter theorem is an improvement of Theorem 2.2.4 in the case of $2 < p \leq 4$. In fact it is not only an improvement in this interval but it gives also the best possible rate for all $p > 2$. The fact that this is indeed true follows from

Theorem 2.6.4. Let $p \geq 2$. Then

(2.6.5) $$\overline{\lim_{n \to \infty}} \frac{S_n - W(n)}{n^{1/p}} \stackrel{a.s.}{=} \infty,$$

for whatever $W(\cdot)$, provided $E|X_1|^p$ does not exist.

This was proved by Breiman (1967), using the method we have also utilized in the proof of Theorem 2.3.2 (cf. (2.3.2)).

A common generalization of Theorems 2.6.1 and 2.6.3 would be achieved if we could answer the

Question. Given a function $H(x) > 0$ ($x \geq 0$) such that $EH(|X_1|) < \infty$, what is the best possible order of the strong invariance principle?

Such a question was first studied by Breiman, who, using the Skorohod Embedding Scheme, proved

Theorem 2.6.5 (Breiman 1967). *Let $H(x) > 0$ ($x \geq 0$) be a continuous function such that $x^{-2} H(x)$ is non-decreasing and for some $r > 0$, $H(x) x^{-4+r}$ is non-increasing. Assume that $EH(|X_1|) < \infty$. Then there exists a Wiener process such that*

(2.6.6) $$\frac{S_n - W(n)}{g(n)} \xrightarrow{a.s.} 0,$$

where $g(n) = \text{inv } H(n) \sqrt{\log \psi(n)}$ and $\psi(n)$ is any non-decreasing function such that for some $m > 0$, $\sum_n (\text{inv } H(n))^{-2} (\psi(n))^{-m} < \infty$.

Remark 2.6.1. This theorem clearly implies Theorem 2.2.1 and (2.2.8) of Theorem 2.2.4, but it does not imply (2.2.9) of the latter. It is also clear that this theorem is not strong enough to imply any of the strong invariance principles of this Section or that of Section 2.5.

Komlós, Major and Tusnády have also contributed greatly towards the solution of the above formulated question, proving

Theorem 2.6.6 (Komlós, Major, Tusnády 1975, 1976 and Major 1976a). *Let $H(x) > 0$ ($x \geq 0$) be a non-decreasing continuous function such that $x^{-2-\gamma} H(x)$ is non-decreasing for some $\gamma > 0$ and $x^{-1} \log H(x)$ is non-increasing. Assume that $EH(|X_1|) < \infty$. Then we have*

(2.6.7) $$S_n - W(n) \stackrel{a.s.}{=} O(\text{inv } H(n)).$$

Remark 2.6.2. Taking $H(x) = e^{tx}$, we immediately get Theorem 2.6.1. As to Theorem 2.6.3, it does not follow so immediately. However, the following two simple lemmas will show that it is also a consequence of the above theorem.

Lemma 2.6.1. Let $H(\cdot)$ be a function as in Theorem 2.6.6 with $EH(|X_1|)<\infty$. Then there exists a function $h(x)\nearrow\infty$ $(x\geq 0)$ such that $H_1(x)=h(x)\cdot H(x)$ again satisfies the conditions imposed on $H(\cdot)$ in Theorem 2.6.6 and $E(H_1(|X_1|))<\infty$.

Lemma 2.6.2. Assume that $H(x)$ satisfies the conditions of Theorem 2.6.6 and $h(x)$ satisfies those of Lemma 2.6.1, i.e., $|\int H_1(x)dF(x)|=|\int H(x)h(x)dF(x)|<\infty$, where $F(x)=P(X_1\leq x)$. Then

(2.6.8) $$\operatorname{inv} H_1(x) = o(\operatorname{inv} H(x))$$

provided

(2.6.9) $$\lim_{x\to\infty} \frac{H(\varepsilon x)}{H(x)} > 0$$

for all $\varepsilon>0$.

The proof of Lemma 2.6.1 is only routine, that of Lemma 2.6.2 is a little harder. Hence we present the

Proof of Lemma 2.6.2. Suppose that (2.6.8) does not hold. Then there exist a $\delta>0$ and a sequence $x_n\nearrow\infty$ such that

$$\operatorname{inv} H_1(x_n) \geq \delta \operatorname{inv} H(x_n).$$

Consider

$$x_n = H_1(\operatorname{inv} H_1(x_n)) = H(\operatorname{inv} H_1(x_n))h(\operatorname{inv} H_1(x_n)).$$

We get

$$H(\operatorname{inv} H_1(x_n)) = o(x_n)$$

and also

$$H(\delta \operatorname{inv} H(x_n)) = o(x_n).$$

This, when replacing x_n by $H(u_n)$, contradicts condition (2.6.9), and proves the Lemma.

Let now $H(x)=x^p$ $(x>0, p>2)$ and let $h(x)$ be as in Lemma 2.6.1. Then, applying Theorem 2.6.6 with $H_1(x)=x^p\cdot h(x)$ we get Theorem 2.6.3.

We also note that Theorem 2.6.6 gives nearly the best possible rate in the sense that if (2.6.7) holds, then

(2.6.10) $$E(H(|X_1|)) < \infty.$$

This can be proved along the lines of the second part of our proof of Theorem 2.3.2. However, one does not know in general for what functions

$H(\cdot)$ of Theorem 2.6.6 would (2.6.10) imply (2.6.7) with $o(\text{inv } H(n))$ instead of $O(\text{inv } H(n))$. We have already seen, for example, that if $H(x)=e^{tx}$, then $O(\cdot)$ of (2.6.7) cannot be replaced by $o(\cdot)$, while this can be done if $H(x)=x^p$, $p>2$. An example of a case when it is not known whether $o(\cdot)$ can replace $O(\cdot)$ in (2.6.7) is $H(x)=e^{x^r}$, $0<r<1$.

It is also of interest to find an analogue of Theorem 2.6.2 when (2.6.10) holds for functions $H(\cdot)$ of Theorem 2.6.6. In this regard we have

Theorem 2.6.7 (Komlós, Major, Tusnády 1975, 1976). *Assume the conditions of Theorem 2.6.6. Then*

$$(2.6.11) \qquad P\{\max_{1 \leq k \leq n} |S_k - W(k)| > x_n\} \leq C_2 \frac{n}{H(ax_n)},$$

provided $\text{inv } H(n) < x_n < C_1 (n \log n)^{1/2}$, *where* C_1, C_2 *and* a *are positive constants depending only on the distribution function of* X_1.

It was an advantage of Theorem 2.6.2 that it implied Theorem 2.6.1. Unfortunately the corresponding statement is not true in the case of Theorem 2.6.7; in fact it does not imply Theorem 2.6.6 and not even Theorem 2.6.3.

It is clear from our discussion so far that we would have to prove only Theorems 2.6.2, 2.6.6 and 2.6.7 in order to complete the treatment of the Komlós, Major, Tusnády theorems of this Section. The details of the proofs of these theorems are very complicated and would take up a lot of space to give them here. The proof of Theorem 2.6.3 in the case of $2<p\leq3$ is relatively simple. However this special case was not covered by the original paper of Komlós, Major, Tusnády (1975, 1976), who treated only the case $p>3$. The case $2<p\leq3$ was settled by Major (1976a), whose proof is based on the same construction as that of the proof of Theorem 2.5.1 but, instead of Theorem 2.5.2, the following moderate deviation theorem is applied.

Theorem 2.6.8 (Major 1976a). *Put*

$$X_i' = \begin{cases} X_i & \text{if} \quad |X_i| < c(n \log n)^{1/2}, \\ 0, & \text{otherwise}, \end{cases}$$

$$\tilde{X}_i = \frac{X_i' - EX_i'}{\sqrt{\text{Var } X_i'}},$$

$$F_n(x) = P\left(\frac{\tilde{X}_1 + \ldots + \tilde{X}_n}{\sqrt{n}} \leq x\right).$$

Then for any $0<\varepsilon<p-2$ and $0<x<\dfrac{\varepsilon}{8c}\sqrt{\log n}$ we have
$$1-F_n(x) = (1-\Phi(x))(1+O(n^{\frac{2+\varepsilon-p}{2}}))$$
and
$$F_n(-x) = \Phi(-x)(1+O(n^{\frac{2+\varepsilon-p}{2}})),$$
provided the conditions of Theorem 2.6.3 hold with $2<p\leq 3$.

As mentioned already, the proof of Theorem 2.6.2 is very complicated. Instead of its details, we will sketch its new ideas as compared to the proof of Theorem 2.5.1. First we recall that in the proof of the latter we constructed a Wiener process in terms of the sequence of partial sums $\{S_k\}$ given there. When proving Theorem 2.6.2 one does the opposite of this. Namely, the sequence $\{S_k\}$ is constructed in terms of a given Wiener process W. Also, in Theorem 2.5.1 first we constructed our Wiener process only at the sequence of points $n_k=[k^\alpha]$, $\alpha=\tfrac{3}{2}$, and then we joined these constructed points of W randomly, so that after this randomization we should still have a Wiener process. In the proof of Theorem 2.6.2 one first constructs the sequence $\{S_n\}$ at the points $n_k=2^k$, but then these points are not joined randomly. Rather we continue *constructing* also the values of S_n for $n_k<n<n_{k+1}$. For example, when constructing $S_{3N/2}=S_{(n_k+n_{k+1})/2}=$ $=\tfrac{1}{2}(S_N+S_{2N})+(S_{3N/2}-\tfrac{1}{2}(S_N+S_{2N}))$, $N=n_k$, it appears that we should construct the r.v. $S_{3N/2}-\tfrac{1}{2}(S_N+S_{2N})$ in terms of $W(3N/2)-\tfrac{1}{2}(W(N)+W(2N))$. However, there is a bit of difficulty. Namely, while the r.v. $W(3N/2)-\tfrac{1}{2}(W(N)+W(2N))$ is independent of the r.v. $W(N)$ and $W(2N)$, the r.v. $S_{3N/2}-\tfrac{1}{2}(S_N+S_{2N})$ is not independent of the already constructed r.v. S_N and S_{2N}.

It appears to be reasonable to believe, however, that the r.v. $S_{3N/2}-\tfrac{1}{2}(S_N+S_{2N})$ should not be effected very much by the already given values of S_N, S_{2N}. Now the transformation

$$\operatorname{inv} F_N\left(G\left(W\left(\frac{3N}{2}\right)-\frac{W(N)+W(2N)}{2}\right)\right),$$

where G is the distribution function of the r.v. $W\left(\dfrac{3N}{2}\right)-\tfrac{1}{2}(W(N)+W(2N))$ and F_N is the distribution function of $S_{3N/2}-\tfrac{1}{2}(S_N+S_{2N})$, gives us a r.v. with the desired distribution F_N. Unfortunately, this latter r.v. is still not independent of S_N and S_{2N}. Whence the joint distribution of $(S_N, S_{3N/2}, S_{2N})$ is not the desired one yet. In order to construct the latter properly, Komlós, Major and Tusnády (1975) introduced the conditional transformation

$$\text{inv } F_N^* \left(G \left(W \left(\frac{3N}{2} \right) - \frac{W(N) + W(2N)}{2} \right) \bigg| S_{2N} - S_N \right),$$

where $F_N^*(x|y) = P\{S_{3N/2} - \frac{1}{2}(S_N + S_{2N}) \leq x | S_{2N} - S_N = y\}$. Continuing along these lines, in the next step one constructs the r.v. $S_{N+N/4}$, $S_{N+3N/4}$, and then, step by step, all the r.v. S_j ($j = N+1, \ldots, 2N$). It is clear, that the joint distribution of the thus constructed sequence is as desired. In order to prove now that the constructed sequence S_j is so close to the given Wiener process $W(j)$ as stated in Theorem 2.6.2, one has to prove that the conditional distribution function $F_N^*(x|y)$ is appropriately near to $G(x)$. This means that, instead of Theorem 2.5.2, one now needs a stronger theorem and that for conditional distributions.

Supplementary Remarks

Section 2.2.

Theorem S. 2.2.1 (Major 1979). *Let a distribution function $F(x)$ be given with $\int x \, dF(x) = 0$, $\int x^2 \, dF(x) = 1$. Define*

$$\sigma_k^2 = \int_{-\sqrt{2^n}}^{\sqrt{2^n}} x^2 \, dF(x) - \left(\int_{-\sqrt{2^n}}^{\sqrt{2^n}} x \, dF(x) \right)^2 \quad \text{if} \quad 2^n \leq k < 2^{n+1}, \ n = 1, 2, \ldots.$$

A sequence of i.i.d.r.v. $X_1, X_2, \ldots$, having the distribution function F as above, and a sequence of independent normal random variables $Y_1, Y_2, \ldots$ with $EY_k = 0$, $EY_k^2 = \sigma_k^2$ can be constructed in such a way that the partial sums $S_n = X_1 + \ldots + X_n$, $T_n = Y_1 + \ldots + Y_n$, $n = 1, 2, \ldots$ ($S_0 = T_0 \equiv 0$) satisfy the relation

(S.2.2.1) $\qquad\qquad |S_n - T_n| \stackrel{\text{a.s.}}{=} o(n^{1/2}).$

Note that in this theorem the partial sum sequence $\{S_n\}$ is approximated by a Gaussian process $\{T_n\}$ and not by a Wiener process $\{W(n)\}$. However, this rate of approximation is better than that of Strassen's theorem (cf. (2.2.2*)). On the other hand, Theorem 2.2.3 states that the rate of (2.2.2*) is the best possible if S_n is to be approximated by a Wiener process. Also, just preceding Theorem 2.2.2, we complained that Donsker's theorem (cf. (0.5)) could not follow from a strong invariance principle if S_n were to be estimated by a Wiener process. We are happy now to be able to say that Theorem S.2.2.1 is that strong invariance principle which implies Donsker's theorem and also that of Strassen. In order to see this, define $T_n(t)$ à la $S_n(t)$ of (0.3), i.e.,

$$T_n(t) = n^{-1/2}\{T_{[nt]}+Y_{[nt]+1}(nt-[nt])\},$$

and a Wiener process such that

$$W(n) = \sum_{i=1}^{n} \frac{Y_i}{\sigma_i}, \quad n = 1, 2, \ldots.$$

Since $\sigma_n \to 1$, we have

(S.2.2.2) $$\sup_{0 \le t \le 1} |T_n(t) - n^{-1/2} W(nt)| \xrightarrow{P} 0,$$

and, by an appropriate law of iterated logarithm (cf. e.g. Feller 1943)

(S.2.2.3) $$\lim_{n \to \infty} \frac{|T_n - W(n)|}{\sqrt{n \log \log n}} \stackrel{\text{a.s.}}{=} 0.$$

Clearly, (S.2.2.2) implies (0.5) and (S.2.2.3) implies (2.2.2*).

Section 2.4. Concerning the problem of longest runs of pure heads there is a gap between the statements of Theorems 2.4.1 and 2.4.2. In a recent paper Erdős and Révész (1976) proved some further related results. One of them states: Let X_i be i.i.d.r.v. with $P(X_1=0)=P(X_1=1)=\frac{1}{2}$, $S_n = X_1 + \ldots + X_n$ and $\alpha_n(k) = [\log n - \log \log \log n + \log \log e - k]$, where the basis of log is two. Then

(i) for any $\varepsilon > 0$ and almost all ω there exists $n_0 = n_0(\omega, \varepsilon)$ such that

$$\max_{1 \le k \le n - \alpha_n(2+\varepsilon)} (S_{k+\alpha_n(2+\varepsilon)} - S_k) = \alpha_n(2+\varepsilon),$$

provided $n \ge n_0$;

(ii) for any $\varepsilon > 0$ and almost all ω there exists a sequence $\{n_j(\omega, \varepsilon)\}$ such that

$$\max_{1 \le k \le n_j - \alpha_{n_j}(1-\varepsilon)} (S_{k+\alpha_{n_j}(1-\varepsilon)} - S_k) < \alpha_{n_j}(1-\varepsilon).$$

Theorem 2.4.3 is somewhat stronger than the original result of Erdős and Rényi (1970). This form as well as its proof is due to P. Bártfai.

For further developments concerning Erdős–Rényi laws (Theorem 2.4.3) we refer to Komlós and Tusnády (1975), Book (1976) and S. Csörgő (1979). The result in Theorem 2.4.3 remains true if we require the finiteness of the moment generating function only in some right hand side neighbourhood of the origin (in this case EX_1 can even be $-\infty$).

Section 2.5. In dealing with the stochastic geyser problem, Bártfai (1966) used a technique similar to that of the proof of Theorem 2.5.1. Reformulating his result in strong invariance context, one gets a somewhat weaker form of Strassen's result (2.2.9). His work was done independently in about the same time as Strassen's. The relatedness of their results, however, was only realized later on.

The form of Theorem 2.5.1 quoted here, differs slightly from that of Csörgő and Révész (1975a). This version achieves the same goal in giving a counterexample to Strassen's conjecture (1965b) and also provides a more straightforward illustration of our method.

3. A Study of Partial Sums with the Help of Strong Approximation Methods

3.0. Introduction

The fundamental aim of Chapter 1 was to study the one dimensional Wiener process, keeping a special eye on properties which might be inherited by sums of r.v. In this chapter we intend to realize the ones which are indeed inherited. This goal is going to be achieved with the help of the invariance principles covered in Chapter 2. It is clear from these invariance principles that those partial sums whose summands have many moments will inherit more than those whose summands have fewer moments.

3.1. How big are the increments of partial sums of I.I.D.R.V. when the moment generating function exists?

The message of this section is summarized in the following

Theorem 3.1.1. Let $X_1, X_2, \ldots$ be a sequence of i.i.d.r.v. with mean zero and variance one, satisfying also the condition

(3.1.1) there exists a $t_0 > 0$ such that $E e^{tX_1}$ is finite if $|t| < t_0$.

Let $\{a_N\}$ be a monotonically non-decreasing sequence of integers satisfying the conditions (i), (ii) of Theorem 1.2.1 and assume also that

(3.1.2) $\quad\quad\quad\quad\quad a_N/\log N \to \infty \quad \text{as} \quad N \to \infty.$

Then for the sums $S_n = X_1 + X_2 + \ldots + X_n$ we have

(3.1.3) $\quad\quad \varlimsup_{N\to\infty} \max_{1 \le n \le N-a_N} \max_{1 \le k \le a_N} \beta_N |S_{n+k} - S_n| \stackrel{\text{a.s.}}{=} 1,$

(3.1.4) $\quad\quad \varlimsup_{N\to\infty} \max_{1 \le n \le N-a_N} \beta_N |S_{n+a_N} - S_n| \stackrel{\text{a.s.}}{=} 1,$

(3.1.5) $\quad\quad \varlimsup_{N\to\infty} \max_{1 \le k \le a_N} \beta_N |S_{N+k} - S_N| \stackrel{\text{a.s.}}{=} 1,$

(3.1.6) $\quad\quad \varlimsup_{N\to\infty} \beta_N |S_{N+a_N} - S_N| \stackrel{\text{a.s.}}{=} 1$

where

$$\beta_N = \left(2a_N\left[\log\frac{N}{a_N}+\log\log N\right]\right)^{-1/2}.$$

If condition (iii) *of Theorem 1.2.1 is also satisfied, then*

(3.1.7) $$\lim_{N\to\infty}\max_{1\le n\le N-a_N}\beta_N|S_{n+a_N}-S_n|\stackrel{a.s.}{=}1$$

and

(3.1.8) $$\lim_{N\to\infty}\max_{1\le n\le N-a_N}\max_{1\le k\le a_N}\beta_N|S_{n+k}-S_n|\stackrel{a.s.}{=}1.$$

Proof. Combining Theorems 1.2.1 and 2.6.1 and noticing that (3.1.2) implies that $\beta_N\log N\to 0$ as $N\to\infty$, the result follows.

Remark 3.1.1. Remark 1.2.1 is also applicable here. That is to say the absolute value signs can be omitted from formulas (3.1.3)–(3.1.8), and changing the lim sup resp. max to lim inf resp. min, the $+1$ on the right hand sides of (3.1.3)–(3.1.8) will be changed to -1.

Remark 3.1.2. It is clear that this theorem plays the same role for partial sums of i.i.d.r.v. having moment generating function as Theorem 1.2.1 does for Wiener processes. However, the extra assumption that $a_N/\log N\to\infty$ is absolutely crucial here in the sense that if $a_N=c\log N$, $c>0$, then, by Theorem 2.4.3, the left-hand side limit of (3.1.7) still exists, but its value will depend on c and the distribution function of X_1, that is to say this latter case cannot be explained from any invariance principle. Thus, Theorems 3.1.1 and 2.4.3 together completely describe the almost sure behaviour of the increments of length $c\log N\le a_N\le N$ of partial sum sequences when the moment generating function of the summands exists. The case of $a_N=o(\log N)$ does not appear to be known in general; it is clear, however, that this case also cannot be treated from invariance-principle-like considerations.

Remark 3.1.3. Besides condition (3.1.2), condition (3.1.1) is also necessary in Theorem 3.1.1 in the following sense: *Let $F(x)$ be a distribution function with mean 0 and variance 1 not satisfying condition (3.1.1). Then there exists a sequence $\{a_N\}$ (depending on F) satisfying conditions* (i), (ii) *of Theorem 1.2.1 (as well as condition (3.1.2) of Theorem 3.1.1) such that none of the statements (3.1.3)–(3.1.8) hold true.* This fact can be seen by going through the following simple steps.

1) If for every $\varepsilon > 0$ $\int_{-\infty}^{+\infty} e^{\varepsilon x} dF(x) = \infty$ (that is (3.1.1) does not hold true), then there exists a positive non-increasing function $\varepsilon(x) \to 0$ such that $\int_{-\infty}^{+\infty} e^{x\varepsilon(x)} dF(x) = \infty$.

2) Let $H(x) = e^{x\varepsilon(x)}$. Then the inverse function of $H(x)$ can be written as inv $H(x) = v(x) \log x$ where $v(x)$ is a non-decreasing function, tending to $+\infty$.

3) Since
$$\sum_{k=1}^{\infty} P\{|X_k| \geq \text{inv } H(k)\} = \infty$$
infinitely many of the events $\{|X_k| \geq v(k) \log k\}$ will occur with probability 1.

Choosing $a_k = v(k) \log k$, we see immediately that the statements (3.1.3), (3.1.5) and (3.1.8) cannot be true. As to the others not being true, let $n_1 = n_1(\omega) < n_2 = n_2(\omega) < \dots$ be a sequence such that $|X_{n_k}| \geq v(n_k) \log n_k$ and consider the sequence $S_{n_1-1} - S_{n_1-a_{n_1}}, S_{n_2-1} - S_{n_2-a_{n_2}}, \dots$. The independence of $S_{n_k-1} - S_{n_k-a_{n_k}}$ and X_{n_k} for each k now implies our claim for (3.1.4), (3.1.6) and (3.1.7).

3.2. How big are the increments of partial sums of I.I.D.R.V. when the moment generating function does not exist?

The aim of this section is to find a theorem which should play the role of Theorem 3.1.1 when we assume only the existence of a finite number of moments instead of that of the moment generating function. An answer to this question is summarized in the following

Theorem 3.2.1. *Let $X_1, X_2, \dots$ be a sequence of i.i.d.r.v. with mean zero and variance one and let $H(x)$, $x > 0$, be a non-decreasing continuous function for which the following assumptions hold:*

(3.2.1) $\qquad EH(|X_1|) < \infty,$

(3.2.2) $\qquad \lim_{x \to \infty} \dfrac{H(\varepsilon x)}{H(x)} > 0$ *for every* $\varepsilon > 0,$

(3.2.3) $x^{-(2+\varepsilon)} H(x)$ *is an increasing function of x for some $\varepsilon > 0$,*

(3.2.4) $x^{-1} \log H(x)$ *is non-increasing.*

Let $\{a_N\}$ be a non-decreasing sequence of integers, satisfying conditions (i), (ii) of Theorem 1.2.1 and also assume that there exists a $C>0$ such that

(3.2.5) $$a_N \geq C \frac{(\text{inv } H(N))^2}{\log N}.$$

Then the statements (3.1.3)–(3.1.6) of Theorem 3.1.1 are true. If we also assume that condition (iii) of Theorem 1.2.1 is satisfied then the statements (3.1.7) and (3.1.8) are also true.

Proof. Combining Theorems 1.2.1, 2.6.6 and Lemma 2.6.2 the result follows, since $\varlimsup\limits_{N\to\infty} \beta_N \text{ inv } H(N) < \infty$.

Remark 3.2.1. The condition (3.2.2) is satisfied by all functions varying slowly (cf. Feller, 1966, p. 269).

Remark 3.2.2. An important special case of Theorem 3.2.1 is when we take $H(x)=x^p$, $p>2$, that is when $E|X_1|^p$, $p>2$ is finite. Then conditions (3.2.2), (3.2.3) and (3.2.4) are satisfied and the assumption (3.2.5) is equivalent to $a_N \geq CN^{2/p}/\log N$. Hence, for all these latter increments a_N, Theorem 3.2.1 holds. Similarly, if we take $H(x)=x^p/(\log x)^{2/p}$, $p>2$, then Theorem 3.2.1 is true for $a_N \geq N^{2/p}$. This latter case was directly studied by Lai (1974) who proved (3.1.5) and (3.1.6) in the special case when $a_N = N^\alpha$, $\alpha \geq 2p^{-1}$, $p>2$.

Another interesting special case of Theorem 3.2.1 is the function $H(x)=e^{x^\alpha}$, $0<\alpha<1$, i.e. we assume that $Ee^{|X_1|^\alpha}<\infty$. Then all the moments of X_1 exist but not the moment generating function. Again conditions (3.2.2), (3.2.3) and (3.2.4) are satisfied and the assumption (3.2.5) is equivalent to $a_N \geq C(\log N)^{2/\alpha-1}$. Hence, for all these latter increments a_N, Theorem 3.2.1 holds.

Remark 3.2.3. Comparing our Theorem 3.1.1 to Theorem 3.2.1, we see that the essential difference between them concerns the length a_N of the increments. That is to say, when we only assume the existence of a finite number of moments instead of that of the moment generating function of X_1, then we impose a stronger restriction on a_N, namely the condition (3.2.5) instead of (3.1.2). This, of course, is technically due to the fact that in the case when only a finite number of moments exist the rate of convergence of the strong invariance principle is weaker than in the case when moment generating function exists (cf. Theorems 2.6.6 and 2.6.1). On the other hand, the assumption (3.2.1) is the best possible one in the following sense: *if there exists a function $H(x)$ satisfying conditions (3.2.2), (3.2.3), (3.2.4) and $EH(|X_1|)=\infty$ then there exists a $C>0$ (depending on H) such that*

$$a_N = C \frac{(\operatorname{inv} H(N))^2}{\log N}$$

and none of the statements (3.1.3)–(3.1.8) *holds true.* In order to see this remark, observe that in this case

$$\beta_N \approx \left(2(\operatorname{inv} H(n))^2 \frac{\log (N(\operatorname{inv} H(N))^{-2})}{\log N}\right)^{1/2} \geqq C(\operatorname{inv} H(N))^{-1}$$

and

$$\sum_{n=1}^{\infty} P\{|X_n| \geqq \operatorname{inv} H(n)\} = \infty$$

which, in turn, contradicts the statements (3.1.3)–(3.1.8) (cf. Remark 3.1.3).

Remark 3.2.4. The case which we have not mentioned so far is that of $H(x)=x^2$. In order to handle this latter one we have to combine the first strong invariance principle, namely that of Strassen (Theorem 2.2.1), with Theorem 1.2.1. This way we get, with $a_N=cN$, $0<c\leqq 1$, that

(3.2.6) $$\varlimsup_{N\to\infty} \max_{0\leqq n\leqq N-cN} \frac{|S_{n+cN}-S_n|}{\sqrt{2cN\log\log N}} \overset{\text{a.s.}}{=} 1$$

which, in the case of $c=1$, reduces to the classical law of iterated logarithm (Hartman–Wintner 1941). We note that Strassen's law of iterated logarithm (Theorem 1.3.2) holds also for partial sums. Indeed, applying Theorem 2.2.1 it follows from Theorem 1.3.2 that

Theorem 3.2.2 (Strassen 1964). *Let $X_1, X_2, \ldots$ be i.i.d.r.v. with mean zero and variance one. Let $S_n(t)$ be defined as in (0.3). Define*

(3.2.7) $$\gamma_n(t) = \frac{S_n(t)}{\sqrt{2\log\log n}}.$$

Then the sequence $\{\gamma_n(t)\}$ is relatively compact in $C(0, 1)$ with probability one, and the set of its limit points is $\mathscr{S}$, where $\mathscr{S}$ is as in Theorem 1.3.2.

3.3. How small are the increments of partial sums of I.I.D.R.V.?

In Section 2.4. we investigated the length of the longest head run in a coin tossing sequence of size N. As we have already noted (cf. Supplementary Remarks 2.4), the length of the longest head run in N experiments is more than

$$\alpha_N = [\log N - \log\log\log N + \log\log e - 2 - \varepsilon]$$

(for any $\varepsilon > 0$, if N is large enough) and it is less than

$$[\log N - \log\log\log N + \log\log e - 1 + \varepsilon]$$

for infinitely many N, where the base of log is 2.

Call a run regular if each tail in it is followed by a head and vice versa (that is a regular run goes like HTHTHT... or THTHT...). In a sense a converse of the problem of T. Varga (cf. Section 2.4) is to ask for the length of the longest regular run. It is not hard to see that the answer to this question is exactly the same as that to the original one.

This remark suggests the following classroom joke. Describe the experiment of T. Varga to the students, and do the experiment in your class only after this information had been given. Clearly now, head runs of length seven will occur in both groups. For your selection of students doing the experiment randomly, pick the ones whose tally would contain at least one regular run of length seven.

An analogue of the above described result on the length of longest regular runs can be formulated for partial sums of i.i.d.r.v. as follows:

Theorem 3.3.1. *Let* $X_1, X_2, \ldots$ *be a sequence of i.i.d.r.v. with mean* 0 *and variance* 1, *satisfying also the condition* (3.1.1). *Let* $\{a_n\}$ *be a non-decreasing sequences of integers satisfying the conditions* (i), (ii) *of Theorem 1.7.1 and assume also that*

(3.3.1) $$a_n (\log n)^{-3} \to \infty \quad \text{as} \quad n \to \infty.$$

Then for the sum $S_n = X_1 + X_2 + \ldots + X_n$ *we have*

(3.3.2) $$\lim_{N \to \infty} \min_{1 \leq n \leq N - a_N} \max_{1 \leq k \leq a_N} \gamma_N |S_{n+k} - S_n| \stackrel{a.s.}{=} 1,$$

where

$$\gamma_N = \left(\frac{8}{\pi^2} \frac{\log N/a_N + \log\log N}{a_N} \right)^{1/2}.$$

If condition (iii) of Theorem 1.7.1 is also satisfied, then
$$\lim_{N\to\infty} \min_{1\le n\le N-a_N} \max_{1\le k\le a_N} \gamma_N |S_{n+k}-S_n| \stackrel{\text{a.s.}}{=} 1. \qquad (3.3.3)$$

Proof. Notice that (3.3.1) implies $\gamma_N \log N \to 0$ as $N \to \infty$ and apply Theorems 1.7.1 and 2.6.1.

In case $a_N \le C \log^3 N$, Theorem 1.7.1 cannot be extended to partial sums of i.i.d.r.v. via the invariance principle. However, applying a small deviation theorem of A. A. Mogul'skiĭ (1974) one can generalize Theorem 3.3.1 for a wider class of sequences $\{a_N\}$. The mentioned result of Mogul'skiĭ is

Theorem 3.3.2 (Mogul'skiĭ 1974). *Let $X_1, X_2, \ldots$ be a sequence of i.i.d.r.v. with mean 0 and variance 1. Let $\{t_n\}$ be a sequence of positive numbers, for which*
$$t_n \to \infty, \quad n^{-1/2} t_n \to 0.$$
Then
$$\log P\{\max_{1\le k\le n} |S_k| \le t_n\} \approx -\frac{\pi^2}{8} \frac{n}{t_n^2}.$$

Applying this theorem with
$$t_n = \left(\frac{\pi^2}{8} \frac{n}{\log\log n}\right)^{1/2}$$
and repeating the proof of Theorem 1.7.1, one gets the following sharper form of Theorem 3.3.1.

Theorem 3.3.1*. *Let $X_1, X_2, \ldots$ be a sequence of i.i.d.r.v. with mean 0 and variance 1. Let $\{a_n\}$ be a non-decreasing sequence of integers satisfying conditions (i), (ii) of Theorem 1.7.1 and assume also that $a_n(\log n)^{-1} \to \infty$. Then we have (3.3.2). If condition (iii) of Theorem 1.7.1 is also satisfied then we have also (3.3.3).*

Comparing Theorems 3.1.1, 3.2.1, Remarks 3.1.2 and 3.2.3 on one side and Theorem 3.3.1* on the other side, one should observe that, when investigating big increments of partial sums of i.i.d.r.v., stronger restrictions are required for the sequence $\{a_n\}$ when less is assumed about the existence of moments, but, in case of small increments, the restriction $a_n(\log n)^{-1} \to \infty$ is already sufficient if only two moments exist. In case of $a_n = c \log n$ ($c > 0$), Theorem 2.4.3 solves the problem of big increments. The problem of small increments in this case is yet unsolved. For the latter we formulate the following

Conjecture. Let $F(x)$ be a distribution function with $\int x dF = 0$, $\int x^2 dF = 1$. Let $X_1, X_2, \ldots$ be a sequence of i.i.d.r.v. with $P(X_1 \leq t) = F(t)$. Then

$$\lim_{N \to \infty} \min_{1 \leq n \leq N - a_N} \max_{1 \leq k \leq a_N} |S_{n+k} - S_n| \stackrel{\text{a.s.}}{=} r(c),$$

where $a_N = [c \log N]$, and $r(c)$ is a function which uniquely determines F.

It is interesting to consider the special case of Theorem 3.3.1* when $a_n = n$. In this case we have

(3.3.4) $$\lim_{n \to \infty} \left(\frac{8 \log \log n}{\pi^2 n} \right)^{1/2} \max_{1 \leq k \leq n} |S_k| \stackrel{\text{a.s.}}{=} 1.$$

This result was proved at first by Chung (1948) under somewhat stronger moment restrictions. Jain and Pruitt (1975) were the first to prove that (3.3.4) is still true if only the existence of two moments is assumed.

3.4. A summary

In the above sections we investigated how the strong laws proved in Chapter 1 can be inherited by partial sums of i.i.d.r.v. applying the strong invariance principles of Chapter 2. We saw that in many cases strong invariance techniques produced the best results also for partial sums. In some cases, however, direct methods produced better ones. When using the technique of strong invariance, we always had to pay special attention to moment conditions. The question of inheritance is much simpler when talking about weak convergence. Indeed, we can simply say that once Donsker's theorem (cf. (0.5)) is proved, distribution results for given functionals of a Wiener process (cf. e.g. (1.5.1), (1.5.2) and Theorem 1.5.4) are inherited by the same functionals of $\{S_k\}$ in the limit. As to inheriting Theorem 1.5.4 we also refer to Erdős, Kac (1947).

Supplementary remarks

Section 3.1. In connection with the Erdős–Rényi law of large numbers (Theorem 2.4.3) Komlós and Tusnády (1975) studied the question how frequently the event

$$\frac{S_{k+[c \log n]} - S_k}{[c \log n]} \geq x \quad (k = 0, 1, 2, \ldots, n - [c \log n]; \, x < \alpha(c))$$

will occur. They proved that the sequence of indices $0 \leq k_1 < k_2 < \ldots$
$\ldots < k_v \leq n - [c \log n]$ for which the above inequality holds true forms
a Poisson process in the limit as $n \to \infty$. For further related results we refer
to Book (1976), S. Csörgő (1979), Révész (1978), Steinebach (1979), Csörgő,
Steinebach (1980) and Guibas, Odlyzko (1980).

Section 3.2. In connection with the Hartman–Wintner law of iterated
logarithm ((3.2.6) with $c=1$) it is of interest to investigate the case $EX_1^2 = \infty$.
This was done by Strassen (1966) who proved that

$$\varlimsup_{n \to \infty} \frac{|X_1 + X_2 + \ldots + X_n|}{\sqrt{2n \log \log n}} \stackrel{\text{a.s.}}{=} \infty,$$

if $X_1, X_2, \ldots$ are i.i.d.r.v. with $EX_1 = 0$, $EX_1^2 = \infty$. Later I. Berkes (1972)
has shown that this result of Strassen is the strongest possible one in the
following sense: *for any function $f(n)$ with $\lim_{n \to \infty} f(n) = \infty$ there exists
a sequence $X_1, X_2, \ldots$ of i.i.d.r.v. for which $EX_1 = 0$, $EX_1^2 = \infty$ and*

$$\lim_{n \to \infty} \frac{|X_1 + X_2 + \ldots + X_n|}{f(n) \sqrt{n \log \log n}} \stackrel{\text{a.s.}}{=} 0.$$

For further results in this connection we refer to Klass (1976, 1977),
who evaluated the normalizing factor $\mathcal{K}_n$ for which

$$\varlimsup_{n \to \infty} \frac{|X_1 + X_2 + \ldots + X_n|}{\mathcal{K}_n} \stackrel{\text{a.s.}}{=} 1,$$

where $EX_1 = 0$, $EX_1^2 = \infty$. This normalizing factor $\mathcal{K}_n$ depends on the
distribution function $F(t) = P(X_1 \leq t)$.

A more detailed characterization of the behaviour of the partial sum
sequence $\{S_n\}$ can be given using the concept of upper and lower classes
introduced by P. Lévy. Lévy says that a function $f(n)$ belongs to the
upper class if the inequality $|S_n| \leq f(n)$ holds with probability 1 for all
but finitely many n and $f(n)$ belongs to the lower class if the inequality
$|S_n| > f(n)$ holds infinitely often with probability 1. Using this terminology
the law of iterated logarithm says that for any $\varepsilon > 0$ the function $f(n) =$
$= ((2+\varepsilon) n \log \log n)^{1/2}$ belongs to the upper class and $((2-\varepsilon) n \log \log n)^{1/2}$
belongs to the lower class.

Feller (1943) gave a test to decide whether a function belongs to the
upper or lower class in the case when the moment condition $E(X_1^2 |\log|X_1||) <$
$< \infty$ holds true. His theorem states that a function $f(n) = n^{1/2} \Psi(n)$, with

$\Psi(n)$ non-decreasing, belongs to the upper class if the series

$$\sum_{n=1}^{\infty} n^{-1} \Psi(n) e^{-\Psi^2(n)/2}$$

converges and $f(n)$ belongs to the lower class if it diverges. For example for any $\varepsilon > 0$ the function $f(n) = (n(2\log\log n + (3+\varepsilon)\log\log\log n))^{1/2}$ belongs to the upper class and $f(n) = (n(2\log\log n + 3\log\log\log n))^{1/2}$ belongs to the lower class. In a sense this result gives the rate of convergence for the law of iterated logarithm. In fact the original form of the law of iterated logarithm implies that for any $\varepsilon > 0$ and for all but finitely many n we have $(2n\log\log n)^{-1/2}|S_n| \in (-1-\varepsilon, 1+\varepsilon)$ with probability one. Our last statement implies that $(2n\log\log n)^{-1/2}|S_n| \in (-1-\varepsilon_n, 1+\varepsilon_n)$ with probability 1, for all but finitely many n, where

$$\varepsilon_n = \frac{(3+\varepsilon)\log\log\log n}{4\log\log n} \quad (\varepsilon > 0).$$

The question of how a rate of convergence for Strassen's law of iterated logarithm (Theorem 3.2.2) could be produced was investigated by Bolthausen (1978). It is not known whether his result is best possible or not.

We should also mention that the above formulated test of Feller could also be proved by the invariance principle (Theorem 2.6.6) with the stronger moment condition $E|X_1|^{2+\varepsilon} < \infty$ ($\varepsilon > 0$), if we had the same test for a Wiener process.

Section 3.3. In his original paper Chung (1948) assumes the existence of the third moments but he proves a stronger result than (3.3.4). In fact he gets the following upper-lower class type result:

$$P\{\max_{1 \leq k \leq n} |S_k| < a_n n^{1/2} \text{ i.o.}\} = \begin{cases} 1 & \text{if } \sum_{n=1}^{\infty} \frac{1}{na_n^2} e^{-\frac{\pi^2}{8a_n^2}} = \infty, \\ 0 & \text{otherwise.} \end{cases}$$

Hirsch (1965) investigated the properties of the sequence $\{\max_{1 \leq k \leq n} S_k = M_n^+\}$ and proved

$$P\{M_n^+ < a_n n^{1/2} \text{ i.o.}\} = \begin{cases} 1 & \text{if } \sum_{n=1}^{\infty} a_n/n = \infty, \\ 0 & \text{otherwise,} \end{cases}$$

when the third moment exists. Replacing M_n^+ by $M_n^- = -\min_{1 \leq k \leq n} S_k$ in the formula given above, it clearly remains true.

Since $M_n = \max_{1 \leq k \leq n} |S_k| = \max(M_n^+, M_n^-)$, formula (3.3.4) implies that with probability 1 there are only finitely many n for which

$$M_n^+ < (1-\varepsilon)\left(\frac{\pi^2 n}{8 \log \log n}\right)^{1/2} \quad \text{and} \quad M_n^- < (1-\varepsilon)\left(\frac{\pi^2 n}{8 \log \log n}\right)^{1/2}.$$

Roughly speaking this means that if M_n^+ is small (smaller than $(1-\varepsilon) \cdot \left(\frac{\pi^2 n}{8 \log \log n}\right)^{1/2}$) then M_n^- is big (bigger than $(1-\varepsilon)\left(\frac{\pi^2 n}{8 \log \log n}\right)^{1/2}$) provided that n is big enough. Reasoning along this line of thought, one can expect that if M_n^+ is much smaller than $(1-\varepsilon) \cdot \left(\frac{\pi^2 n}{8 \log \log n}\right)^{1/2}$ then M_n^- must be much larger. Csáki (1978) proved that this is indeed true. Let $a(t) > 0$, $b(t) > 0$ be non-increasing functions such that $a(t) > 0$ and $b(t) > 0$ are increasing. Then, assuming the existence of two moments,

$$P\{M_n^+ \leq a(n)\sqrt{n} \text{ and } M_n^- \leq b(n)\sqrt{n} \text{ i.o.}\} = \begin{cases} 1 & \text{if } \sum_{n=1}^{\infty} \frac{a(n)}{nc^3(n)} e^{-\frac{\pi^2}{2c^2(n)}} = \infty, \\ 0 & \text{otherwise}, \end{cases}$$

where $c(n) = a(n) + b(n)$.

The special case $a(n) = b(n)$ of this theorem also gives Chung's law of iterated logarithm (3.3.4). Formally this theorem does not contain Hirsch's law of iterated logarithm, but Csáki also proved the validity of Hirsch's theorem in the case when only two moments exist.

In order to illustrate what Csáki's theorem is all about we present here an

Example. Put
$$a(n) = C(\log \log n)^{-1/2} \quad (0 < C < \sqrt{\pi^2/8})$$
and
$$b(n) = D(\log \log n)^{-1/2} \quad (D > 0).$$
Then the series
$$\sum_{n=1}^{\infty} \frac{a(n)}{nc^3(n)} e^{-\frac{\pi^2}{2c^2(n)}}$$
is convergent if $D < \pi/\sqrt{2} - C$ and divergent if $D \geq \pi/\sqrt{2} - C$. Applying Csáki's theorem this fact implies that the events

$$M^+(n) < C\left(\frac{n}{\log \log n}\right)^{1/2} \quad \text{and} \quad M^-(n) < D\left(\frac{n}{n \log \log}\right)^{1/2}$$

occur infinitely often with probability 1 if $0 < C < \sqrt{\pi^2/8}$ and $D \geq \pi/\sqrt{2} - C$.

However it is not so if $D < \pi/\sqrt{2} - C$. That is to say if n is big enough and
$$M^+(n) < C\left(\frac{n}{\log\log n}\right)^{1/2} \quad (0 < C < \sqrt{\pi^2/8}),$$
then it follows that
$$M^-(n) \geq D\left(\frac{n}{\log\log n}\right)^{1/2} \quad (0 < D < \pi/\sqrt{2} - C).$$

Csáki also proved an analogue of Strassen's converse law of iterated logarithm. In fact he proved that if (3.3.4) holds true then the second moment of the r.v. X_1 exists.

As we mentioned already, Csáki's theorem states that if one of the r.v. $M^+(n)$ and $M^-(n)$ is very small then the other one must be big. It is interesting to ask what happens if one of the r.v. $M^+(n)$ and $M^-(n)$ is very big. Strassen's law of iterated logarithm (Theorem 3.2.2) easily implies that for any $\varepsilon > 0$ the events

$$M^+(n) \geq \frac{1-\varepsilon}{3}(2n \log\log n)^{1/2} \quad \text{and} \quad M^-(n) \geq \frac{1-\varepsilon}{3}(2n \log\log n)^{1/2}$$

occur infinitely often with probability 1 but of the events

$$M^+(n) \geq \frac{1+\varepsilon}{3}(2n \log\log n)^{1/2} \quad \text{and} \quad M^-(n) \geq \frac{1+\varepsilon}{3}(2n \log\log n)^{1/2}$$

only finitely many can occur with probability 1. That is to say if n is big enough and
$$M^+(n) \geq \frac{1+\varepsilon}{3}(2n \log\log n)^{1/2},$$
then
$$M^-(n) \leq \frac{1+\varepsilon}{3}(2n \log\log n)^{1/2}.$$

In general, one can say that for any $\varepsilon > 0$ and $\frac{1}{3} \leq q < 1$ the events

$$M^+(n) \geq (1-\varepsilon)q(2n \log\log n)^{1/2} \quad \text{and} \quad M^-(n) \geq (1-\varepsilon)\frac{1-q}{2}(2n \log\log n)^{1/2}$$

occur infinitely often with probability 1 but of the events

$$M^+(n) \geq (1+\varepsilon)q(2n \log\log n)^{1/2} \quad \text{and} \quad M^-(n) \geq (1+\varepsilon)\frac{1-q}{2}(2n \log\log n)^{1/2}$$

only finitely many occur with probability 1.

4. Strong Approximations of Empirical Processes by Gaussian Processes

4.1. Some classical results

The role and development of the various invariance principles in the areas of partial sum and empirical processes are similar. Studying the latter, this chapter plays the same role as Chapter 2 does in the study of the former.

In this introduction we are going to introduce some of the classical definitions and results. Let $X_1, X_2, \ldots$ be i.i.d.r.v. with distribution function $F(\cdot)$ and define *the empirical distribution function* of the sample $X_1, \ldots, X_n$ by

$$F_n(x) = \frac{1}{n} \sum_{i=1}^{n} I_{(-\infty, x]}(X_i).$$

One can give an equivalent definition of the empirical distribution function $F_n(\cdot)$ in terms of the order statistics $X_1^{(n)} \leq X_2^{(n)} \leq \ldots \leq X_n^{(n)}$ of the random sample $X_1, \ldots, X_n$ as follows:

$$F_n(x) = \begin{cases} 0, & X_1^{(n)} > x, \\ \frac{k}{n}, & X_k^{(n)} \leq x < X_{k+1}^{(n)}, \quad k = 1, 2, \ldots, n-1, \\ 1, & X_n^{(n)} \leq x. \end{cases}$$

Clearly, for every fixed x, $F_n(x)$ is the relative frequency of successes in a Bernoulli sequence of trials with $EF_n(x) = F(x)$ and $\operatorname{Var} F_n(x) = \frac{1}{n} F(x)(1-F(x))$. Consequently, by the classical strong law of large numbers,

(4.1.1) $\qquad F_n(x) \xrightarrow{\text{a.s.}} F(x), \quad \text{for } x \text{ fixed.}$

Hence, using the language of statistics, $F_n(x)$ is an unbiased and strongly consistent estimator of $F(x)$ for each fixed x.

Viewing $\{F_n(x); -\infty < x < \infty\}$ as a stochastic process, its sample functions are distribution functions and it is of great importance that $F(x)$ can be uniformly estimated by this process with probability one. This is formulated in

Theorem 4.1.1 (Cantelli 1917 and Glivenko 1933).

(4.1.2) $$\sup_{-\infty < x < \infty} |F_n(x) - F(x)| \xrightarrow{\text{a.s.}} 0.$$

This theorem is rightly called the fundamental theorem of mathematical statistics. It tells us that, sampling ad infinitum, $F(x)$ can be uniquely determined with probability one. In fact, (4.1.2) is a simple consequence of (4.1.1).

From a practical point of view it is also of interest to study the rate of convergence in (4.1.2). Towards this end we define *the empirical process*

$$\beta_n(x) = \sqrt{n}(F_n(x) - F(x)), \quad -\infty < x < \infty.$$

The pointwise behaviour of $\beta_n(x)$ is quite simple. One immediately has the central limit theorem for each fixed x:

(4.1.3) $$\beta_n(x) \xrightarrow{\mathcal{D}} \mathcal{N}(0, F(x)(1-F(x))).$$

As to a rate of convergence in (4.1.2) we have

Theorem 4.1.2 (Kolmogorov 1933 and Smirnov 1939). *If $F(x)$ is a continuous distribution function, then*

(4.1.4) $$P\{\sup_{-\infty < x < \infty} |\beta_n(x)| \leq y\} \to K(y)$$

where

$$K(y) = \begin{cases} \sum_{k=-\infty}^{\infty} (-1)^k e^{-2k^2 y^2}, & y > 0 \\ 0, & \text{otherwise,} \end{cases}$$

and

(4.1.5) $$P\{\sup_{-\infty < x < \infty} \beta_n(x) \leq y\} \to S(y)$$

where

$$S(y) = \begin{cases} 1 - e^{-2y^2}, & y > 0, \\ 0, & \text{otherwise.} \end{cases}$$

A large deviation type result for the rate of convergence in (4.1.2) is also available:

Theorem 4.1.3 (Dvoretzky, Kiefer, Wolfowitz 1956). *There exists a universal constant C such that, for all $n > 0$ and $r > 0$,*

(4.1.6) $$P\{\sup_{-\infty < x < \infty} |\beta_n(x)| \geq r\} \leq C e^{-2r^2}.$$

4.2. Why should the empirical process behave like a Brownian bridge?

As we have seen, the study of $\beta_n(x)$, for x fixed, can be done on traditional grounds, as a result of its binomial distribution. The global statements (4.1.4), (4.1.5) and (4.1.6), however, cannot be handled easily. The original proofs of Kolmogorov and Smirnov, for example, are very complicated (cf. however Feller 1948).

As mentioned already in the Introduction, Doob (1949) suggested a novel approach for the study of the process $\{\beta_n(x); -\infty < x < \infty\}$. In order to describe Doob's idea we let $U_i = F(X_i)$, $i=1, 2, \ldots$. Then the U_i are $U(0, 1)$ r.v. provided $F(\cdot)$ is continuous. Let now $E_n(y)$ be the empirical distribution function of the sample $U_1, \ldots, U_n$ and denote the resulting empirical process in this case by

$$\alpha_n(y) = \sqrt{n}\,(E_n(y) - y), \quad 0 \leq y \leq 1.$$

Then $E\alpha_n(y) = 0$ and the covariance function of the process $\{\alpha_n(y); 0 \leq y \leq 1\}$ is

$$\varrho(s, t) = E\alpha_n(s)\alpha_n(t) = s \wedge t - st,$$

which coincides with that of a Brownian bridge $\{B(y); 0 \leq y \leq 1\}$. Using a multivariate version of the central limit theorem one can immediately say that

(4.2.1) $\qquad (\alpha_n(y_1), \ldots, \alpha_n(y_k)) \xrightarrow{\mathscr{D}} (B(y_1), \ldots, B(y_k)),$

for any fixed sequence $0 \leq y_1 < y_2 < \ldots < y_k \leq 1$. This should then suggest that the distributional properties of $\{\alpha_n(y); 0 \leq y \leq 1\}$ should coincide with those of $\{B(y); 0 \leq y \leq 1\}$ as $n \to \infty$. Indeed, Doob (1949), on this heuristic basis, said: "...in calculating asymptotic $\alpha_n(y)$ process distributions when $n \to \infty$ we may simply replace the $\alpha_n(y)$ process by the $B(y)$ process."

At this stage we call attention to the fact that this idea is essentially the same as the one which led us to Donsker's theorem (cf. (0.4)). Indeed, Donsker (1952) was again the first one who justified Doob's heuristic approach.

In order to formulate his result, we give a continuous version $F_n^*(\cdot)$ of the empirical distribution function:

$$F_n^*(x) = F_n(X_k^{(n)}) + 2\left(x - \frac{X_k^{(n)} + X_{k+1}^{(n)}}{2}\right)\left(\frac{F_n(X_{k+1}^{(n)}) - F_n(X_k^{(n)})}{X_{k+2}^{(n)} - X_k^{(n)}}\right),$$

if
$$\frac{X_k^{(n)} + X_{k+1}^{(n)}}{2} \leq x \leq \frac{X_{k+1}^{(n)} + X_{k+2}^{(n)}}{2}.$$

It is clear that
$$\sup_{-\infty < x < \infty} |F_n(x) - F_n^*(x)| \leq \frac{1}{n}.$$

The corresponding continuous versions $\beta_n^*(x)$ and $\alpha_n^*(y)$ of the empirical process then become

$$\beta_n^*(x) = \sqrt{n}(F_n^*(x) - F(x)), \quad -\infty < x < \infty,$$

$$\alpha_n^*(y) = \sqrt{n}(E_n^*(y) - y), \quad 0 \leq y \leq 1.$$

Theorem 4.2.1 (Donsker 1952). *We have*

(4.2.2) $$h(\alpha_n^*(y)) \xrightarrow{\mathscr{D}} h(B(y)),$$

for every continuous functional $h: C(0, 1) \to R^1$.

We are not going to prove this theorem here. It will, however, follow from Theorem 4.4.1.

We also note already here that Theorem 4.2.1 and Theorem 1.5.1 together imply Theorem 4.1.2. This is true because, for every $\omega \in \Omega$, $\alpha_n(F(x)) = \beta_n(x)$ and $\sup_{0 \leq y \leq 1} \alpha_n(y) = \sup_{-\infty < x < \infty} \alpha_n(F(x)) = \sup_{-\infty < x < \infty} \beta_n(x)$, provided F is continuous (note: $\sup_{0 \leq y \leq 1}$ is a continuous functional).

4.3. The first strong approximations of the empirical process

In the light of Strassen's strong approximation result (2.2.2) it was natural to look for analogous approximations concerning the empirical process. Indeed, the first one of these is directly based on (2.2.2) and is due to Brillinger.

Theorem 4.3.1 (Brillinger 1969). *Given independent $U(0, 1)$ r.v. $U_1, U_2, \ldots$, there exists a probability space with sequences of Brownian bridges $\{B_n(y); 0 \leq y \leq 1\}$ and processes $\{\tilde{\alpha}_n(y); 0 \leq y \leq 1\}$ such that*

(4.3.1) $$\{\tilde{\alpha}_n(y); 0 \leq y \leq 1\} \stackrel{\mathscr{D}}{=} \{\alpha_n(y); 0 \leq y \leq 1\}$$

for each $n=1, 2, \ldots$, *and*

(4.3.2) $\quad\sup\limits_{0\leq y\leq 1} |\tilde{\alpha}_n(y) - B_n(y)| \stackrel{a.s.}{=} O(n^{-1/4}(\log n)^{1/2}(\log\log n)^{1/4}).$

The fact that such a theorem can be directly based on (2.2.2) is due to the construction of the process $\{\tilde{\alpha}_n(y); 0\leq y\leq 1\}$, which hinges on the observation that for each $n=1, 2, \ldots$

(4.3.3) $\quad\{U_k^{(n)}; 1\leq k\leq n\} \stackrel{\mathscr{D}}{=} \left\{\dfrac{S_k}{S_{n+1}}; 1\leq k\leq n\right\},$

where $S_k = E_1 + \ldots + E_k$ and the E_i are independent exponential r.v. with mean one. Consequently, for each n, one can define an empirical distribution function $\tilde{E}_n(y)$ in terms of $\left\{\dfrac{S_k}{S_{n+1}}; 1\leq k\leq n\right\}$. Then the process $\tilde{\alpha}_n(y) = \sqrt{n}(\tilde{E}_n(y) - y)$ clearly satisfies (4.3.1) and it is also not difficult to arrive at (4.3.2) (e.g. via Theorem 2.2.1). Further details of proof are omitted, for Theorem 4.3.1 will follow from Theorem 4.4.1.

Remark 4.3.1. In the sequel we will not emphasize that a new probability space should be introduced. That is to say, in an appropriate sense, Remark 2.2.1 should be applied in this Chapter too.

Remark 4.3.2. Naturally, one can replace (2.2.2) by Theorem 2.6.1 in the above sketched proof of Brillinger's theorem. His result however will not be improved by doing so. In fact Theorem 2.6.1 will imply that $\tilde{\alpha}_n\left(\dfrac{S_k}{S_{n+1}}\right)$ is close to $B_n\left(\dfrac{k}{n+1}\right)$, but then we would still have to estimate $\sup\limits_{1\leq k\leq n}\left|\tilde{\alpha}_n\left(\dfrac{S_k}{S_{n+1}}\right) - \tilde{\alpha}_n\left(\dfrac{k}{n}\right)\right|$. One possible way of doing this is via Theorem 5.2.1. However, this approach will destroy the gain obtained via Theorem 2.6.1.

It is clear that Theorem 4.3.1 immediately implies Theorem 4.2.1. Also, it can be immediately generalized to the case of the empirical process $\beta_n(x)$ with an arbitrary continuous distribution function $F(\cdot)$ via applying again the equality $\alpha_n(F(x)) = \beta_n(x)$. Moreover, we also have the following result with a not necessarily continuous F.

Theorem 4.3.1*. *Given i.i.d.r.v.* $X_1, X_2, \ldots$ *with distribution function* $F(\cdot)$, *there exists a sequence of Brownian bridges* $\{B_n(y); 0\leq y\leq 1\}$ *such that*

(4.3.4) $\quad\sup\limits_{-\infty < x < \infty} |\beta_n(x) - B_n(F(x))| \stackrel{a.s.}{=} O(n^{-1/4}(\log n)^{1/2}(\log\log n)^{1/2}).$

Proof. Let $U_1, \ldots, U_n$ be a $U(0, 1)$ random sample, and let $\alpha_n(y)$ and $B_n(y)$ be as in Theorem 4.3.1. Then, for any $U(0, 1)$ r.v. U, $\operatorname{inv} F(U)$ has distribution function $F(x)$. Define $X_k = \operatorname{inv} F(U_k)$, $k=1, 2, \ldots, n$, and let $F_n(x)$ be the empirical distribution function of these r.v. Since $P\{X_k \le x\} = F(x)$, $k=1, 2, \ldots, n$, we have $F_n(x) = E_n(F(x))$ for every $\omega \in \Omega$. Hence $\beta_n(x) = \sqrt{n}(F_n(x) - F(x)) = \sqrt{n}(E_n(F(x)) - F(x))$ for every $\omega \in \Omega$, and $\sup_{-\infty < x < \infty} |\beta_n(x) - B_n(F(x))| = \sup_{-\infty < x < \infty} |\sqrt{n}(E_n(F(x)) - F(x)) - B_n(F(x))| \le$
$\le \sup_{0 \le y \le 1} |\sqrt{n}(E_n(y) - y) - B_n(y)| = \sup_{0 \le y \le 1} |\alpha_n(y) - B_n(y)|$; for the latter (4.3.2) holds and whence Theorem 4.3.1* is also true.

We should like to call attention to the fact that though Theorem 4.3.1 might appear as a strong approximation theorem, it is not at all like Theorems 2.2.4. and 2.6.1 are in the area of partial sum approximation in Chapter 2. It is rather like Lemma 2.5.1, that is in both of them one has only an approximation for each n. More precisely, in the present case we have not succeeded in bringing together the stochastic processes $\{\alpha_n(y); 0 \le y \le 1, n=1, 2, \ldots\}$ and $\{B_n(y); 0 \le y \le 1, n=1, 2, \ldots\}$. Thus no strong law type behaviour of the process $\alpha_n(y)$, like, e.g., the Smirnov (1944), Chung (1949) law of iterated logarithm, can be deduced from Theorem 4.3.1.

Kiefer was the first one to call attention to the desirability of viewing the empirical process $\alpha_n(y)$ as a two-parameter process and that it should be approximated by an appropriate two-dimensional Gaussian process. He also gave a solution to this problem by proving

Theorem 4.3.2 (Kiefer 1972). *Given independent $U(0, 1)$ r.v. $U_1, U_2, \ldots$, there exists a Kiefer process $\{K(y, t); 0 \le y \le 1, 0 \le t < \infty\}$ such that*

(4.3.5) $$\sup_{0 \le y \le 1} |\sqrt{n}\alpha_n(y) - K(y, n)| \stackrel{\text{a.s.}}{=} O(n^{1/3}(\log n)^{2/3}).$$

A Theorem 4.3.1* type analogue of this theorem is immediate.

The above result of Kiefer is the first two-dimensional strong approximation theorem. Consequently, he had to develop a new technique. He did this, generalizing Skorohod's embedding scheme to vectors. An elementary proof of Theorem 4.3.2 was given in Csörgő, Révész (1975b), using the technique of Theorem 2.5.1 instead of Skorohod's scheme. Subsequently Komlós, Major, Tusnády (1975) have succeeded in applying their one dimensional technique to improve Kiefer's above theorem, and the next section is devoted to their corresponding results.

4.4. Best strong approximations of the empirical process

First we formulate a result, which improves the rate in (4.3.2), giving also the best possible one.

Theorem 4.4.1 (Komlós, Major, Tusnády 1975). *Given independent $U(0, 1)$ r.v. $U_1, U_2, \ldots$, there exists a sequence of Brownian bridges $\{B_n(y); 0 \leq y \leq 1\}$ such that for all n and x we have*

$$(4.4.1) \quad P\{\sup_{0 \leq y \leq 1} |\alpha_n(y) - B_n(y)| > n^{-1/2}(C \log n + x)\} \leq L e^{-\lambda x},$$

where C, L, λ are positive absolute constants. (For example they can be chosen as $C = 100$, $L = 10$, $\lambda = 1/50$.) Consequently

$$(4.4.2) \quad \sup_{0 \leq y \leq 1} |\alpha_n(y) - B_n(y)| \stackrel{a.s.}{=} O(n^{-1/2} \log n).$$

The following three lemmas play a crucial role in the proof of Theorem 4.4.1.

Lemma 4.4.1 (Tusnády 1977b). *Let $G(x) \in \mathcal{N}\left(\dfrac{n}{2}, \dfrac{n}{4}\right)$, i.e.,*

$$(4.4.3) \quad G(x) = \left(\frac{2}{\pi n}\right)^{1/2} \int_{-\infty}^{x} \exp\left(-\frac{2\left(u - \dfrac{n}{2}\right)^2}{n}\right) du.$$

Then

$$(4.4.4) \quad G\left((2kn)^{1/2} - \frac{n}{2}\right) \leq \sum_{j=0}^{k} \binom{n}{j} 2^{-n} \leq G(k+1), \quad \text{if} \quad k \leq n/2.$$

Although the proof of this inequality is elementary, it is not at all simple. It will not be given here however.

Lemma 4.4.2 (Tusnády 1977b). *Let $Y \in \mathcal{N}\left(\dfrac{n}{2}, \dfrac{n}{4}\right)$. There exists then a r.v. $X \in \mathcal{B}(n, \tfrac{1}{2})$ such that*

$$(4.4.5) \quad Y - 1 \leq X \leq Y + \frac{\left(Y - \dfrac{n}{2}\right)^2}{2n} + 1, \quad \text{if} \quad Y \leq n/2,$$

and

$$(4.4.6) \quad Y - \frac{\left(Y - \dfrac{n}{2}\right)^2}{2n} - 1 \leq X \leq Y + 1, \quad \text{if} \quad Y \geq n/2.$$

Combining these two inequalities we get

(4.4.7) $$|X-Y| \leq \frac{1}{8} Z^2 + 1$$

and

(4.4.8) $$\left| X - \frac{n}{2} \right| \leq \left| Y - \frac{n}{2} \right| + 1,$$

where $Z = 2n^{-1/2}\left(Y - \frac{n}{2}\right)$, i.e. the standardized version of the r.v. Y.

The role of (4.4.7) will be the same in the sequel as that of (2.5.3) in the proof of Theorem 2.5.1.

Proof. We will detail only (4.4.5). With G of (4.4.3), define the sequence of real numbers $c_1 < c_2 < ... < c_n$ by

$$G(c_k) = \sum_{j=0}^{k} \binom{n}{j} 2^{-n}.$$

Next, we define the r.v. X as follows: X equals to k when the event $A_k = \{\omega : c_{k-1} < Y(\omega) \leq c_k\}$ occurs. Since by Lemma 4.4.1

$$G(c_k) = \sum_{j=0}^{k} \binom{n}{j} 2^{-n} \leq G(k+1),$$

on the set A_k ($k \leq n/2$) we have

$$Y - 1 \leq c_k + 1 \leq k = X,$$

and the left hand side of (4.4.5) is proved.

In order to see the right-hand side of (4.4.5), we observe that if $Y \leq -n/2$ then $X \leq 1$ and $Y + \frac{(Y-n/2)^2}{2n} + 1 \geq 1$. Hence we can assume that $Y \geq -n/2$. By Lemma 4.4.1 for $k \leq n/2$ we have

$$G\left((2kn)^{1/2} - \frac{n}{2}\right) \leq \sum_{j=0}^{k} \binom{n}{j} 2^{-n} = G(c_k).$$

Hence

$$(2(k-1)n)^{1/2} - \frac{n}{2} \leq c_{k-1}$$

and so

$$k - 1 \leq \frac{\left(c_{k-1} + \frac{n}{2}\right)^2}{2n} = c_{k-1} + \frac{\left(c_{k-1} - \frac{n}{2}\right)^2}{2n}.$$

The latter inequality implies that on the set A_k ($k \leq n/2$) we have

$$X = k \leq c_{k-1} + \frac{\left(c_{k-1} - \frac{n}{2}\right)^2}{2n} + 1 \leq Y + \frac{\left(Y - \frac{n}{2}\right)^2}{2n} + 1,$$

and the proof of (4.4.5) is now complete.

Lemma 4.4.3. *Let $X_{n,k}$ ($k=1, 2, \ldots, 2^n$; $n=1, 2, \ldots$) be a triangular array of independent $\mathcal{N}(0,1)$ r.v. Define, for any k, n ($1 \leq k \leq 2^n$),*

$$S_{n,k}^2 = X_{n,k_0}^2 + X_{n-1,k_1}^2 + \ldots + X_{2,k_{n-2}}^2 + X_{1,k_{n-1}}^2,$$

where $k_0 = k$ and $k_{i+1} = \left[\frac{k_i + 1}{2}\right]$, $i = 0, 1, \ldots, n-2$. Then there exists a $C > 0$ such that

$$\sum_{n=1}^{\infty} P\{n^{-1} \max_{1 \leq k \leq 2^n} S_{n,k}^2 > C\} < \infty.$$

Proof. By Theorem 2.4.4 there exists an absolute constant $C > 0$ such that

$$P\left\{\frac{|S_{n,k}^2 - n|}{\sqrt{n}} \geq C\sqrt{n}\right\} \leq e^{-n}, \quad k = 1, 2, \ldots, 2^n; \; n = 1, 2, \ldots.$$

Hence

$$P\left\{\max_{1 \leq k \leq 2^n} \frac{|S_{n,k}^2 - n|}{\sqrt{n}} \geq C\sqrt{n}\right\} \leq \left(\frac{2}{e}\right)^n,$$

and Lemma 4.4.3 is proved.

Proof of (4.4.2). Let us assume that we are given a sequence of Brownian bridges $\{B_n(y); 0 \leq y \leq 1\}$. We define, for each $n = 1, 2, \ldots$,

$$b_n(y) = ny + \sqrt{n} B_n(y).$$

Then $b_n(\tfrac{1}{2}) \in \mathcal{N}(n/2, n/4)$ and, by (4.4.7), one can define also a binomial r.v. $a_n(\tfrac{1}{2}) \in \mathcal{B}(n, \tfrac{1}{2})$ such that

$$|b_n(\tfrac{1}{2}) - a_n(\tfrac{1}{2})| \leq \tfrac{1}{2} B_n^2(\tfrac{1}{2}) + 1.$$

In terms of the thus defined binomial r.v. $a_n(\tfrac{1}{2})$, we can now define the r.v.

$$\alpha_n(\tfrac{1}{2}) = n^{-1/2}\left(a_n(\tfrac{1}{2}) - \frac{n}{2}\right),$$

whose distribution, for each n, is the same as that of a uniform empirical process at $y = \tfrac{1}{2}$ and, in addition to this, we have

$$n^{1/2}|\alpha_n(\tfrac{1}{2}) - B_n(\tfrac{1}{2})| \leq \tfrac{1}{2} B_n^2(\tfrac{1}{2}) + 1.$$

Given this initial step, we define the r.v.
$$\{\alpha_n((2k+1)2^{-(m+1)}); \ k = 0, 1, 2, \ldots, 2^m - 1\}$$
on the basis of
$$\{B_n(k2^{-(m+1)}); \ k = 0, 1, 2, \ldots, 2^{m+1}\},$$
assuming that the r.v.
$$\{\alpha_n(k2^{-m}), \ k = 1, 2, \ldots, 2^m\}$$
are already constructed.

Before doing this, we put
$$a_n(y) = ny + \sqrt{n}\,\alpha_n(y), \quad 0 \leq y \leq 1,$$
and observe

(4.4.9) $\quad E\left(a_n\left(\dfrac{2k+1}{2^{m+1}}\right) \Big| a_n\left(\dfrac{l}{2^m}\right), \ l = 0, 1, \ldots, 2^m\right) = \dfrac{a_n\left(\dfrac{k}{2^m}\right) + a_n\left(\dfrac{k+1}{2^m}\right)}{2},$

(4.4.10) $\quad E\left(\left(a_n\left(\dfrac{2k+1}{2^{m+1}}\right) - \dfrac{a_n\left(\dfrac{k}{2^m}\right) + a_n\left(\dfrac{k+1}{2^m}\right)}{2}\right)^2 \Big| a_n\left(\dfrac{l}{2^m}\right), \ l = 0, 1, \ldots, 2^m\right)$

$$= \dfrac{1}{4}\left(a_n\left(\dfrac{k+1}{2^m}\right) - a_n\left(\dfrac{k}{2^m}\right)\right), \quad k = 0, 1, \ldots, 2^m - 1,$$

(4.4.11) $\quad \Gamma_n(k, m) = B_n\left(\dfrac{2k+1}{2^{m+1}}\right) - \dfrac{B_n\left(\dfrac{k}{2^m}\right) + B_n\left(\dfrac{k+1}{2^m}\right)}{2}$

for each $\quad k = 0, 1, \ldots, 2^m - 1$

is independent of the vector of r.v. $\left\{B_n\left(\dfrac{l}{2^m}\right); \ l = 0, \ldots, 2^m\right\}$.

Given $a_n\left(\dfrac{l}{2^m}\right)$, $l = 0, \ldots, 2^m$, we have

(4.4.12)
$$Y_n(k, m) = \sqrt{2^{m+2}} \sqrt{\dfrac{1}{4}\left(a_n\left(\dfrac{k+1}{2^m}\right) - a_n\left(\dfrac{k}{2^m}\right)\right)} \Gamma_n(k, m) + \dfrac{a_n\left(\dfrac{k+1}{2^m}\right) - a_n\left(\dfrac{k}{2^m}\right)}{2}$$

$$\in \mathcal{N}\left(\dfrac{a_n\left(\dfrac{k+1}{2^m}\right) - a_n\left(\dfrac{k}{2^m}\right)}{2}, \ \dfrac{a_n\left(\dfrac{k+1}{2^m}\right) - a_n\left(\dfrac{k}{2^m}\right)}{4}\right),$$

$k = 0, 1, \ldots, 2^m - 1.$

Now we construct

$$(4.4.13) \quad X_n(k, m) = a_n\left(\frac{2k+1}{2^{m+1}}\right) - a_n\left(\frac{k}{2^m}\right) \in \mathscr{B}\left(a_n\left(\frac{k+1}{2^m}\right) - a_n\left(\frac{k}{2^m}\right), \frac{1}{2}\right)$$

in terms of $Y_n(k, m)$ of (4.4.12) via Lemma 4.4.2, and then we put

$$(4.4.14) \quad \alpha_n\left(\frac{2k+1}{2^{m+1}}\right) = \sqrt{n}\left[\frac{a_n\left(\frac{2k+1}{2^{m+1}}\right)}{n} - \frac{2k+1}{2^{m+1}}\right]$$

$$= \sqrt{n}\left[\frac{X_n(k, m) + a_n\left(\frac{k}{2^m}\right)}{n} - \frac{2k+1}{2^{m+1}}\right]$$

for $k=0, 1, \ldots, 2^m-1$, $m=0, 1, 2, \ldots, M$, where M is an integer to be picked later on.

So far we have constructed the process $\alpha_n(y)$ at $y=l/2^m$ ($l=0, 1, \ldots, 2^m$; $m=0, 1, 2, \ldots, M$) and our construction (4.4.14) shows that the distribution of the array of r.v. $\left\{\alpha_n\left(\frac{l}{2^m}\right); l=0, 1, \ldots, 2^m; m=0, 1, 2, \ldots, M\right\}$ is as desired. Next we show that this array is as close to the given array $\left\{B_n\left(\frac{l}{2^m}\right); l=0, 1, \ldots, 2^m; m=1, 2, \ldots, M\right\}$ as claimed. Towards this end we first consider

$$(4.4.15)$$

$$\left| B_n\left(\frac{2k+1}{2^{m+1}}\right) - \frac{B_n\left(\frac{k}{2^m}\right) + B_n\left(\frac{k+1}{2^m}\right)}{2} - \left(\alpha_n\left(\frac{2k+1}{2^{m+1}}\right) - \frac{\alpha_n\left(\frac{k}{2^m}\right) + \alpha_n\left(\frac{k+1}{2^m}\right)}{2}\right)\right|$$

$$\leq \left|\Gamma_n(k, m) - \sqrt{\frac{2^{m+2}}{n}}\sqrt{\frac{1}{4}\left(a_n\left(\frac{k+1}{2^m}\right) - a_n\left(\frac{k}{2^m}\right)\right)}\,\Gamma_n(k, m)\right|$$

$$+ \left|\sqrt{\frac{2^{m+1}}{n}}\sqrt{\frac{1}{4}\left(a_n\left(\frac{k+1}{2^m}\right) - a_n\left(\frac{k}{2^m}\right)\right)}\,\Gamma_n(k, m)\right.$$

$$\left. - \left(\alpha_n\left(\frac{2k+1}{2^{m+1}}\right) - \frac{\alpha_n\left(\frac{k}{2^m}\right) + \alpha_n\left(\frac{k+1}{2^m}\right)}{2}\right)\right| = J_1 + J_2,$$

where

$$J_2 = \left| \frac{Y_n(k,m)}{\sqrt{n}} - \left(\alpha_n\left(\frac{2k+1}{2^{m+1}}\right) - \left(\frac{\alpha_n\left(\frac{k}{2^m}\right) + \alpha_n\left(\frac{k+1}{2^m}\right)}{2} \right) \right) - \frac{a_n\left(\frac{k+1}{2^m}\right) - a_n\left(\frac{k}{2^m}\right)}{2\sqrt{n}} \right|$$
(4.4.16)

$$= \left| \frac{X_n(k,m) - Y_n(k,m)}{\sqrt{n}} \right|$$

$$\leq n^{-1/2}\left(\frac{1}{8} 2^{m+2}(\Gamma_n(k,m))^2 + 1\right),$$

and

(4.4.17) $\quad J_1 = |\Gamma_n(k,m)| \cdot \left| 1 - \sqrt{\frac{2^{m+2}}{n}} \sqrt{\frac{1}{4}\left(a_n\left(\frac{k+1}{2^m}\right) - a_n\left(\frac{k}{2^m}\right)\right)} \right|$

$$= |\Gamma_n(k,m)| \cdot \left| 1 - \sqrt{1 + \frac{2^m\left(\alpha_n\left(\frac{k+1}{2^m}\right) - \alpha_n\left(\frac{k}{2^m}\right)\right)}{\sqrt{n}}} \right|$$

$$\leq \frac{2^{m-1}}{\sqrt{n}} |\Gamma_n(k,m)| \cdot \left| \alpha_n\left(\frac{k+1}{2^m}\right) - \alpha_n\left(\frac{k}{2^m}\right) \right|$$

$$\leq \frac{2^{m-2}}{\sqrt{n}} \Gamma_n^2(k,m) + \frac{2^{m-2}}{\sqrt{n}} \left(\alpha_n\left(\frac{k+1}{2^m}\right) - \alpha_n\left(\frac{k}{2^m}\right) \right)^2.$$

We note that the inequality of (4.4.16) is by (4.4.7). Whence we have

(4.4.18) $\quad J_1 + J_2 \leq n^{-1/2}\left(3(2^{m-2})\Gamma_n^2(k,m) + 2^{m-2}\left(\alpha_n\left(\frac{k+1}{2^m}\right) - \alpha_n\left(\frac{k}{2^m}\right) \right)^2 + 1 \right)$

$$\leq n^{-1/2}(5(2^{m-2})\Gamma_n^2(k,m) + 3), \quad m = 0, 1, 2, \ldots, M,$$

where the last inequality is by (4.4.8). The right-hand side term of (4.4.18) estimates the error incurred in our construction at the mth step for each $k = 0, 1, \ldots, 2^m - 1$. The total error of construction up to M can be estimated by the above error terms. Applying now Lemma 4.4.3 we get

(4.4.19) $\quad \displaystyle\sum_{M=1}^{\infty} P\left\{ \frac{\sqrt{n} \sup_{0 \leq l \leq 2^M} \left| B_n\left(\frac{l}{2^M}\right) - \alpha_n\left(\frac{l}{2^M}\right) \right|}{M} > C \right\} < \infty$

for $n = 1, 2, \ldots$. Choosing $M = \log n$, we have, so far, proved (4.4.2) at the indicated points. As to filling out the gaps in between, we do it via observing that

$$\sum_{n=1}^{\infty} P\left\{\sup_{0\leq x\leq 1-1/n}\sup_{0\leq s\leq 1/n}\left|\frac{B_n(x+s)-B_n(x)}{\sqrt{n^{-1}\log n}}\right|>C\right\}<\infty$$

and

$$\sum_{n=1}^{\infty} P\left\{\sup_{0\leq x\leq 1-1/n}\left|\alpha_n\left(x+\frac{1}{n}\right)-\alpha_n(x)\right|>C\log n\right\}<\infty,$$

where the first convergence is by Lemma 1.1.1 and the second one follows from elementary calculations. The proof of (4.4.2) is now almost complete (cf. Remark 4.4.2).

Remark 4.4.1. We note that, in the above proof, we actually proved a bit more than (4.4.2). Namely we showed that there exists an absolute positive constant C such that

(4.4.20) $$\sum_{n=1}^{\infty} P\{\sup_{0\leq y\leq 1}\sqrt{n}|\alpha_n(y)-B_n(y)|>C\log n\}<\infty.$$

While this is indeed more than the statement of (4.4.2), it is considerably less than that of (4.4.1). In order to prove the latter, one would use the same construction, but instead of Lemma 4.4.3, we would have to estimate the large deviation distribution of

$$\max_{1\leq k\leq 2^n}|S_{n,k}^2-n|/\sqrt{n}$$

a bit more exactly.

Remark 4.4.2. We note also that what we have proved above is somewhat different than claimed in the statement of Theorem 4.4.1. Namely, there we claimed the existence of sequences $\{U_n\}$ and $\{B_n(y), 0\leq y\leq 1\}$ satisfying (4.4.1) and (4.4.2). On the other hand, in our proof of (4.4.2) we started out with a given sequence of Brownian bridges and constructed, for each $n=1, 2, \ldots$, a vector $\{U_{n,j}: 1\leq j\leq n\}$ of independent $U(0, 1)$ r.v. such that the empirical process α_n of these satisfied (4.4.1) and (4.4.2). However, we did not say anything about the joint distribution of the triangular array of vectors $\{U_{n,j}: 1\leq j\leq n\}$, $n=1, 2, \ldots$. Consequently, we cannot really say that we have obtained the sequence $\{U_n\}$ of Theorem 4.4.1 as desired. Nevertheless (4.4.20) enables us to conclude also what we really wanted to say, as a result of the following

Lemma 4.4.4. *Let $\mu(\cdot)$ be a probability measure defined on the Borel sets of the Banach space $D(0, 1) \times D(0, 1)$, and let ξ (resp. η) be $D(0, 1)$ valued r.v. defined on $(\Omega_1, \mathcal{A}_1, P_1)$ (resp. on $(\Omega_2, \mathcal{A}_2, P_2)$) with*

$$P_1\{\xi \in A\} = \mu(A \times D(0, 1)) \quad \text{resp.} \quad P_2\{\eta \in A\} = \mu(D(0, 1) \times A)$$

for any Borel set A of $D(0, 1)$. Then there exists a probability measure P defined on $(\Omega_1 \times \Omega_2, \mathcal{A}_1 \times \mathcal{A}_2)$ such that

$$P\{(\omega_1, \omega_2) \in \Omega_1 \times \Omega_2 : (\xi(\omega_1), \eta(\omega_2)) \in B\} = \mu(B),$$

for any Borel set B of $D(0, 1) \times D(0, 1)$.

The proof of this lemma, which is based on standard measure theoretic methods, will not be given here.

Applying now Lemma 4.4.4 with $\mu = \mu_n$, $n = 1, 2, \ldots$, being the joint distribution of (α_n, B_n), (where α_n and B_n are the processes constructed in our proof), and $\xi = \alpha_n^*$ being the empirical process defined in terms of $U_1, U_2, \ldots, U_n$ of Theorem 4.4.1, via (4.4.20) and the Kolmogorov extension theorem we get

$$(4.4.20^*) \qquad \sum_{n=1}^{\infty} P\{\sup_{0 \leq y \leq 1} \sqrt{n}|\alpha_n^*(y) - B_n^*(y)| > C \log n\} < \infty,$$

where, for $n = 1, 2, \ldots, B_n^* = \eta$ of Lemma 4.4.4.

At the beginning of this paragraph we mentioned that the rate of (4.4.2) is the best possible one. Now we show that this is indeed so.

Theorem 4.4.2 (Komlós, Major, Tusnády 1975). *For any sequence of Brownian bridges $\{B_n(y), 0 \leq y \leq 1\}$ on the probability space of our empirical process $\{\alpha_n(y); 0 \leq y \leq 1\}$ we have*

$$P\{\sup_{0 \leq y \leq 1} |\alpha_n(y) - B_n(y)| \geq \frac{1}{6} n^{-1/2} \log n\} \to 1.$$

Proof. Recall (4.3.3). Then, for any $c > 0$

$$(4.4.21) \qquad \sup_{1 \leq k \leq n - [c \log n]} n^{1/2} \frac{\alpha_n(U_{k+[c \log n]}^{(n)}) - \alpha_n(U_k^{(n)})}{[c \log n]}$$

$$\stackrel{\mathscr{D}}{=} \sup_{1 \leq k \leq n - [c \log n]} \frac{-1}{[c \log n]} \frac{n}{S_{n+1}} \left\{ (S_{k+[c \log n]} - S_k - [c \log n]) - \frac{[c \log n]}{n}(S_{n+1} - n) \right\}.$$

Now Theorem 2.4.3 implies that the limit of the right-hand side of (4.4.21) is almost surely equal to

$$\lim_{n\to\infty} \sup_{1\le k\le n-[c\log n]} (-1)\frac{S_{k+[c\log n]}-S_k-[c\log n]}{[c\log n]} = \alpha(c),$$

where $\alpha(c)$ is as in (2.4.5). Consequently, (4.4.21) implies

$$n^{1/2} \sup_{1\le k\le n-[c\log n]} \frac{\alpha_n(U^{(n)}_{k+[c\log n]})-\alpha_n(U^{(n)}_k)}{[c\log n]} \xrightarrow{P} \alpha(c).$$

Similarly, one can also show that

$$n^{1/2} \sup_{1\le k\le n-[c\log n]} \frac{B_n(U^{(n)}_{k+[c\log n]})-B_n(U^{(n)}_k)}{[c\log n]} \xrightarrow{P} \alpha^*(c),$$

where $\alpha^*(c)$ is as in (2.4.5). We know that $\alpha(c)$ and $\alpha^*(c)$ cannot be identical. Studying the difference $\alpha(c)-\alpha^*(c)$ one finds that for any appropriate c we will have that $\frac{c}{2}(\alpha(c)-\alpha^*(c))$ is greater than or equal to $\frac{1}{6}$. This also completes the proof of Theorem 4.4.2.

As to the two dimensional Kiefer type approximation of the empirical process we have:

Theorem 4.4.3 (Komlós, Major, Tusnády 1975). *Given independent $U(0,1)$ r.v. $U_1, U_2, \ldots$, there exists a Kiefer process $\{K(y,t); 0\le y\le 1, 0\le t<\infty\}$ such that*

(4.4.22) $\quad P\{\sup_{1\le k\le n}\sup_{0\le y\le 1} |k^{1/2}\alpha_k(y)-K(y,k)| > (C\log n+x)\log n\} < Le^{-\lambda x}$

for all x and n, where C, L, λ are positive absolute constants. Consequently,

(4.4.23) $\quad \sup_{0\le y\le 1} |n^{1/2}\alpha_n(y)-K(y,n)| \stackrel{a.s.}{=} O(\log^2 n).$

It is clear from Theorem 4.4.2 that the rate of (4.4.23) cannot be improved beyond $O(\log n)$. However, the best possible rate is not yet known in this case.

Due to its length and complexity, the proof of Theorem 4.4.3 is omitted here.

Remark 4.4.3. A Theorem 4.3.1* type analogue of Theorems 4.4.1 and 4.4.3 is also true.

Corollary 4.4.1. *Theorems 4.4.1 and 4.4.3 imply*

(4.4.24) $$\left|\sup_{0\leq y\leq 1}|\alpha_n(y)|-\sup_{0\leq y\leq 1}|B_n(y)|\right|\stackrel{a.s.}{=} O(n^{-1/2}\log n)$$

$$\left|\sup_{0\leq y\leq 1}|\alpha_n(y)|-\sup_{0\leq y\leq 1}n^{-1/2}|K(y,n)|\right|\stackrel{a.s.}{=} O(n^{-1/2}\log^2 n),$$

and

(4.4.25) $$\left|\int_0^1 \alpha_n^2(y)\,dy-\int_0^1 B_n^2(y)\,dy\right|\stackrel{a.s.}{=} O(n^{-1/2}\log n(\log\log n)^{1/2})$$

$$\left|\int_0^1 \alpha_n^2(y)\,dy-n^{-1}\int_0^1 K^2(y,n)\,dy\right|\stackrel{a.s.}{=} O(n^{-1/2}\log^2 n(\log\log n)^{1/2}).$$

Proof. The two statements of (4.4.24) follow directly from Theorems 4.4.1 and 4.4.3. As to (4.4.25), we have

$$\left|\int_0^1 (\alpha_n^2(y)-n^{-1}K^2(y,n))\,dy\right|$$

$$=\left|\int_0^1 (\alpha_n(y)-n^{-1/2}K(y,n))(\alpha_n(y)+n^{-1/2}K(y,n))\,dy\right|$$

$$\leq \sup_{0\leq y\leq 1}|\alpha_n(y)-n^{-1/2}K(y,n)|\sup_{0\leq y\leq 1}|\alpha_n(y)+n^{-1/2}K(y,n)|$$

$$\stackrel{a.s.}{=} O(n^{-1/2}\log^2 n)O((\log\log n)^{1/2}),$$

by (4.4.23) and by applying the law of iterated logarithm twice (cf. Corollary 1.15.1 and Theorem 5.1.1). As to the first statement of (4.4.25), the left-hand side of the latter is bounded above by

$$\sup_{0\leq y\leq 1}|\alpha_n(y)-B_n(y)|\sup_{0\leq y\leq 1}|\alpha_n(y)+B_n(y)|$$

$$\leq \sup_{0\leq y\leq 1}|\alpha_n(y)-B_n(y)|\left(\sup_{0\leq y\leq 1}|B_n(y)-\alpha_n(y)|+2\sup_{0\leq y\leq 1}|\alpha_n(y)|\right)$$

$$\stackrel{a.s.}{=} O(n^{-1/2}\log n)(O(n^{-1/2}\log n)+O((\log\log n)^{1/2}))$$

$$\stackrel{a.s.}{=} O(n^{-1/2}\log n(\log\log n)^{1/2}),$$

on applying Theorem 4.4.1 twice to $\sup_{0\leq y\leq 1}|\alpha_n(y)-B_n(y)|$ and the law of iterated logarithm once to $\sup_{0\leq y\leq 1}|\alpha_n(y)|$ (cf. Theorem 5.1.1).

4.5. Strong approximations of the quantile process

In dealing with the empirical process in the first four sections of this chapter, so far we have emphasized that all the results proved for the uniform case have an immediate form for the general case with an arbitrary continuous distribution function $F(\cdot)$ simply by noticing that $\alpha_n(F(x)) = \beta_n(x)$. Such a substitution will not work immediately for the quantile process and we will have to deal with the general case separately. First some definitions.

Let $U_1, U_2, \ldots$ be a sequence of independent $U(0,1)$ r.v. and, for each $n \geq 1$, let $0 = U_0^{(n)} \leq U_1^{(n)} \leq \ldots \leq U_n^{(n)} \leq U_{n+1}^{(n)} = 1$ denote the order statistics of the random sample $U_1, U_2, \ldots, U_n$. Define *the uniform quantile function*

$$U_n(y) = \begin{cases} U_k^{(n)} & \text{if } \frac{k-1}{n} < y \leq \frac{k}{n}, \quad k = 1, 2, \ldots, n. \\ 0 & \text{if } y = 0 \end{cases}$$

and *the uniform sample quantile process*

$$u_n(y) = n^{1/2}(U_n(y) - y), \quad 0 \leq y \leq 1.$$

Let $X_1, X_2, \ldots$ be a sequence of i.i.d.r.v. with a continuous distribution function $F(\cdot)$ and let $X_1^{(n)} \leq \ldots \leq X_n^{(n)}$ denote the order statistics of the random sample $X_1, X_2, \ldots, X_n$. Define *the quantile function*

$$Q_n(y) = X_k^{(n)} \quad \text{if } \frac{k-1}{n} < y \leq \frac{k}{n}, \quad k = 1, 2, \ldots, n,$$

and *the sample quantile process*

$$q_n(y) = n^{1/2}(Q_n(y) - \text{inv } F(y)), \quad 0 < y \leq 1.$$

This latter process sometimes is called the inverse empirical process.

The pointwise properties of the empirical process (cf. (4.1.1), (4.1.3)) are quite simple and follow from well-known properties of partial sum sequences of i.i.d.r.v. The corresponding properties of the quantile process are not so immediate. An analogue of (4.1.3) is

Theorem 4.5.1 (cf., e.g., Rényi 1970, p. 490). *Suppose that $F(x)$ is absolutely continuous in an interval around* $\text{inv } F(y_0)$. *Then, with $f = F'$, we have*

(4.5.1) $\quad P\left\{ f(\text{inv } F(y_0)) \frac{q_n(y_0)}{(y_0(1-y_0))^{1/2}} < t \right\} \to \varphi(t) \quad (n \to \infty)$

provided $f(\text{inv } F(y_0))$ is positive and continuous at y_0.

The proof of this theorem is elementary and will not be presented here. Under somewhat stronger restrictions we will approximate the quantile process by Brownian bridges (Theorem 4.5.6), and the latter approximation, in turn, will imply the validity of (4.5.1).

In order to give an analogue of (4.1.1), it is easily seen that

$$\lim_{n\to\infty} [Q_n(y_0) - \mathrm{inv}\, F(y_0)] \stackrel{\mathrm{a.s.}}{=} 0,$$

provided inv $F(\cdot)$ is continuous at y_0. Otherwise, the above statement cannot be true.

Towards an analogue of Theorem 4.1.1, we first note that

$$P\{\lim_{n\to\infty} \sup_{0<y<1} |Q_n(y) - \mathrm{inv}\, F(y)| = \infty\} = 1,$$

unless F has finite support. Given Theorem 4.5.1, it appears to be natural to consider under what conditions should the statement

$$\lim_{n\to\infty} \sup_{0<y<1} f(\mathrm{inv}\, F(y)) |Q_n(y) - \mathrm{inv}\, F(y)| \stackrel{\mathrm{a.s.}}{=} 0$$

be true. We are not going to study this question directly. However, a law of iterated logarithm will be proved for the quantile process in Chapter 5.

Now we turn to the problem of approximating the quantile process by a sequence of Brownian bridges and first we prove an analogue of Theorem 4.4.1.

Theorem 4.5.2 (Csörgő, Révész 1975c, 1978b). *Given independent $U(0, 1)$ r.v. $U_1, U_2, \ldots$, there exists a sequence of Brownian bridges $\{B_n(y); 0 \leq y \leq 1\}$ such that for each $n = 1, 2, \ldots$, and for all $|z| \leq c\sqrt{n}$ and $c > 0$ we have*

(4.5.2) $\quad P\{\sup_{0 \leq y \leq 1} |u_n(y) - B_n(y)| > n^{-1/2}(A \log n + z)\} \leq Be^{-Cz},$

where A, B, C, c are positive absolute constants. Whence we also have

(4.5.3) $\quad \sup_{0 \leq y \leq 1} |u_n(y) - B_n(y)| \stackrel{\mathrm{a.s.}}{=} O(n^{-1/2} \log n).$

Proof. Put $E_k = \log(1/U_k)$, $k = 1, 2, \ldots$, $S_0 = 0$, $S_k = \sum_{j=1}^{k} E_j$, $k = 1, 2, \ldots$, and

$$\tilde{U}_n(y) = \begin{cases} S_k/S_{n+1} & \text{if } \frac{k-1}{n} < y \leq \frac{k}{n}, \quad k = 1, 2, \ldots, n, \\ 0 & \text{if } y = 0, \end{cases}$$

$$\tilde{u}_n(y) = n^{1/2}(\tilde{U}_n(y) - y), \quad 0 \leq y \leq 1.$$

Then the E_k are independent exponential r.v. with mean value one and, for each $n=1, 2, \ldots$ (cf. (4.3.3))

$$\{\tilde{U}_n(y);\ 0 \leq y \leq 1\} \stackrel{\mathscr{D}}{=} \{U_n(y);\ 0 \leq y \leq 1\}.$$

Whence

(4.5.4) $\quad \{\tilde{u}_n(y);\ 0 \leq y \leq 1\} \stackrel{\mathscr{D}}{=} \{u_n(y);\ 0 \leq y \leq 1\}, \quad n = 1, 2, \ldots.$

A simple calculation yields

$$\tilde{u}_n\left(\frac{k}{n}\right) = n^{1/2}\left(\frac{S_k}{S_{n+1}} - \frac{k}{n}\right) = n^{-1/2}\frac{n}{S_{n+1}}\left[(S_k - k) - \frac{k}{n}(S_{n+1} - n)\right].$$

Let $W(t)$ be a Wiener process approximating the sequence $\{S_n - n\}$ at the rate $O(\log n)$ (cf. Theorem 2.6.1). Define

(4.5.5) $\quad B_n(y) = n^{-1/2}(W(ny) - yW(n)), \quad 0 \leq y \leq 1,$

and put $1 + \varepsilon_n = n/S_{n+1}$. Now consider

(4.5.6)

$$\tilde{u}_n\left(\frac{k}{n}\right) - B_n\left(\frac{k}{n}\right) = n^{-1/2}\left[((S_k - k) - W(k)) - \frac{k}{n}((S_n - n) - W(n)) - \frac{k}{n}E_{n+1}\right.$$
$$\left. + \varepsilon_n\left((S_k - k) - \frac{k}{n}(S_{n+1} - n)\right)\right].$$

We have for all $z > 0$

$$P\left\{\sup_{1 \leq k \leq n} n^{-1/2}|(S_k - k) - W(k)| \geq \left(\frac{A \log n + z}{5}\right)n^{-1/2}\right\} \leq Be^{-Cz},$$

$$P\left\{n^{-1/2}|(S_n - n) - W(n)| \geq \left(\frac{A \log n + z}{5}\right)n^{-1/2}\right\} \leq Be^{-Cz},$$

$$P\left\{n^{-1/2}E_{n+1} \geq \left(\frac{A \log n + z}{5}\right)n^{-1/2}\right\} \leq Be^{-Cz},$$

$$P\left\{\sup_{1 \leq k \leq n} n^{-1/2}|S_k - k| \geq \left(\frac{A \log n + z}{5}\right)^{1/2}\right\} \leq Be^{-Cz},$$

$$P\left\{|\varepsilon_n| \geq \left(\frac{A \log n + z}{5n}\right)^{1/2}\right\} \leq Be^{-Cz},$$

on choosing A, B, C appropriately, where in the last two inequalities we applied Theorem 2.4.4. Consequently, (4.5.6) and the above inequalities imply

$$P\left\{\sup_{1\leq k\leq n}\left|\tilde{u}_n\left(\frac{k}{n}\right)-B_n\left(\frac{k}{n}\right)\right|>n^{-1/2}(A\log n+z)\right\}\leq 6Be^{-Cz},$$

which by Theorem 1.5.1, in turn, gives

(4.5.7) $\quad P\{\sup_{0\leq y\leq 1}|\tilde{u}_n(y)-B_n(y)|>n^{-1/2}(A\log n+z)\}\leq 6Be^{-Cz}.$

Since (4.5.4) holds and the $B_n(y)$ of (4.5.5) is a Brownian bridge for each n, (4.5.7) also completes the proof with applying again Lemma 4.4.4.

Our next theorem is an analogue statement of (4.4.23) for the uniform quantile process $u_n(y)$.

Theorem 4.5.3 (Csörgő, Révész 1975c). *Given independent $U(0, 1)$ r.v. $U_1, U_2, \ldots$, there exists a Kiefer process $\{K(y, t); 0\leq y\leq 1, t\geq 0\}$ such that*

(4.5.8) $\quad \sup_{0\leq y\leq 1}|u_n(y)-n^{-1/2}K(y,n)| \stackrel{a.s.}{=} O(n^{-1/4}(\log\log n)^{1/4}(\log n)^{1/2}).$

The proof of Theorem 4.5.3 hinges on Theorem 1.15.2 and the following

Lemma 4.5.1. *Let $U_1, U_2, \ldots$ be i.i.d. $U(0, 1)$ r.v. and let $\{K(y, t); 0\leq y\leq 1, t\geq 0\}$ be a Kiefer process on the same probability space. Then*

$$\sup_{1\leq k\leq n}\left|K(U_k^{(n)}, n)-K\left(\frac{k}{n}, n\right)\right| \stackrel{a.s.}{=} O((n\log\log n)^{1/4}(\log n)^{1/2}).$$

Proof. Since $\alpha_n(U_k^{(n)})=-u_n\left(\frac{k}{n}\right)=\sqrt{n}\left(\frac{k}{n}-U_k^{(n)}\right)$, by the Chung–Smirnov law of iterated logarithm (cf. Theorem 5.1.1) we have

$$\varlimsup_{n\to\infty}\frac{n^{1/2}\sup_{1\leq k\leq n}\left|U_k^{(n)}-\frac{k}{n}\right|}{(\log\log n)^{1/2}} \stackrel{a.s.}{=} 2^{-1/2}.$$

Whence, with $a_n=\sup_{1\leq k\leq n}\left|U_k^{(n)}-\frac{k}{n}\right|$, we have

$$\sup_{1\leq k\leq n}\sup_{0\leq s\leq a_n}\left|K\left(\frac{k}{n}+s,n\right)-K\left(\frac{k}{n},n\right)\right|$$

$$\leq \sup_{1\leq k\leq n}\sup_{0\leq s\leq O(1)\left(\frac{\log\log n}{n}\right)^{1/2}}\left|K\left(\frac{k}{n}+s,n\right)-K\left(\frac{k}{n},n\right)\right|$$

almost surely, for all but a finite number of n. Lemma 4.5.1 now follows upon taking $h_n = O(1) \left(\frac{\log \log n}{n} \right)^{1/2}$ in Theorem 1.15.2.

Proof of Theorem 4.5.3. First we note that $\alpha_n(U_k^{(n)}) = -u_n\left(\frac{k}{n}\right)$. Hence (4.4.23) of Theorem 4.4.3 implies

$$\sup_{0 \leq k \leq n} \left| u_n\left(\frac{k}{n}\right) - n^{-1/2} K(U_k^{(n)}, n) \right| \stackrel{a.s.}{=} O(n^{-1/2} \log^2 n).$$

Combining now this latter statement with that of Lemma 4.5.1 we get

$$\sup_{0 \leq k \leq n} \left| u_n\left(\frac{k}{n}\right) - n^{-1/2} K\left(\frac{k}{n}, n\right) \right| \stackrel{a.s.}{=} O(n^{-1/4} (\log \log n)^{1/4} (\log n)^{1/2}),$$

and Theorem 4.5.3 follows, since $\left| u_n(y) - u_n\left(\frac{k}{n}\right) \right| \leq n^{-1/2}$ for $\frac{k-1}{n} < y < \frac{k}{n}$ and, by Theorem 1.15.2

$$\varlimsup_{n \to \infty} \sup_{1 \leq k \leq n} \sup_{0 \leq s \leq \frac{1}{n}} \frac{\left| K\left(\frac{k}{n}+s, n\right) - K\left(\frac{k}{n}, n\right) \right|}{\left(2nh_n \log \frac{1}{h_n} \right)^{1/2}} \leq 1 \quad \text{a.s.}$$

with $h_n = 1/n$.

Remark 4.5.1. Observing again that $\alpha_n(U_k^{(n)}) = -u_n\left(\frac{k}{n}\right)$, it follows from the proof of Theorem 4.4.2 that the rate of (4.5.3) is best possible. As to the rate of embedding of (4.5.8), it is probably very far from being best in spite of the fact that the nearness of Lemma 4.5.1 is best possible if K is the Kiefer process of Theorem 4.4.3 (cf. Theorem 5.2.1).

Our aim now is to prove an analogous statement of (4.5.3) and (4.5.8) for the general quantile process $q_n(y)$. As we have already mentioned at the beginning of this section, there is no such immediate handle in this case like simply replacing y by $F(x)$ in Theorem 4.5.1. However, the distance between $q_n(y)$ and $u_n(y)$, respectively defined in terms of $X_k^{(n)}$ and $U_k^{(n)} = F(X_k^{(n)})$, can be computed accurately enough, so that Theorems 4.5.2 and 4.5.3 can actually be used to obtain strong approximations also for the quantile process $q_n(y)$.

In order to be able to estimate the distance between $q_n(y)$ and $u_n(y)$ we still have to study the latter for a little while. Csáki (1977) investigated the limes superior of the sequence

$$\sup_{\varepsilon_n \leq y \leq 1-\varepsilon_n} (y(1-y) \log \log n)^{-1/2} |\alpha_n(y)|,$$

and succeeded in evaluating this lim sup for a wide class of sequences $\{\varepsilon_n\}$, $\varepsilon_n \searrow 0$. Here we mention only one special case of his many results for later use.

Theorem 4.5.4 (Csáki 1977). *With $\varepsilon_n = dn^{-1} \log \log n$ and $d \geq 0.236...$ we have*

$$\varlimsup_{n \to \infty} \sup_{\varepsilon_n \leq y \leq 1-\varepsilon_n} (y(1-y) \log \log n)^{-1/2} |\alpha_n(y)| \stackrel{a.s.}{=} 2.$$

For further details we refer to Theorem 5.1.6 and Remark 5.1.1.

Our next step is to prove an analogue of this theorem for $u_n(y)$. Actually, this analogue is going to be weaker than Theorem 4.5.4. Applying Csáki's method, however, it does not seem to be too difficult to get complete analogues. The herewith presented one suffices for our purposes in the sequel.

Theorem 4.5.5 (Csörgő, Révész 1978b). *With $\delta_n = 25n^{-1} \log \log n$ we have*

$$\varlimsup_{n \to \infty} \sup_{\delta_n \leq y \leq 1-\delta_n} (y(1-y) \log \log n)^{-1/2} |u_n(y)| \leq 4 \quad a.s.$$

Proof. Let
$$x_1 = x_1(y) = y - 4(y(1-y)n^{-1} \log \log n)^{1/2},$$
$$x_2 = x_2(y) = y + 4(y(1-y)n^{-1} \log \log n)^{1/2}.$$

Then, for $n \geq 3$

(4.5.9) $\qquad \varepsilon_n \leq x_1 < x_2 \leq 1-\varepsilon_n, \quad \text{provided} \quad \delta_n \leq y \leq 1-\delta_n,$

where again $\varepsilon_n = dn^{-1} \log \log n$ ($d=0.236...$). In order to see that (4.5.9) holds for $n \geq 3$, we note that

$$x_1 - \varepsilon_n \geq \left(\frac{y}{25} - \varepsilon_n\right) + \left(\frac{24}{25} y^{1/2} - 4(n^{-1} \log \log n)^{1/2}\right) y^{1/2} \geq 0,$$

and similarly, $1 - x_2 - \varepsilon_n \geq 0$. Hence for x_1 as defined at the beginning of this proof and n large, Theorem 4.5.4 gives

$$nF_n(x_1) \leq nx_1 + 2(x_1(1-x_1)n \log \log n)^{1/2}(1+o(1))$$
$$= ny - 4(y(1-y)n \log \log n)^{1/2}$$
$$\quad + 2(x_1(1-x_1)n \log \log n)^{1/2}(1+o(1)) \leq ny$$

almost surely if n is large enough, where, for n large, the last inequality follows from the inequality $x_1(1-x_1)(1+o(1)) \leq 4y(1-y)$. The latter, in turn, is true since $x_1 < y$ and

$$(1+o(1))\frac{1-x_1}{1-y} = (1+o(1))\left(1+4\left(\frac{y}{1-y}n^{-1}\log\log n\right)^{1/2}\right) \le$$
$$\le (1+o(1))(1+4(1/25)^{1/2}) < 4$$

if n is large enough. Similarly we have $nF_n(x_2) \ge ny$, that is to say we now have $nF_n(x_1) \le ny \le nF_n(x_2)$. Hence, for n large, $x_1 \le U_n(y) \le x_2$ whenever $\delta_n \le y \le 1-\delta_n$, since $F_n(\cdot)$ is monotone non-decreasing. This, in turn, is the statement of Theorem 4.5.5.

Now we are in the position to estimate the distance between $q_n(y)$ and $u_n(y)$. Namely we have

Theorem 4.5.6 (Csörgő, Révész 1978b). *Let* $X_1, X_2, \ldots$ *be i.i.d.r.v. with a continuous distribution function* F *which is also twice differentiable on* (a, b), *where* $-\infty \le a = \sup\{x: F(x)=0\}$, $+\infty \ge b = \inf\{x: F(x)=1\}$ *and* $F' = f \ne 0$ *on* (a, b). *Let the quantile process* $q_n(y)$ *resp.* $u_n(y)$ *be defined in terms of* $X_k^{(n)}$ *resp.* $U_k^{(n)} = F(X_k^{(n)})$. *Assume that for some* $\gamma > 0$,

(4.5.10) $$\sup_{a<x<b} F(x)(1-F(x))\left|\frac{f'(x)}{f^2(x)}\right| \le \gamma.$$

Then, with δ_n *as in Theorem 4.5.5*

(4.5.11) $$\varlimsup_{n\to\infty} \frac{n^{1/2}}{\log\log n} \sup_{\delta_n \le y \le 1-\delta_n} |f(\operatorname{inv} F(y))q_n(y) - u_n(y)| \le K \quad a.s.,$$

where $K = 289\gamma \, 72^\gamma$.

If, in addition to (4.5.10), *we also assume that*

(4.5.12) $$\begin{cases} f \text{ is non-decreasing (non-increasing) on an interval} \\ \text{to the right of } a \text{ (to the left of } b), \end{cases}$$

then

(4.5.13)
$$\sup_{0<y<1} |f(\operatorname{inv} F(y))q_n(y) - u_n(y)| \stackrel{a.s.}{=} \begin{cases} O(n^{-1/2}\log\log n) & \text{if } \gamma < 1 \\ O(n^{-1/2}(\log\log n)^2) & \text{if } \gamma = 1 \\ O(n^{-1/2}(\log\log n)^\gamma (\log n)^{(1+\varepsilon)(\gamma-1)}) \\ & \text{if } \gamma > 1, \end{cases}$$

where $\varepsilon > 0$ *is arbitrary. The respective constants of the* $\stackrel{a.s.}{=} O(\cdot)$ *of* (4.5.13) *may be taken to be:* $\left(46\sqrt{25}\frac{2^\gamma}{1-\gamma}\right)2 + K$ *if* $\gamma < 1$, 51 *if* $\gamma = 1$ *and arbitrarily small if* $\gamma > 1$ *and* $\varepsilon > 0$ *is fixed*.

The following Lemma is going to be useful in the sequel.

Lemma 4.5.2. *Under condition* (4.5.10) *of Theorem 4.5.5 we have*

$$(4.5.14) \qquad \frac{f(\operatorname{inv} F(y_1))}{f(\operatorname{inv} F(y_2))} \leq \left\{ \frac{y_1 \vee y_2}{y_1 \wedge y_2} \cdot \frac{1-(y_1 \wedge y_2)}{1-(y_1 \vee y_2)} \right\}^\gamma,$$

for every pair $y_1, y_2 \in (0, 1)$.

Proof. (4.5.10) implies

$$\left| \frac{d}{dy} \log f(\operatorname{inv} F(y)) \right| \leq \gamma (y(1-y))^{-1} = \gamma \frac{d}{dy} \log \frac{y}{1-y}.$$

Whence, if $y_1 > y_2$, then

$$\log \frac{f(\operatorname{inv} F(y_1))}{f(\operatorname{inv} F(y_2))} \leq \gamma \log \frac{y_1}{1-y_1} - \gamma \log \frac{y_2}{1-y_2} = \gamma \log \frac{y_1}{y_2} \frac{1-y_2}{1-y_1}$$

and, if $y_1 < y_2$, then

$$\log \frac{f(\operatorname{inv} F(y_1))}{f(\operatorname{inv} F(y_2))} \leq \gamma \log \frac{y_2}{1-y_2} - \gamma \log \frac{y_1}{1-y_1} = \gamma \log \frac{y_2}{y_1} \frac{1-y_1}{1-y_2}.$$

Hence (4.5.14) is proved.

Proof of Theorem 4.5.6. For $\frac{k-1}{n} < y \leq \frac{k}{n}$,

$$f(\operatorname{inv} F(y)) q_n(y) = n^{1/2} f(\operatorname{inv} F(y))(\operatorname{inv} F(U_k^{(n)}) - \operatorname{inv} F(y))$$
$$= n^{1/2} f(\operatorname{inv} F(y))(\operatorname{inv} F(y + n^{-1/2} u_n(y)) - \operatorname{inv} F(y))$$
$$= u_n(y) - \tfrac{1}{2} n^{-1/2} u_n^2(y) f(\operatorname{inv} F(y)) \frac{f'(\operatorname{inv} F(\xi))}{f^3(\operatorname{inv} F(\xi))},$$

where ξ is between y and $U_k^{(n)} = y + n^{-1/2} u_n(y)$, i.e. $|\xi - y| \leq n^{-1/2} |u_n(y)|$. Hence

$$(4.5.15) \quad |f(\operatorname{inv} F(y)) q_n(y) - u_n(y)| \leq \tfrac{1}{2} n^{-1/2} u_n^2(y) f(\operatorname{inv} F(y)) \left| \frac{f'(\operatorname{inv} F(\xi))}{f^3(\operatorname{inv} F(\xi))} \right|.$$

Theorem 4.5.5 implies that, uniformly for $\delta_n \leq y \leq 1 - \delta_n$, the right-hand side of the above inequality is almost surely majorized for large n by

$$(4.5.16) \quad 8n^{-1/2} (\log \log n) y(1-y) f(\operatorname{inv} F(y)) \frac{|f'(\operatorname{inv} F(\xi))|}{f^3(\operatorname{inv} F(\xi))} (1 + o(1))$$

$$= 8n^{-1/2} (\log \log n) \left[\frac{y(1-y)}{\xi(1-\xi)} \right] \left[\xi(1-\xi) \frac{|f'(\operatorname{inv} F(\xi))|}{f^2(\operatorname{inv} F(\xi))} \right] \left[\frac{f(\operatorname{inv} F(y))}{f(\operatorname{inv} F(\xi))} \right] (1 + o(1))$$

with $|\xi - y| \leq 4(y(1-y) n^{-1} \log \log n)^{1/2} (1 + o(1))$.

First we estimate $\dfrac{y(1-y)}{\xi(1-\xi)}$. Since $\xi > y - 4(y(1-y)n^{-1}\log\log n)^{1/2}(1+o(1))$ and $y \geq \delta_n$, for n large enough we have

$$\frac{y}{\xi} \leq 1 + \frac{4(y(1-y)n^{-1}\log\log n)^{1/2}(1+o(1))}{y - 4(y(1-y)n^{-1}\log\log n)^{1/2}(1+o(1))}$$

$$= 1 + \frac{4(y^{-1}(1-y)n^{-1}\log\log n)^{1/2}(1+o(1))}{1 - 4(y^{-1}(1-y)n^{-1}\log\log n)^{1/2}(1+o(1))} \leq 1 + \frac{4/5(1+o(1))}{1 - 4/5(1+o(1))} \leq 6.$$

Now applying the inequality $\xi < y + 4(y(1-y)n^{-1}\log\log n)^{1/2}(1+o(1))$, where $y \leq 1 - \delta_n$, a similar computation yields that for n large $\dfrac{1-y}{1-\xi} \leq 6$. Hence for large enough n

$$y(1-y)/\xi(1-\xi) \leq 36,$$

and by condition (4.5.10)

$$\xi(1-\xi)|f'(\operatorname{inv} F(\xi))/f^2(\operatorname{inv} F(\xi))| \leq \gamma.$$

Finally by Lemma 4.5.2

$$\frac{f(\operatorname{inv} F(y))}{f(\operatorname{inv} F(\xi))} \leq \left[\frac{\xi(1-y)}{y(1-\xi)} + \frac{y(1-\xi)}{\xi(1-y)}\right]^{\gamma} \leq 72^{\gamma}.$$

From these statements and from (4.5.16) it follows that for large n the left-hand side of (4.5.15) is bounded above by $289\gamma \, 72^{\gamma} \, n^{-1/2}(\log\log n)$, and (4.5.11) follows.

In order to prove (4.5.13), it suffices to show that

$$\sup_{0 < y \leq \delta_n} |f(\operatorname{inv} F(y))q_n(y) - u_n(y)| \quad \text{and} \quad \sup_{1-\delta_n \leq y < 1} |f(\operatorname{inv} F(y))q_n(y) - u_n(y)|$$

are $\overset{\text{a.s.}}{=} O(\cdot)$ as indicated on the right-hand side of (4.5.13). We demonstrate this only for the first one of these sups since for the second one a similar argument holds. First of all we show that for n large

(4.5.17) $\qquad \sup_{0 \leq y \leq \delta_n} |u_n(y)| \leq 46 n^{-1/2} \log\log n \quad \text{a.s.}$

The proof of (4.5.17) is as follows: for $0 \leq y \leq \delta_n$

(4.5.18) $\qquad |u_n(y)| = \sqrt{n}\,|U_n(y) - y| \leq \sqrt{n}\,y \leq 25 n^{-1/2} \log\log n,$

whenever $y \geq U_n(y)$, and $|u_n(y)| = \sqrt{n}\,|U_n(y) - y| \leq \sqrt{n}\,U_n(y) \leq \sqrt{n}\,U_{[n\delta_n]}^{(n)}$, whenever $y \leq U_n(y)$.

In the latter case we consider

(4.5.19)
$$n^{1/2}(U^{(n)}_{[n\delta_n]}-\delta_n)+n^{1/2}\delta_n \leq (1+o(1))4(\delta_n \log\log n)^{1/2}+25n^{-1/2}\log\log n$$
$$\leq 46n^{-1/2}\log\log n \quad \text{a.s. for } n \text{ large,}$$

where the above a.s. inequality follows from Theorem 4.5.5. Now (4.5.18) and (4.5.19) combined, imply (4.5.17).

Restricting attention then to the region $0<y\leq\delta_n$, we assume that $f(\operatorname{inv} F(y))$ is non-decreasing on an interval to the right of a (cf. (4.5.12)). Let $\frac{k-1}{n}<y\leq\frac{k}{n}$. If $U^{(n)}_k \geq y$,

(4.5.20)
$$|f(\operatorname{inv} F(y))q_n(y)| = n^{1/2}\int_y^{U^{(n)}_k} \frac{f(\operatorname{inv} F(y))}{f(\operatorname{inv} F(u))}\,du \leq u_n(y),$$

where the inequality on the right-hand side results from the assumption that $f(\operatorname{inv} F(y))$ is non-decreasing on an interval to the right of a. If $U^{(n)}_k < y$, then

(4.5.21)
$$|f(\operatorname{inv} F(y))q_n(y)| = n^{1/2}\int_{U^{(n)}_k}^y \frac{f(\operatorname{inv} F(y))}{f(\operatorname{inv} F(u))}\,du$$
$$\leq n^{1/2}\int_{U^{(n)}_k}^y \left(\frac{y(1-u)}{u(1-y)}\right)^\gamma du, \quad \text{by (4.5.14)}$$
$$\leq 2^\gamma n^{1/2}\int_{U^{(n)}_k}^y \left(\frac{y}{u}\right)^\gamma du$$
$$\leq \begin{cases} \dfrac{2^\gamma}{1-\gamma} n^{1/2} y & \text{if } \gamma < 1 \\ \dfrac{2^\gamma}{\gamma-1} n^{1/2} y^\gamma (U^{(n)}_k)^{-(\gamma-1)} & \text{if } \gamma > 1 \\ 2n^{1/2} y \log \dfrac{y}{U^{(n)}_k} & \text{if } \gamma = 1. \end{cases}$$

Hence (4.5.20) (with the help of (4.5.17)) and (4.5.21) (via $0<y\leq\delta_n$ and in view of $\lim_{n\to\infty} U^{(n)}_1 \cdot n(\log n)^{1+\varepsilon} = \infty$ for every $\varepsilon>0$) together imply (4.5.13). This also completes the proof of Theorem 4.5.6.

A careful investigation of the proof of Theorem 4.5.6 also shows that (4.5.11) gives the best possible result in the following sense:

Assume that the conditions of Theorem 4.5.6 hold and also assume that f is twice differentiable with $|f''| \leq C$ and $|f'| \geq \varrho > 0$ on a finite interval $(\bar{a}, \bar{b}) \subset (a, b)$. Then we have

$$\varlimsup_{n \to \infty} \frac{n^{1/2}}{\log \log n} \sup_{\bar{a} \leq y \leq \bar{b}} |f(\operatorname{inv} F(y)) q_n(y) - u_n(y)| > K > 0 \quad \text{a.s.,}$$

where the constant K depends on f.

Remark 4.5.2. Using Theorems 4.5.5 and 4.5.6, a Theorem 4.5.5-type result could be proved also for the general quantile process $f(\operatorname{inv} F(y)) q_n(y)$. It would be more desirable, however, first to produce complete analogues for the uniform quantile process $u_n(y)$ according to Csáki (1977) and then to use these exact analogues, instead of our Theorem 4.5.5, in combination with Theorem 4.5.6 to prove the same complete Csáki-type analogues for the general quantile process $f(\operatorname{inv} F(y)) q_n(y)$. That is to say Theorem 4.5.6 may be viewed and can be used as a strong invariance theorem for studying the problem of what kind of a.s., in-probability and in-distribution properties of $u_n(y)$ should be inherited by $f(\operatorname{inv} F(y)) q_n(y)$. For example, it follows from (4.5.3) and (4.5.13) that $f(\operatorname{inv} F(y)) q_n(y) \xrightarrow{\mathscr{D}} B(y)$, a Brownian bridge on $[0, 1]$, given the conditions (4.5.10) and (4.5.12). Some further examples are given in Chapter 5, and our next theorem is also of this nature.

Having now Theorem 4.5.6 at our disposal, our desired analogue of (4.5.3) and (4.5.8) for the quantile process $q_n(y)$ is immediately at hand as follows:

Theorem 4.5.7 (Csörgő, Révész 1978b). *Let $X_1, X_2, \ldots$ be i.i.d.r.v. with a continuous distribution function F which is also twice differentiable on (a, b), where $-\infty \leq a = \sup\{x: F(x) = 0\}$, $+\infty \geq b = \inf\{x: F(x) = 1\}$ and $F' = f \neq 0$ on (a, b). One can then define a Brownian bridge $\{B_n(y); 0 \leq y \leq 1\}$ for each n, and a Kiefer process $\{K(y, t); 0 \leq y \leq 1, 0 \leq t\}$ such that if condition (4.5.10) of Theorem 4.5.6 is assumed then*

(4.5.22) $\quad \sup\limits_{\delta_n \leq y \leq 1 - \delta_n} |f(\operatorname{inv} F(y)) q_n(y) - B_n(y)| \stackrel{\text{a.s.}}{=} O(n^{-1/2} \log n)$

and

(4.5.23)

$\quad \sup\limits_{\delta_n \leq y \leq 1 - \delta_n} |n^{1/2} f(\operatorname{inv} F(y)) q_n(y) - K(y, n)| \stackrel{\text{a.s.}}{=} O((n \log \log n)^{1/4} (\log n)^{1/2}),$

where δ_n is as in Theorem 4.5.5.

If, in addition to (4.5.10), *condition* (4.5.12) *of Theorem 4.5.6 is also assumed, then*

(4.5.24)
$$\sup_{0<y<1} \left|f(\text{inv } F(y))q_n(y) - B_n(y)\right| \stackrel{a.s.}{=} \begin{cases} O(n^{-1/2}\log n) & \text{if } \gamma < 2 \\ O\left(n^{-1/2}(\log\log n)^\gamma (\log n)^{(1+\varepsilon)(\gamma-1)}\right) \\ & \text{if } \gamma \geq 2, \end{cases}$$

where γ is as in (4.5.10) *and $\varepsilon > 0$ is arbitrary; also*

(4.5.25) $\sup_{0<y<1} \left|n^{1/2}f(\text{inv } F(y))q_n(y) - K(y,n)\right| \stackrel{a.s.}{=} O\left((n\log\log n)^{1/4}(\log n)^{1/2}\right).$

Proof. Let $U_k^{(n)} = F(X_k^{(n)})$ and define $u_n(y)$ in terms of these uniform order statistics. Let $B_n(y)$ resp. $K(y,t)$ be as in Theorem 4.5.2 resp. in Theorem 4.5.3. Then Theorem 4.5.2 resp. Theorem 4.5.3 holds for the thus defined $u_n(y)$ and combining them with Theorem 4.5.6 we get Theorem 4.5.7. The $\stackrel{a.s.}{=} O(n^{-1/2}\log n)$ rate of (4.5.24) for $\gamma < 2$ holds because of the first two $\stackrel{a.s.}{=} O(\cdot)$ rates of (4.5.13), and by taking $\varepsilon < \dfrac{1}{\gamma-1} - 1$ (if $1 < \gamma < 2$) in the $\stackrel{a.s.}{=} O\left(n^{-1/2}(\log\log n)^\gamma (\log n)^{(1+\varepsilon)(\gamma-1)}\right)$ rate of (4.5.13).

Supplementary remarks

Section 4.2. Donsker's original formulation is slightly different from the one given in (4.2.2) in that he works on what is called the $D(0,1)$ space today.

The idea of studying the empirical process via appropriate Gaussian processes can be also found in the papers of Kac (1949), Bartlett (1949) and Kendall's remark to the latter. The just quoted paper of Kac is also discussed in Chapter 7.

Section 4.4. The herewith presented proof of Theorem 4.4.1 is different from the original proof of Komlós, Major and Tusnády (1975). This proof is based on the one given by Tusnády in his dissertation (1977b).

We note also that Theorem 4.4.1 when combined with the Erdős–Rényi law (cf. Komlós, Major and Tusnády 1975a and Tusnády 1977) gives the rate $O(\log n/n^{1/2})$ for the Prohorov distance of probability measures generated by $\alpha_n(y)$ and $B_n(y) \overset{\mathscr{D}}{=} B(y)$, and that the latter rate of $\log n/n^{1/2}$ turns out to be also best possible in this case.

Section 4.5. The non-uniform quantile process was also studied by Shorack (1972a, 1972b). Under somewhat different conditions than ours he proved a number of results. The sharpest one of them reads as follows (Shorack 1972b, Corollary 1):

$$\sup_{n^{-1} \leq y \leq 1-n^{-1}} \frac{|f(\text{inv } F(y))q_n(y) - B_n(y)|}{g(y)} = o(1)$$

for certain functions g.

In a recent paper (Csörgő, Révész 1980a) we noted that the condition (4.5.12) of Theorems 4.5.6 and 4.5.7 can be replaced by the following conditions:

(S.4.5.1) $\qquad A = \lim_{x \downarrow a} f(x) < \infty, \quad B = \lim_{x \uparrow b} f(x) < \infty,$

(S.4.5.2) one of the following conditions hold
  (a) $\min(A, B) > 0$,
  (b) if $A = 0$ (resp. $B = 0$) then f is non-decreasing (resp. non-increasing) on an interval to the right of a (resp. to the left of b).

In the sequel we will often refer to Theorems 4.5.6 and 4.5.7 as well as to (4.5.12). Whenever we do so, the just mentioned two conditions are to replace (4.5.12).

5. A Study of Empirical and Quantile Processes with the Help of Strong Approximation Methods

5.0. Introduction

The role of this chapter in relation to Chapters 1 and 4 is similar to that of Chapter 3 to Chapters 1 and 2. This is why, in addition to studying the Wiener process, in Chapter 1 we also studied certain distributional and almost sure fluctuational properties of Brownian bridges and the Kiefer process with the aim that they might be directly inherited by the empirical and quantile processes via the invariance principles covered in Chapter 4. A look at these strong invariance principles makes it clear that, unlike in the case of sums of r.v., the number of finite moments of the original r.v. sequence does not play a role as to what might be inherited by the empirical and quantile processes themselves. Thus, in the latter sense, our job of sorting out almost sure inheritance is somewhat easier here. On the other hand, the job of sorting out in distribution-type inheritance is made somewhat more difficult by the numerous statistical-type questions one can ask and answer in terms of these processes. Indeed, apart from the first three sections, this chapter is entirely devoted to tackling these latter problems.

5.1. The law of iterated logarithm for the empirical process

Let $X_1, X_2, \ldots$ be a sequence of i.i.d.r.v. with distribution function F. A trivial consequence of the simplest form of the law of iterated logarithm (Theorem 3.2.2 or Remark 3.2.4) is:

Let x be any real number for which $0 < F(x) < 1$. Then

$$(5.1.1) \qquad \varlimsup_{n\to\infty} \frac{n^{1/2}|F_n(x)-F(x)|}{(2F(x)(1-F(x))\log\log n)^{1/2}} \stackrel{a.s.}{=} 1.$$

This implies:

Suppose that F is a distribution function for which there exists a real m such that $F(m) = \frac{1}{2}$. Then

$$(5.1.2) \qquad \varlimsup_{n\to\infty} \sup_{-\infty < x < \infty} \frac{n^{1/2}|F_n(x) - F(x)|}{(\log\log n)^{1/2}} \geq 2^{-1/2} \quad a.s.$$

Formulas (5.1.1) and (5.1.2) suggest the question: how can the rate of convergence of $\sup_x |F_n(x) - F(x)|$ to 0 (cf. Theorem 4.1.1) be estimated? An answer to this question was obtained by Smirnov (1944) and, independently, also by Chung (1949). Their result says:

Theorem 5.1.1 (Smirnov 1944, Chung 1949). *Suppose that F is continuous. Then*

$$(5.1.3) \qquad \varlimsup_{n\to\infty} \left(\frac{n}{\log\log n}\right)^{1/2} \sup_{-\infty < x < +\infty} |F_n(x) - F(x)| \stackrel{a.s.}{=} 2^{-1/2}.$$

(This Theorem will be a consequence of Theorem 5.1.2 and is also a consequence of Theorem 4.4.3 and Corollary 1.15.1.)

As we saw in Chapter 3, having the law of iterated logarithm, a natural next step was to look for the accumulation points of the functions $\gamma_n(t)$ (of Theorem 3.2.2). In the case of the empirical process this was done by Finkelstein (1971), who proved:

Theorem 5.1.2 (Finkelstein 1971). *Let F be a continuous distribution over the real line. Then the sequence*

$$\tilde{\beta}_n(y) = \frac{\alpha_n(\mathrm{inv}\, F(y))}{(2\log\log n)^{1/2}} = \left(\frac{n}{2\log\log n}\right)^{1/2} (F_n(\mathrm{inv}\, F(y)) - y), \quad 0 \leq y \leq 1,$$

is relatively compact with probability 1, and the set of its limit points (with respect to the sup norm) is the set $\mathscr{F} \subset C(0,1)$ of absolutely continuous functions, defined in Theorem 1.15.1.

Proof. This theorem is a straight consequence of Theorems 1.15.1 and 4.4.3.

Studying the properties of the set $\mathscr{F}$ by standard calculus of variation methods, Finkelstein proves:

Lemma 5.1.1. *We have*

(5.1.4) $$\sup_{f \in \mathscr{F}} \sup_{0 \leq x \leq 1} |f(x)| = \tfrac{1}{2}$$

and

(5.1.5) $$\sup_{f \in \mathscr{F}} \int_0^1 f^2(x)\,dx = \pi^{-2}.$$

(5.1.4) and Theorem 5.1.2 clearly imply Theorem 5.1.1. Similarly (5.1.5) and Theorem 5.1.2 together imply

Theorem 5.1.3. *Let F be an arbitrary continuous distribution function. Then*

$$\varlimsup_{n \to \infty} \frac{n}{2 \log \log n} \int_{-\infty}^{+\infty} (F_n - F)^2 \, dF \overset{\text{a.s.}}{=} \pi^{-2}.$$

Relation (5.1.1) suggests that, having studied $\tilde{\beta}_n(y)$ (cf. Theorem 5.1.2), we should also study the behaviour of the sequence of processes

$$\beta_n^*(x) = \frac{n^{1/2}(F_n(x) - F(x))}{(2(F(x)(1-F(x))\log \log n)^{1/2}}.$$

As a straight consequence of (1.15.2) and Theorem 4.4.3, we get

Theorem 5.1.4. *Let F be a continuous distribution function. Then for any $0 < \varepsilon < \tfrac{1}{2}$ we have*

(5.1.6) $$\varlimsup_{n \to \infty} \sup_{\varepsilon < F(x) < 1-\varepsilon} |\beta_n^*(x)| \overset{\text{a.s.}}{=} 1.$$

The latter suggests that we should also be interested in the properties of the sequence

$$\sup_{\varepsilon_n \leq F(x) \leq 1-\varepsilon_n} |\beta_n^*(x)|,$$

where $0 < \varepsilon_n < \tfrac{1}{2}$ is a sequence tending to 0 in a given order. Applying (1.15.4) as well as Theorem 4.4.3, we get

Theorem 5.1.5. *Let F be a continuous distribution function and let $\varepsilon_n = \varepsilon \exp\left(-(\log n)^c\right)$ ($0 \leq c < 1$, $0 < \varepsilon < \tfrac{1}{2}$). Then*

$$\varlimsup_{n \to \infty} \sup_{\varepsilon_n \leq F(x) \leq 1-\varepsilon_n} |\beta_n^*(x)| \overset{\text{a.s.}}{=} (c+1)^{1/2}.$$

In case $c = 0$, we obtain Theorem 5.1.4 as a special case.

When proving Theorem 5.1.5, in the application of Theorem 4.4.3 we needed to apply the relation $\varepsilon_n n (\log n)^{-3} \to \infty$. The case of $\varepsilon_n \leq n^{-1}(\log n)^3$

cannot be treated via strong approximation methods, since they do not imply that the processes

$$\frac{K(F(x), n)}{(F(x)(1-F(x)))^{1/2}} \quad \text{and} \quad \frac{n(F_n(x)-F(x))}{(F(x)(1-F(x)))^{1/2}}$$

are near enough to each other when $F(x) \leq \varepsilon_n = o(n^{-1} \log^3 n)$. Surprisingly enough, in spite of this fact, the limiting behaviour of the suprema of these processes are the same if $d_0 n^{-1} \log \log n \leq F(x) \leq 1 - d_0 n^{-1} \log \log n$, where $d_0 = 0.236...$; they behave differently when $F(x)$ is outside this interval. In fact Csáki (1977) proved the following:

Theorem 5.1.6 (Csáki 1977). *If F is a continuous distribution function and $\varepsilon_n = d n^{-1} \log \log n$, then*

(5.1.7) $$\varlimsup_{n \to \infty} \sup_{\varepsilon_n \leq F(x) \leq 1-\varepsilon_n} |\beta_n^*(x)| \stackrel{\text{a.s.}}{=} \max \left\{ 2^{1/2}, \left(\frac{d}{2}\right)^{1/2} (b_d - 1) \right\}$$

where $b_d > 1$ is the solution of the equation

$$b_d (\log b_d - 1) = d^{-1}(1-d).$$

Remark 5.1.1. One gets by simple numerical methods that $d^{1/2}(b_d - 1) = 2$ if $d = d_0 = 0.236...$, which means that the limiting behaviour of the Kiefer process and that of the empirical process is the same if $\varepsilon_n \geq d_0 n^{-1} \log \log n$, and different if $\varepsilon_n \leq (d_0 - \varepsilon) n^{-1} \log \log n$ ($0 < \varepsilon < d_0$) (cf. also (1.15.5)).

In Chapter 3, besides the law of iterated logarithm (Remark 3.2.4), we also presented Chung's theorem (cf. (3.3.4)). An analogue of the latter one for empirical processes was found by Mogul'skiĭ (1977), who proved

Theorem 5.1.7 (Mogul'skiĭ 1977). *Let F be a continuous distribution function. Then*

(5.1.8) $$\varliminf_{n \to \infty} (n \log \log n)^{1/2} \sup_x |F_n(x) - F(x)| = 8^{-1/2} \pi$$

and

(5.1.9) $$\varliminf_{n \to \infty} (n \log \log n)^{1/2} \left(\int_0^1 (F_n(x) - F(x))^2 \, dF(x) \right)^{1/2} \stackrel{\text{a.s.}}{=} 8^{-1/2}.$$

The proof will not be presented here.

5.2. The distance between the empirical and the quantile processes

Bahadur (1966) was the first to investigate the distance between the empirical and quantile processes in the case when the sample is coming from the uniform $U(0, 1)$ distribution. The best result, concerning this problem, is due to Kiefer (1970). He proved

Theorem 5.2.1 (Kiefer 1970). *Let $X_1, X_2, \ldots$ be i.i.d.r.v. with a twice differentiable distribution function F on the unit interval. Let $f=F'$,*

$$R_n = \sup_{0 \leq y \leq 1} |(F_n(\operatorname{inv} F(y)) - y) - (\operatorname{inv} F(y) - Q_n(y)) f(\operatorname{inv} F(y))|$$

and assume that $\inf_{0 \leq x \leq 1} f(x) > 0$ *and* $\sup_{0 \leq x \leq 1} |f'(x)| < \infty$. *Then*

(5.2.1) $$\varlimsup_{n \to \infty} \frac{n^{3/4}}{(\log n)^{1/2}(\log \log n)^{1/4}} R_n \overset{\text{a.s.}}{=} 2^{-1/4}.$$

Applying this theorem in the uniform case and our Theorem 4.5.6, we immediately get the following extension of the former.

Theorem 5.2.2. *Let $X_1, X_2, \ldots$ be i.i.d.r.v. with a continuous distribution function F which is also twice differentiable on (a, b), where $-\infty \leq a = \sup_x \{x: F(x) = 0\}$, $+\infty \geq b = \inf \{x: F(x) = 1\}$ and $F' = f \neq 0$ on (a, b). Assume that F also satisfies conditions (4.5.10) and (4.5.12) of Theorem 4.5.6. Then the statement of (5.2.1) is still true.*

Remark 5.2.1. We wish to emphasize here that the conditions (4.5.10) and (4.5.12) of Theorem 4.5.6 are much weaker than those of Theorem 5.2.1. Especially it is not assumed here that a and b are necessarily finite. We should, however, also emphasize that Theorem 5.2.1 in the uniform case is applied in the proof of Theorem 5.2.2.

Studying the properties of R_n, Kiefer (1970) also obtained the limit theorem:

Theorem 5.2.3 (Kiefer 1970). *Under the conditions of Theorem 5.2.1 we have*

(5.2.2) $$\lim_{n \to \infty} P\{n^{3/4}(\log n)^{-1/2} R_n > t\} = 2 \sum_{m=1}^{\infty} (-1)^{m+1} e^{-2m^2 t^4} \quad (t > 0).$$

As a matter of fact, (5.2.2) states that, under the conditions of Theorem 5.2.1, $n^{3/4}(\log n)^{-1/2} R_n$ has the same limiting distribution as the square root of the Kolmogorov–Smirnov statistic $D_n = n^{1/2} \sup_x |F_n(x) - F(x)|$ (cf. Theorem 4.1.2). Indeed, Kiefer proved (5.2.2) via the more fundamental

Theorem 5.2.4 (Kiefer 1970). *Under the conditions of Theorem 5.2.1, as* $n \to \infty$,

$$\text{(5.2.3)} \qquad \frac{n^{3/4} R_n}{(D_n \log n)^{1/2}} \xrightarrow{P} 1.$$

The latter theorem implies (5.2.2) at once. Kiefer (1970) also noted that (5.2.3) was actually true with probability one, but did not publish his proof.

Applying this theorem in the uniform case and our Theorem 4.5.6, we can extend the former again to the case when only the weaker conditions of the latter are assumed. However this extension is not so immediate as in the case of Theorem 5.2.1. In fact Theorem 5.1.7 will be also applied in the proof of

Theorem 5.2.5. *Under the conditions of Theorem 5.2.2 on F, the statement of (5.2.3) and hence also that of (5.2.2) are still true.*

Proof. Let $U_k^{(n)} = F(X_k^{(n)})$ and define $U_n(y)$ in terms of these uniform order statistics. Put $k_n = n^{3/4}(D_n \log n)^{-1/2}$. Then, applying Theorems 4.5.6 and 5.1.7, we get

$$\text{(5.2.4)} \qquad \lim_{n \to \infty} \sup_{0 < y < 1} \left| (y - U_n(y)) - (\operatorname{inv} F(y) - Q_n(y)) f(\operatorname{inv} F(y)) \right| k_n$$

$$\leq \frac{\varlimsup_{n \to \infty} n^{1/2} \sup_{0 < y < 1} \left| (y - U_n(y)) - (\operatorname{inv} F(y) - Q_n(y)) f(\operatorname{inv} F(y)) \right|}{(\varlimsup_{n \to \infty} (\log \log n)^{1/2} D_n)^{1/2}}$$

$$\cdot n^{1/4} (\log n)^{-1/2} (\log \log n)^{1/4} \stackrel{\text{a.s.}}{=} 0.$$

For $k_n R_n$, we have the following estimation

$$\sup_{0 < y < 1} \left| (F_n(\operatorname{inv} F(y)) - y) - (y - U_n(y)) \right| k_n$$

$$- \sup_{0 < y < 1} \left| (y - U_n(y)) - (\operatorname{inv} F(y) - Q_n(y)) f(\operatorname{inv} F(y)) \right| k_n \leq k_n R_n$$

$$\leq \sup_{0 < y < 1} \left| (F_n(\operatorname{inv} F(y)) - y) - (y - U_n(y)) \right| k_n$$

$$+ \sup_{0 < y < 1} \left| (y - U_n(y)) - (\operatorname{inv} F(y) - Q_n(y)) f(\operatorname{inv} F(y)) \right| k_n.$$

Taking the in-probability limit as $n \to \infty$, and applying Theorem 5.2.4 in the uniform case and the above (5.2.4), we get Theorem 5.2.5.

Remark 5.2.1. An application of the almost sure version of (5.2.3) produces an almost sure version of Theorem 5.2.5.

5.3. The law of iterated logarithm for the quantile process

In Section 4.5 we called attention to the fact that the analogue of the Glivenko–Cantelli theorem for the quantile process does not hold true without further restrictions. In this section we intend to point out that most of the strong laws proved for the empirical process are also true for the quantile process if we assume the conditions (4.5.10) and (4.5.12) of Theorem 4.5.6. We can do this in three ways: (i) apply Theorem 4.5.7 saying that the quantile process is near to a Kiefer process and deduce that theorems proved for a Kiefer process are also true for the quantile process; (ii) apply Theorem 5.2.2 saying that the quantile process is near to an empirical process and deduce that theorems proved for empirical processes are also true for quantile processes; (iii) apply Theorem 4.5.6 saying that a quantile process is near to a suitable uniform quantile process, hence theorems for the uniform quantile process extend to more general quantile processes. We followed the latter approach in the proofs of Theorems 5.2.2 and 5.2.5.

Applying any of the methods (i) and (ii) one gets (5.3.1)–(5.3.5) immediately, while method (iii) together with Theorem 4.5.6 implies (5.3.6). Thus we have:

Theorem 5.3.1. *Under the conditions of Theorem 5.2.2 on F we have*

(5.3.1) $\quad \varlimsup_{n\to\infty} (\log\log n)^{-1/2} \sup_{0<y<1} f(\text{inv } F(y))|q_n(y)| \stackrel{\text{a.s.}}{=} 2^{-1/2},$

(5.3.2) $\varlimsup_{n\to\infty} (2\log\log n)^{-1} \int_0^1 f^2(\text{inv } F(y)) q_n^2(y)\, dy \stackrel{\text{a.s.}}{=} \pi^{-2},$ (cf. Theorem 5.1.3),

(5.3.3) $\quad \varliminf_{n\to\infty} (\log\log n)^{1/2} \sup_{0<y<1} f(\text{inv } F(y))|q_n(y)| \stackrel{\text{a.s.}}{=} 8^{-1/2}\pi,$ (cf. (5.1.8))

(5.3.4) $\quad \varliminf_{n\to\infty} (\log\log n)^{1/2} \Big(\int_0^1 f^2(\text{inv } F(y)) q_n^2(y)\, dy\Big)^{1/2} \stackrel{\text{a.s.}}{=} 8^{-1/2},$ (cf. (5.1.9)),

(5.3.5) $\quad \varlimsup_{n\to\infty} (2\log\log n)^{-1/2} \sup_{\varepsilon<y<1-\varepsilon} (y(1-y))^{-1/2} f(\text{inv } F(y))|q_n(y)| \stackrel{\text{a.s.}}{=} 1.$

Assume that conditions (4.5.10), (4.5.12) *with* $\gamma<1$ *hold true. Then there exists a* $C>0$ *such that*

(5.3.6) $\quad \varlimsup_{n\to\infty} \sup_{\delta_n<y<1-\delta_n} (y(1-y)\log\log n)^{-1/2} |f(\text{inv } F(y)) q_n(y)| \leq C \quad a.s.$

where $\delta_n = 25 n^{-1} \log\log n$.

The version of Theorem 5.1.2 in terms of $f(\text{inv } F(y))q_n(y)$ is also straightforward. However the generalization of Theorem 5.1.5 resp. 5.1.6 to the quantile process is not immediate at all. The only such version we have at present is that of (5.3.6).

5.4. Asymptotic distribution results for some classical functionals of the empirical process

We have already seen that Donsker's theorem for the empirical process (Theorem 4.2.1) is a direct consequence of any one of the Brillinger and/or Kiefer type approximation theorems of Sections 4.3 and 4.4, and hence that, for example, Theorem 4.1.2 is implied by (1.5.3) and (1.5.4) of Theorem 1.5.1. Some further Corollary 4.4.1 type results follow here.

Corollary 5.4.1. *Consider the empirical process* $\beta_n(x) = \alpha_n(F(x))$. *Let* $y = F(x)$ *be a continuous distribution function and let* $g(y) \neq 0$ *be a real valued function for which we also have*

$$(5.4.1) \qquad \sup_y |g(y)| < \infty.$$

There exist then a sequence of Brownian bridges $\{B_n(y); 0 \leq y \leq 1\}$ *and a Kiefer process* $\{K(y, t); 0 \leq y \leq 1, 0 \leq t < \infty\}$ *such that*

$$(5.4.2) \quad \left| \sup_{-\infty < x < \infty} \beta_n(x) g(F(x)) - \sup_{0 \leq y \leq 1} B_n(y) g(y) \right| \stackrel{a.s.}{=} O(n^{-1/2} \log n),$$

$$\left| \sup_{-\infty < x < \infty} \beta_n(x) g(F(x)) - \sup_{0 \leq y \leq 1} n^{-1/2} K(y, n) g(y) \right| \stackrel{a.s.}{=} O(n^{-1/2} \log^2 n),$$

$$(5.4.3) \left| \sup_{-\infty < x < \infty} |\beta_n(x) g(F(x))| - \sup_{0 \leq y \leq 1} |B_n(y) g(y)| \right| \stackrel{a.s.}{=} O(n^{-1/2} \log n),$$

$$\left| \sup_{-\infty < x < \infty} |\beta_n(x) g(F(x))| - \sup_{0 \leq y \leq 1} |n^{-1/2} K(y, n) g(y)| \right| \stackrel{a.s.}{=} O(n^{-1/2} \log^2 n),$$

$$(5.4.4) \qquad \left| \int_{-\infty}^{+\infty} \beta_n^2(x) g^2(F(x)) \, dF(x) - \int_0^1 B_n^2(y) g^2(y) \, dy \right|$$

$$\stackrel{a.s.}{=} O(n^{-1/2} \log n \, (\log \log n)^{1/2}),$$

$$\left| \int_{-\infty}^{+\infty} \beta_n^2(x) g^2(F(x)) \, dF(x) - \int_0^1 n^{-1} K^2(y, n) g^2(y) \, dy \right|$$

$$\stackrel{a.s.}{=} O(n^{-1/2} \log^2 n (\log \log n)^{1/2})$$

and

$$(5.4.5) \quad \left|n^{1/2}R_n - \left(\sup_y B_n(y)g(y) - \inf_y B_n(y)g(y)\right)\right| \stackrel{a.s.}{=} O(n^{-1/2}\log n),$$

$$\left|n^{1/2}R_n - \left(\sup_y n^{-1/2}K(y,n)g(y) - \inf_y n^{-1/2}K(y,n)g(y)\right)\right| \stackrel{a.s.}{=} O(n^{-1/2}\log^2 n),$$

where

$$R_n = D_n^+ + D_n^- = \sup_x \left(F_n(x) - F(x)\right)g(F(x)) - \inf_x \left(F_n(x) - F(x)\right)g(F(x)).$$

Since $B_n(y) \stackrel{\mathscr{D}}{=} B(y)$, and also $n^{-1/2}K(y,n) \stackrel{\mathscr{D}}{=} K(y,1) \stackrel{\mathscr{D}}{=} B(y)$, a Brownian bridge for each n, we see that the limit distributions of the above-presented functionals of the empirical process agree with the distributions of the corresponding functionals of a Brownian bridge. For example, (5.4.2) and (5.4.3) with $g(y) = 1$ give the asymptotic distribution of the classical Kolmogorov–Smirnov statistics (cf. Theorem 4.1.2) via (1.5.3) and (1.5.4). Again with $g(y)=1$ and applying (5.4.4), we get the limit distribution of the Cramér–von Mises statistic. Namely, Theorem 1.5.2 implies

$$(5.4.6) \quad \lim_{n\to\infty} P\left\{ \int_{-\infty}^{+\infty} \beta_n^2(x)\,dF(x) \leq u \right\} = P\{\omega^2 \leq u\}, \quad u \geq 0,$$

where the latter distribution function is given in Theorem 1.5.2. Another classical result, the distribution of the Kuiper (1960) statistic, can be obtained from (5.4.5) with $g(y)=1$ by Theorem 1.5.3. Namely we have

$$\lim_{n\to\infty} P\{n^{1/2}R_n \leq u\} = 1 - \sum_{j=1}^{\infty} 2(4(ju)^2 - 1)e^{-2j^2 u^2}, \quad u \geq 0.$$

Taking now for example

$$(5.4.7) \quad g_\varepsilon(y) = \begin{cases} 0 & \text{if } 0 \leq y \leq \varepsilon \\ y^{-1} & \text{if } \varepsilon < y \leq 1, \end{cases}$$

$$g_\delta(y) = \begin{cases} (1-y)^{-1} & \text{if } 0 \leq y \leq \delta \\ 0 & \text{if } \delta < y \leq 1, \end{cases}$$

respectively

$$(5.4.8) \quad g_{\varepsilon,\delta}(y) = \begin{cases} (y(1-y))^{-1/2} & \text{if } 0 < \varepsilon \leq y \leq \delta < 1 \\ 0 & \text{otherwise,} \end{cases}$$

in (5.4.2) and (5.4.3), we conclude that the limit distributions of the Rényi (1953) statistics, respectively those of the Anderson–Darling (1952) statistics can be evaluated via the distributions of the corresponding functionals of a Brownian bridge. We mention only two typical results obtainable this way. The first one of these is

(5.4.9) $$\lim_{n\to\infty} P\left\{\sup_{\varepsilon\leq F(x)} \frac{\beta_n(x)}{F(x)} \leq u\right\} = P\left\{\sup_{\varepsilon\leq y} \frac{B(y)}{y} \leq u\right\}$$
$$= 2\Phi\left(u\left(\frac{\varepsilon}{1-\varepsilon}\right)^{1/2}\right) - 1 \quad u \geq 0, \ \varepsilon > 0.$$

The latter equality is true, since

$$\left\{\frac{B(y)}{y};\ 0 < y \leq 1\right\} \stackrel{\mathscr{D}}{=} \left\{W\left(\frac{1-y}{y}\right);\ 0 < y \leq 1\right\},$$

and hence

$$P\left\{\sup_{\varepsilon\leq y} \frac{B(y)}{y} \leq u\right\} = P\left\{\sup_{\varepsilon\leq y} W\left(\frac{1-y}{y}\right) \leq u\right\}$$
$$= P\{\sup_{0<t\leq\frac{1-\varepsilon}{\varepsilon}} W(t) \leq u\},$$

which by (1.5.1) now gives the desired result. The second one we have in mind is

(5.4.10) $$\lim_{n\to\infty} P\left\{\sup_{\varepsilon\leq F(x)} \frac{|\beta_n(x)|}{F(x)} \leq u\right\} = P\left\{\sup_{\varepsilon\leq y} \frac{|B(y)|}{y} \leq u\right\}$$
$$= \frac{4}{\pi} \sum_{k=0}^{\infty} \frac{(-1)^k}{(2k+1)} \exp\{-(2k+1)^2 \pi^2 (1-\varepsilon)/8\varepsilon u^2\}, \quad u > 0, \ \varepsilon > 0.$$

The above two results ((5.4.9) and (5.4.10)) were first given by Rényi (1953) via classical limiting arguments. For a proof of these and some further similar ones along these lines via the invariance principle, we refer to Csörgő (1966, 1967).

So far we have seen how strong invariance principles (cf. Corollary 5.4.1) can be used to prove asymptotic distribution result like e.g., (5.4.6), (5.4.9), (5.4.10). In proving these results, we have not utilized the rates of approximation of Corollary 5.4.1 at all, and did not say anything about the problem of how fast these distribution functions themselves converged to their limits. The next result gives an answer to this problem.

Corollary 5.4.2 (Komlós, Major, Tusnády 1975a). *Let B_n, α_n be as in Theorem 4.4.1, and let ψ be a functional defined on $D(0, 1)$, satisfying the Lipschitz condition*

(5.4.11) $$|\psi(u) - \psi(v)| \leq L \sup_{0\leq y\leq 1} |u(y) - v(y)|$$

with some positive constant L. Assume further that the distribution of the random variable $\psi(B(y)) \stackrel{\mathscr{D}}{=} \psi(B_n(y))$ $(n=1, 2, \ldots)$ has a bounded density with respect to Lebesgue measure. Then

$$(5.4.12) \qquad \sup_{-\infty < x < \infty} \left| P\{\psi(\alpha_n(y)) \leq x\} - P\{\psi(B_n(y)) \leq x\} \right| = O\left(\frac{\log n}{n^{1/2}}\right).$$

The proof of this corollary is similar to that of Corollary 5.4.3 whose proof is given below.

As to the nearness of the processes α_n and B_n, Theorem 4.4.1 gives the best possible rate. It is an open question whether the rate of (5.4.12) is the best possible or not for any given Lipschitzian functional ψ. For example, in the Kolmogorov–Smirnov case (that is when ψ is the sup-functional) the rate of convergence is known to be $O(n^{-1/2})$ (cf. Gnedenko, Korolyuk and Skorohod 1960; Bickel 1974).

The rate in (5.4.12) does not hold necessarily for functionals not satisfying the Lipschitzian condition (5.4.11), but Theorem 4.4.1 might still give us a handle on occasions. The case we have in mind is that of the Cramér–von Mises statistic $\omega_n^2 = \int_0^1 \alpha_n^2(y) dy$. Let $N_n(x) = P\{\omega_n^2 \leq x\}$, $N(x) = P\{\int_0^1 B^2(y) dy \leq x\}$ and put $\Delta_n = \sup_{0 < x < \infty} |N_n(x) - N(x)|$. We have, of course, that $\lim_{n \to \infty} N_n(x) = N(x)$ for every real x (cf. (5.4.6)) and, in addition to this, it can be easily deduced from the first statement of (4.4.25), or from that of (5.4.4) that $\Delta_n = O(n^{-1/2} \log n (\log \log n)^{1/2})$. But, if we are a bit more circumspect, we can actually prove also

Corollary 5.4.3 (S. Csörgő 1976).

$$(5.4.13) \qquad \Delta_n = O(n^{-1/2} \log n).$$

The latter statement is of interest, because it turns out to be a refinement of the best available result of this kind so far for the distribution of ω_n^2. Namely Orlov (1974) proved that for any $\varepsilon > 0$ there exists a positive constant $b(\varepsilon)$ such that for each n $\Delta_n \leq b(\varepsilon) n^\varepsilon / n^{1/2}$. For a complete set of earlier work on this problem we refer to S. Csörgő (1976). We should also remark here that, though the rate of convergence of Δ_n in (5.4.13) is the best available one so far, it is probably far away from the best possible one. Indeed, a complete asymptotic expansion for the Laplace transform of ω_n^2 is given by S. Csörgő (1976) and, on the basis of his work, he conjectures that Δ_n has the order of $1/n$ (concerning this latter problem we refer also to S. Csörgő and Stachó, 1979). For an improved result in this direction we refer to Götze (1979).

Proof of Corollary 5.4.3. Let $\omega^2(n) = \int_0^1 B_n^2(y)\,dy$, where $\{B_n(y); 0 \leq y \leq 1\}$ is the sequence of Brownian bridges for which the statement of Theorem 4.4.1 holds true, and let $\omega^2 = \int_0^1 B^2(y)\,dy$, where $B(y)$ is an arbitrary Brownian bridge. We will use the elementary fact that, if X, Y, Z are arbitrary r.v. such that $P\{|X-Y| > Z\} < \varepsilon$ for some $\varepsilon > 0$, then, for every real x,

(5.4.14) $\quad P\{Y \leq x - Z\} - \varepsilon \leq P\{X \leq x\} \leq P\{Y \leq x + Z\} + \varepsilon.$

In (4.4.1) let $x = \frac{1}{2\lambda} \log n$, $D = C + \frac{1}{2\lambda}$. Then, with the notation $D_n =$
$= \sup\limits_{0 \leq y \leq 1} |\alpha_n(y) - B_n(y)|$ and $\varepsilon_n = D \log n / \sqrt{n}$, we have

$$P\{|\omega_n^2 - \omega^2(n)| > \varepsilon_n^2 + 2\varepsilon_n \omega(n)\}$$
$$= P\left\{\left|\int_0^1 (\alpha_n(y) - B_n(y))(\alpha_n(y) + B_n(y))\,dy\right| > \varepsilon_n^2 + 2\varepsilon_n \omega(n)\right\}$$
$$\leq P\left\{\int_0^1 (\alpha_n(y) - B_n(y))^2\,dy + 2\int_0^1 |\alpha_n(y) - B_n(y)||B_n(y)|\,dy > \varepsilon_n^2 + 2\varepsilon_n \omega(n)\right\}.$$

Let the event of the latter probability statement be denoted by $E(\varepsilon_n)$. Then, from the above inequality,

$$P\{|\omega_n^2 - \omega^2(n)| > \varepsilon_n^2 + 2\varepsilon_n \omega(n)\} \leq P\{E(\varepsilon_n), D_n \leq \varepsilon_n\} + P\{D_n > \varepsilon_n\}$$
$$\leq P\left\{\varepsilon_n^2 + 2\varepsilon_n \int_0^1 |B_n(y)|\,dy > \varepsilon_n^2 + 2\varepsilon_n \omega(n)\right\} + n^{-1/2} L$$
$$= P\left\{\int_0^1 |B_n(y)|\,dy > \omega(n)\right\} + n^{-1/2} L$$
$$\leq P\{\omega(n) > \omega(n)\} + n^{-1/2} L = n^{-1/2} L,$$

where the last inequality follows from that of Schwartz, and L is the positive absolute constant of (4.4.1).

Now we apply (5.4.14) with $X = \omega_n^2$, $Y = \omega(n)$, $Z = \varepsilon_n^2 + 2\varepsilon_n \omega(n)$ and $\varepsilon = n^{-1/2} L$, and get

$$P\{A_n(x)\} - n^{-1/2} L \leq P\{\omega_n^2 \leq x\} \leq P\{G_n(x)\} + n^{-1/2} L,$$

where $A_n(x) = \{\omega^2 \leq x - \varepsilon_n^2 - 2\varepsilon_n \omega\}$, $G_n(x) = \{\omega^2 \leq x + \varepsilon_n^2 + 2\varepsilon_n \omega\}$ and $x > 0$.

Solving the corresponding quadratic inequalities for the events $A_n(x)$ and $G_n(x)$ we find that $P\{A_n(x)\}=0$ if $x \leq \varepsilon_n^2$, while

$$P\{A_n(x)\} = P\{\omega^2 \leq x+\varepsilon_n^2-(4\varepsilon_n^2 x)^{1/2}\} \geq P\{\omega^2 \leq x-\delta_n\}, \quad \text{if} \quad x \geq \varepsilon_n^2,$$

and

$$G_n(x) \subset \{\omega^2 \leq x+\delta_n\} \quad \text{with} \quad \delta_n = 3\varepsilon_n^2 + (8\varepsilon_n^4+4\varepsilon_n^2 x)^{1/2}, \quad x>0.$$

Hence, with $\delta_n = \delta_n(x)$ ($x>0$) as in the preceding line,

$$N(x-\delta_n(x)) - n^{-1/2}L \leq N_n(x) \leq N(x+\delta_n(x)) + n^{-1/2}L,$$

and whence

$$\Delta_n \leq \sup_x P\{x-\delta_n(x) < \omega^2 \leq x+\delta_n(x)\} + n^{-1/2}L$$

$$= \sup_x \int_x^{x+2\delta_n(x)} v(y)\,dy + n^{-1/2}L \leq 2\sup_x \bigl(\delta_n(x) \sup_{x \leq y \leq x+2\delta_n(x)} v(y)\bigr) + n^{-1/2}L$$

$$\leq 2K \sup_x \bigl(\delta_n(x) \sup_{x \leq y \leq x+2\delta_n(x)} (y+1)^{-1/2}\bigr) + n^{-1/2}L$$

$$\leq 2K \sup_x \frac{\delta_n(x)}{(x+1)^{1/2}} + n^{-1/2}L = 2K \sup_x \frac{3\varepsilon_n^2+(8\varepsilon_n^4+4\varepsilon_n^2 x)^{1/2}}{(x+1)^{1/2}} + n^{-1/2}L$$

$$\leq 2K \sup_x \frac{3\varepsilon_n^2(x+1)^{1/2}+(8\varepsilon_n^4(x+1)+4\varepsilon_n^2(x+1))^{1/2}}{(x+1)^{1/2}} + n^{-1/2}L$$

$$= O(\varepsilon_n) = O(n^{-1/2}\log n),$$

where $v(x) = \dfrac{d}{dx}N(x)$, and the third inequality for Δ_n above is by $v(y)(y+1)^{1/2} \leq v(y)(y^{1/2}+1) \leq K$ uniformly in $y \in (0, \infty)$, since it is known that not only the density function $v(x)$ of ω^2 but $v(x)x^{1/2}$ too is bounded on the positive half-line (cf. Lemma 8 in § 5, S. Csörgő 1976).

While Corollary 5.4.1 provides us with weak convergence results for the classical functionals of the empirical process with weight functions $g(y)$ satisfying the condition (5.4.1), the weight function $g(y) = (y(1-y))^{-1/2}$, $0 < y < 1$, which is probably the most natural one, does not fit into its framework. An application of Theorem 4.4.1, however, turns out to be a good initial step also towards this direction of weak convergence problems, which we are going to consider now.

Let $V_n(x) = n^{1/2}(F_n(x)-F(x))/(F(x)(1-F(x)))^{1/2}$, where F is a continuous distribution function and, for $0 \leq \varepsilon < \delta \leq 1$, define $V^n(\varepsilon, \delta) = \sup_{\varepsilon < F(x) < \delta} V_n(x)$ and $W^n(\varepsilon, \delta) = \sup_{\varepsilon < F(x) < \delta} |V_n(x)|$. We note that Anderson and Darling (1952) derived the Laplace transform of the asymptotic distribution of $W^n(\varepsilon, \delta)$ with $0 < \varepsilon < \delta < 1$. It is also natural to ask how one could choose normalizing factors for given sequences ε_n, δ_n so that $V^n(\varepsilon_n, \delta_n)$, $W^n(\varepsilon_n, \delta_n)$, $V^n(0, 1)$ and $W^n(0, 1)$ should have a non-degenerate asymptotic distribution. Indeed this question was asked and answered recently by Jaeschke (1975) and Eicker (1976, 1979) and further studied and developed by Jaeschke (1976, 1979).

For the construction of confidence intervals for F,

$$\hat{V}_n(x) = \begin{cases} 0, & \text{if } F_n(x) = 0 \text{ or } 1 \\ n^{1/2}(F_n(x) - F(x))/(F_n(x)(1 - F_n(x)))^{1/2}, & \text{otherwise,} \end{cases}$$

is more convenient than $V_n(x)$. It is shown by Jaeschke (1976) that for $\hat{V}_n$ the same assertions hold as for V_n, including also the case $\varepsilon_n = 0, \delta_n = 1$. The latter is also a generalization of the earlier results of Eicker (1976). Here we formulate these results with $\varepsilon_n = 0, \delta_n = 1$.

Towards stating these results, let $E_c(t) = \exp(-c \exp(-t))$ $(c \geq 0)$ and $a(\cdot, \cdot)$ be as in Theorem 1.9.1.

Theorem 5.4.1 (Jaeschke 1976). *We have*

(5.4.15) $$\lim_{n \to \infty} P\{V^n(0, 1) \leq a(t, \log n)\} = E_1(t)$$

and

(5.4.16) $$\lim_{n \to \infty} P\{W^n(0, 1) \leq a(t, \log n)\} = E_2(t), \quad -\infty < t < +\infty.$$

Theorem 5.4.2 (Eicker 1976, Jaeschke 1976). *Let*

$$\hat{V}^n(\varepsilon, \delta) = \sup_{\varepsilon < F(x) < \delta} \hat{V}_n(x) \quad \text{and} \quad \hat{W}^n(\varepsilon, \delta) = \sup_{\varepsilon < F(x) < \delta} |\hat{V}_n(x)|.$$

Then

(5.4.17) $$\lim_{n \to \infty} P\{\hat{V}^n(0, 1) \leq a(t, \log n)\} = E_1(t)$$

and

(5.4.18) $$\lim_{n \to \infty} P\{\hat{W}^n(0, 1) \leq a(t, \log n)\} = E_2(t), \quad -\infty < t < +\infty.$$

Just as in the case of the classical empirical process $\beta_n(x) = \alpha_n(F(x))$, we may from now on take $F \in U(0, 1)$, since F is assumed to be continuous. As to the proof of the two theorems formulated above, we give only the main steps, in order to demonstrate how Theorem 4.4.1 can be applied in this situation. Here we follow Jaeschke (1976).

The proof of Theorem 5.4.1 is based on the following lemma:

Lemma 5.4.1 (Jaeschke 1976). *For* $-\infty < t < +\infty$ *we have*

(5.4.19) $\quad \lim_{n \to \infty} P\{W^n(0, \varepsilon_n) \vee W^n(1-\varepsilon_n, 1) \leq a(t, \log n)\} = 1,$

where $\varepsilon_n = n^{-1} \log^3 n$.

For a proof of this lemma we refer to that of Lemma 4 in Jaeschke (1976).

Proof of Theorem 5.4.1. Since by Theorem 4.4.1

$$(2 \log \log n)^{1/2} \sup_{\varepsilon_n < y < 1-\varepsilon_n} |V_n(y) - B_n(y)/(y(1-y))^{1/2}|$$

$$\stackrel{\text{a.s.}}{=} O\left(\left(\frac{\log \log n}{\log n}\right)^{1/2}\right) = o(1),$$

Corollary 1.9.1 and Lemma 5.4.1 imply Theorem 5.4.1.

For a similar proof of the second theorem we need two further lemmas.

Lemma 5.4.2 (Jaeschke 1976). *For* $\varepsilon_n \geq n^{-1} \log n$ *and* $a_n = (2 \log \log n)^{1/2}$ *we have*

(5.4.20) $\quad a_n \hat{V}^n(\varepsilon_n, 1-\varepsilon_n) \stackrel{\text{a.s.}}{=} a_n V^n(\varepsilon_n, 1-\varepsilon_n) + o(1).$

Proof. Due to Theorem 5.1.6 we have

$$\sup_{n^{-1} \log n < y < 1} |1 - y^{-1} F_n(y)| \stackrel{\text{a.s.}}{=} O((\log \log n / \log n)^{1/2})$$

and

$$\sup_{0 < y < 1-n^{-1} \log n} |1 - (1-F_n(y))/(1-y)| \stackrel{\text{a.s.}}{=} O((\log \log n / \log n)^{1/2}).$$

Whence

$$a_n \hat{V}^n(\varepsilon_n, 1-\varepsilon_n) \stackrel{\text{a.s.}}{=} a_n V^n(\varepsilon_n, 1-\varepsilon_n)(1 + O((\log \log n / \log n)^{1/2})).$$

Again Theorem 5.1.6 implies $a_n V^n(\varepsilon_n, 1-\varepsilon_n) \stackrel{\text{a.s.}}{=} O(\log \log n)$, and this is the assertion of (5.4.20).

Lemma 5.4.3 (Eicker 1976, Jaeschke 1976). *With* $\varepsilon_n = n^{-1} \log n$ *and* $-\infty < t < +\infty$, *we have*

(5.4.21) $\quad \lim_{n \to \infty} P\{\hat{W}^n(0, \varepsilon_n) \vee \hat{W}^n(1-\varepsilon_n, 1) \leq a(t, \log n)\} = 1.$

For a proof of this lemma we refer to that of Lemma 6 in Jaeschke (1976).

Proof of Theorem 5.4.2. A combination of Lemmas 5.4.2, 5.4.3 and Theorem 5.4.1 yields (5.4.17) and (5.4.18).

5.5. Asymptotic distribution results for some classical functionals of the quantile process

On the basis of Theorems 4.5.6 and 4.5.7 it is quite immediate to construct a Corollary 5.4.1 type statement for the quantile process $f(\text{inv } F(y))q_n(y)$, provided we assume conditions (4.5.10) and (4.5.12). Such an analogue of Corollary 5.4.1 immediately implies, among others, the following typical statements:

$$(5.5.1) \quad \lim_{n\to\infty} P\{\sup_{0<y<1} f(\text{inv } F(y))q_n(y) \leq u\} = P\{\sup_{0\leq y\leq 1} B(y) \leq u\}$$

$$= 1 - e^{-2u^2}, \quad u \geq 0, \text{ (cf. (1.5.3))},$$

$$(5.5.2) \quad \lim_{n\to\infty} P\{\sup_{0<y<1} |f(\text{inv } F(y))q_n(y)| \leq u\} = P\{\sup_{0\leq y\leq 1} |B(y)| \leq u\}$$

$$= 1 - \sum_{k\neq 0} (-1)^{k+1} e^{-2k^2 u^2}, \quad u \geq 0, \text{ (cf. (1.5.4))},$$

$$(5.5.3) \quad \lim_{n\to\infty} P\{\mathscr{R}_n \leq u\} = P\{\sup_{0<y<1} B(y) - \inf_{0<y<1} B(y) \leq u\}$$

$$= 1 - \sum_{j=1}^{\infty} 2(4(ju)^2 - 1) e^{-2j^2 u^2}, \quad u \geq 0, \text{ (cf. Theorem 1.5.3)},$$

where $\mathscr{R}_n = \sup_{0<y<1} f(\text{inv } F(y))q_n(y) - \inf_{0<y<1} f(\text{inv } F(y))q_n(y),$

$$(5.5.4) \quad \lim_{n\to\infty} P\left\{\sup_{\varepsilon\leq y} \frac{f(\text{inv } F(y))q_n(y)}{y} \leq u\right\} = P\left\{\sup_{\varepsilon\leq y} \frac{B(y)}{y} \leq u\right\}$$

$$= 2\Phi\left(u\left(\frac{\varepsilon}{1-\varepsilon}\right)^{1/2}\right) - 1, \quad u \geq 0, \, \varepsilon > 0, \text{ (cf. (5.4.9))},$$

and

$$(5.5.5) \quad \lim_{n\to\infty} P\left\{\sup_{\varepsilon\leq y} \frac{|f(\text{inv } F(y))q_n(y)|}{y} \leq u\right\} = P\left\{\sup_{\varepsilon\leq y} \frac{|B(y)|}{y} \leq u\right\}$$

$$= \frac{4}{\pi} \sum_{k=0}^{\infty} \frac{(-1)^k}{(2k+1)} \exp\{-(2k+1)^2 \pi^2 (1-\varepsilon)/8\varepsilon u^2\}, \quad u > 0, \, \varepsilon > 0,$$

(cf. (5.4.10)).

We again call attention to the fact that the above asymptotic results hold true, assuming only the reasonably weak conditions (4.5.10) and (4.5.12). An analogue of Corollary 5.4.2 is also immediate.

Corollary 5.5.1. *Let B_n, q_n be as in Theorem 4.5.7, and let ψ be defined on $D(0, 1)$, satisfying the Lipschitz condition of (5.4.11). Assume further* that the distribution of the r.v. $\psi(B(y)) \stackrel{\mathscr{D}}{=} \psi(B_n(y))$ $(n=1, 2, \ldots)$ has a bounded density with respect to Lebesgue measure. Then, under conditions (4.5.10) and (4.5.12) we have

$$(5.5.6) \quad \sup_{-\infty < x < \infty} \left| P\{\psi(f(\operatorname{inv} F(y))q_n(y)) \leq x\} - P\{\psi(B_n(y)) \leq x\} \right|$$

$$= \begin{cases} O(n^{-1/2} \log n) & \text{if } \gamma < 2 \\ O(n^{-1/2} (\log \log n)^\gamma (\log n)^{(1+\varepsilon)(\gamma-1)}) & \text{if } \gamma \geq 2, \end{cases}$$

where γ is as in (4.5.10) and $\varepsilon > 0$ is arbitrary.

Unlike in the case of $\alpha_n(y)$, it does not even appear to be known whether the Kolmogorov–Smirnov functionals of $f(\operatorname{inv} F(y))q_n(y)$ themselves have a limit distribution of rate $1/n^{1/2}$ in (5.5.6) or not. Hence the rates of (5.5.6) appear to be the best available. Now in the case of the uniform quantile process $u_n(y)$, the rate of (5.5.6) is $O(n^{-1/2} \log n)$. It is easy to see that for the Kolmogorov–Smirnov functionals of $u_n(y)$ the rate is $O(1/n^{1/2})$. Hence it is again an open question whether all the Lipschitzian functionals of $u_n(y)$ have a limit distribution of rate $1/n^{1/2}$ or not.

As to an analogue of Corollary 5.4.3, we have:

Corollary 5.5.2. Let $\omega_n^2 = \int_0^1 (f(\operatorname{inv} F(y))q_n(y))^2 dy$. Assume conditions (4.5.10) and (4.5.12). Then

$$(5.5.7) \quad \sup_{0 < x < \infty} \left| P\{\omega_n^2 \leq x\} - P\left\{ \int_0^1 B^2(y) dy \leq x \right\} \right|$$

$$= \begin{cases} O\left(\dfrac{\log n (\log \log n)^{1/2}}{n^{1/2}} \right) & \text{if } \gamma < 2 \\ O\left(\dfrac{(\log \log n)^{\gamma+1/2} (\log n)^{(1+\varepsilon)(\gamma-1)}}{n^{1/2}} \right) & \text{if } \gamma \geq 2, \end{cases}$$

where γ is as in (4.5.10) and $\varepsilon > 0$ is arbitrary.

The proof of this corollary goes along the lines of that of Corollary 5.4.3, but here we use (4.5.24) instead of a possible analogue of (4.4.1) for $f(\operatorname{inv} F(y))q_n(y)$, which, in turn, is not yet known. The non-availability of the latter analogue of (4.4.1) results in the appearance of the extra factor $(\log \log n)^{1/2}$ in the above rates.

We have seen so far that most of the results of Section 5.4 extend immediately to the quantile process. The extension of Theorems 5.4.1 and 5.4.2 is not so immediate and requires some further attention.

Let
$$f(\text{inv } F(y))\hat{q}_n(y) = \begin{cases} f(\text{inv } F(y))q_n(y)/(y(1-y))^{1/2}, & 1/n < n < 1-1/n, \\ 0, & \text{otherwise,} \end{cases}$$

and, for $0<\varepsilon<\delta<1$, define
$$Q^n(\varepsilon, \delta) = \sup_{\varepsilon<y<\delta} f(\text{inv } F(y))\hat{q}_n(y),$$
$$\underset{\sim}{Q}^n(\varepsilon, \delta) = \sup_{\varepsilon<y<\delta} f(\text{inv } F(y))|\hat{q}_n(y)|.$$

Using the notation of Theorems 5.4.1 and 5.4.2 we have

Theorem 5.5.1 (Csörgő, Révész 1979). *Let $X_1, X_2, \ldots$ be i.i.d.r.v. with a continuous distribution function F which is also twice differentiable on (a, b), where $-\infty \le a = \sup\{x: F(x)=0\}$, $+\infty \ge b = \inf\{x: F(x)=1\}$ and $F' = f \ne 0$ on (a, b). Assume condition (4.5.10) of Theorem 4.5.6. Then*

(5.5.8) $\quad \lim_{n\to\infty} P\{Q^n(0, 1) \le a(t, \log n)\} = E_1(t),$

(5.5.9) $\quad \lim_{n\to\infty} P\{\underset{\sim}{Q}^n(0, 1) \le a(t, \log n)\} = E_2(t), \quad -\infty < t < +\infty.$

Towards a proof of this theorem we need four further lemmas.

Lemma 5.5.1. *Let $\varepsilon_n = n^{-1}(\log \log n)^4$. Then for a uniform quantile process $u_n(y)$ we have*

(5.5.10) $\quad \lim_{n\to\infty} P\{\sup_{\varepsilon_n<y<1-\varepsilon_n} u_n(y)/(y(1-y))^{1/2} \le a(t, \log n)\} = E_1(t),$

(5.5.11) $\quad \lim_{n\to\infty} P\{\sup_{\varepsilon_n<y<1-\varepsilon_n} |u_n(y)|/(y(1-y))^{1/2} \le a(t, \log n)\} = E_2(t),$
$$-\infty < t < +\infty.$$

Proof. Put
$$\hat{\alpha}_n(y) = \begin{cases} \alpha_n(y)/(E_n(y)(1-E_n(y)))^{1/2} & \text{if } U_1^{(n)} \le y < U_n^{(n)}, \\ 0 & \text{otherwise,} \end{cases}$$

where $E_n(y)$ is the empirical distribution function of a uniform $(0, 1)$ random sample, and $\alpha_n(y) = n^{1/2}(E_n(y) - y)$. First we note that

(5.5.12) $\quad \sup_{0<y<1} \hat{\alpha}_n(y) = \sup_{1/n \le y \le 1-1/n} -u_n(y)/(y(1-y))^{1/2}.$

For $0 \le \varepsilon < \delta \le 1$ we define, as before,
$$V_n(\varepsilon, \delta) = \sup_{\varepsilon<y<\delta} \hat{\alpha}_n(y) \quad \text{and} \quad W_n(\varepsilon, \delta) = \sup_{\varepsilon<y<\delta} |\hat{\alpha}_n(y)|.$$

Concerning these standardized empirical processes with $\varepsilon_n^* = n^{-1}\log n$, the following statements hold by Corollary 1.9.1, Theorem 4.5.7 and Lemma 5.4.1.

(5.5.13) $\quad\lim_{n\to\infty} P\{V_n(\varepsilon_n^*, 1-\varepsilon_n^*) \leq a(t, \log n)\} = E_1(t),$

(5.5.14) $\quad\lim_{n\to\infty} P\{W_n(\varepsilon_n^*, 1-\varepsilon_n^*) \leq a(t, \log n)\} = E_2(t),$

and

(5.5.15) $\lim_{n\to\infty} P\{W_n(0, \varepsilon_n^*) \vee W_n(1-\varepsilon_n^*, 1) \leq a(t, \log n)\} = 1, \quad -\infty < t < +\infty.$

A combination of (5.5.13), (5.5.14) and (5.5.15) with (5.5.12) now gives Lemma 5.5.1.

Remark 5.5.1. We note that Lemma 5.5.1 holds even with $\varepsilon_n = 1/n$. The only reason we have defined ε_n as in the statement of Lemma 5.5.1 is just for the sake of the method of proof of our next lemma.

Lemma 5.5.2. *Let the quantile processes $q_n(y)$ resp. $u_n(y)$ be defined in terms of $X_k^{(n)}$ resp. $U_k^{(n)} = F(X_k^{(n)})$. Under the conditions of Theorem 5.5.1 and with $\varepsilon_n = n^{-1}(\log\log n)^4$, we have*

(5.5.16) $\quad\overline{\lim_{n\to\infty}} \sup_{\varepsilon_n < y < 1-\varepsilon_n} \left| f(\operatorname{inv} F(y))\hat{q}_n(y) - \frac{u_n(y)}{(y(1-y))^{1/2}} \right| \stackrel{a.s.}{=} o\left((\log\log n)^{-1/2}\right).$

Proof. It follows from (4.5.11) and the definition of ε_n that

$$\sup_{\varepsilon_n < y < 1-\varepsilon_n} \left| f(\operatorname{inv} F(y))\hat{q}_n(y) - \frac{u_n(y)}{(y(1-y))^{1/2}} \right|$$

$$\leq \frac{1}{(\varepsilon_n(1-\varepsilon_n))^{1/2}} \sup_{\delta_n < y < 1-\delta_n} \left| f(\operatorname{inv} F(y)) q_n(y) - u_n(y) \right|$$

$$\stackrel{a.s.}{=} \frac{O(\log\log n / n^{1/2})}{(n^{-1}(\log\log n)^4)^{1/2}} = o((\log\log n)^{-1/2}),$$

where δ_n above is as in Theorem 4.5.5.

Lemma 5.5.3. *Let $\varepsilon_n = n^{-1}(\log\log n)^4$. Then*

$$\lim_{n\to\infty} P\left\{ \sup_{1/n < y \leq \varepsilon_n} \left|\frac{u_n(y)}{\sqrt{y}}\right| > (\log\log n)^{1/4} \right\} = 0.$$

The proof of this lemma is based on (1.9.5) via using the relationship (4.5.4).

Lemma 5.5.4. Let $\varepsilon_n = n^{-1}(\log \log n)^4$. Then

(5.5.17) $\quad \lim_{n \to \infty} P\{\underset{\sim}{Q}^n(0, \varepsilon_n) \vee \underset{\sim}{Q}^n(1-\varepsilon_n, 1) \leq a(t, \log n)\} = 1.$

Proof. Let the quantile process $q_n(y)$ resp. $u_n(y)$ be as in Lemma 5.5.2, and consider $\underset{\sim}{Q}^n(0, \varepsilon_n)$ (for $\underset{\sim}{Q}^n(1-\varepsilon_n, 1)$ a similar argument holds). It follows from the definition of $a(t, \log n)$ that, in order to show $\lim_{n \to \infty} P\{\underset{\sim}{Q}^n(0, \varepsilon_n) \leq a(t, \log n)\} = 1$, it suffices to show

(5.5.18) $\quad \lim_{n \to \infty} P\{\underset{\sim}{Q}(0, \varepsilon_n) > (\log \log n)^{1/2}\} = 0.$

Restricting our attention then to the region $0 < y \leq \varepsilon_n$, it suffices to consider $f(\text{inv } F(y)) q_n(y)/y^{1/2}$ instead of $f(\text{inv } F(y)) q_n(y)/(y(1-y))^{1/2}$. Hence we are going to show that

(5.5.19) $\quad \lim_{n \to \infty} P\{\sup_{\frac{1}{n} < y \leq \varepsilon_n} f(\text{inv } F(y))|q_n(y)|/y^{1/2} > (\log \log n)^{1/2}\} = 0,$

which, in turn, implies (5.5.18).

Now for $\dfrac{k-1}{n} < y \leq \dfrac{k}{n}$, by Lemma 4.5.2, we have

(5.5.20)
$$\left| f(\text{inv } F(y)) q_n(y)/y^{1/2} \right| = n^{1/2} f(\text{inv } F(y)) \left| (\text{inv } F(U_k^{(n)}) - \text{inv } F(y))/y^{1/2} \right|$$
$$= n^{1/2} f(\text{inv } F(y)) \left| (\text{inv } F(y + n^{-1/2} u_n(y)) - \text{inv } F(y))/y^{1/2} \right|$$
$$= \left| \frac{u_n(y)}{y^{1/2}} \right| \left| \frac{f(\text{inv } F(y))}{f(\text{inv } F(\xi))} \right| \leq \left| \frac{u_n(y)}{y^{1/2}} \right| \left\{ \frac{y \vee \xi}{y \wedge \xi} \cdot \frac{1 - (y \wedge \xi)}{1 - (y \vee \xi)} \right\}^\gamma$$
$$\leq \left| \frac{u_n(y)}{y^{1/2}} \right| \left\{ \frac{y(1-\xi)}{\xi(1-y)} + \frac{\xi(1-y)}{y(1-\xi)} \right\}^\gamma \overset{\text{a.s.}}{=} \frac{|u_n(y)|}{y^{1/2}} \left\{ \frac{y}{\xi} + \frac{\xi}{y} \right\}^\gamma O(1),$$

where $|\xi - y| \leq n^{-1/2} |u_n(y)|$ and $y \in (0, \varepsilon_n]$.

Hence by (5.5.19) and (5.5.20) we are to show now that

(5.5.21)
$$\lim_{n \to \infty} P\left\{ \sup_{\frac{1}{n} < y \leq \varepsilon_n} \frac{|u_n(y)|}{y^{1/2}} \left\{ \frac{y}{y - \frac{|u_n(y)|}{n^{1/2}}} + \frac{y + \frac{|u_n(y)|}{n^{1/2}}}{y} \right\}^\gamma > (\log \log n)^{1/2} \right\} = 0.$$

First we observe that $\dfrac{d}{dx} P\left\{ \dfrac{u_n(y)}{\sqrt{y}} \leq x \right\}$ is uniformly bounded in y and n on any bounded interval containing $x=1$. This implies that, as $n \to \infty$, we have

$$(5.5.22) \quad P\left\{\sup_{\frac{1}{n} \leq y \leq \omega_n} \left[\frac{y}{y - \frac{|u_n(y)|}{n^{1/2}}} + \frac{y + \frac{|u_n(y)|}{n^{1/2}}}{y}\right]^\gamma > (\log \log n)^{1/4}\right\} \to 0,$$

where $\omega_n = (\log \log \log n)/n$.

We observe also that, with the same ω_n, we have

$$\sup_{\omega_n \leq y \leq \varepsilon_n} \frac{|u_n(y)|}{n^{1/2} y} \xrightarrow{P} 0, \quad \text{as} \quad n \to \infty.$$

Hence

$$(5.5.23) \quad P\left\{\sup_{\omega_n \leq y \leq \varepsilon_n} \left[\frac{y}{y - \frac{|u_n(y)|}{n^{1/2}}} + \frac{y + \frac{|u_n(y)|}{n^{1/2}}}{y}\right]^\gamma > (\log \log n)^{1/4}\right\} \to 0.$$

Now Lemma 5.5.3 together with (5.5.22) and (5.5.23) implies (5.5.21), and this also completes the proof of (5.5.17).

Proof of Theorem 5.5.1. Combining Lemmas 5.5.1, 5.5.2 and 5.5.4 we get (5.5.8) and (5.5.9).

We note that Theorem 5.5.1 in its present form, and also the weak convergence results of (5.5.1)–(5.5.5), can be used to construct confidence intervals for $f(\text{inv } F(y))q_n(y) = \sqrt{n} f(\text{inv } F(y))(Q_n(y) - \text{inv } F(y))$ and also to test the null hypothesis that X_1 has a given, completely specified density function $f(\cdot)$. For the sake of confidence intervals for $\text{inv } F(y)$ in terms of $Q_n(y)$ one should estimate the factor $f(\text{inv } F(y))$ of $f(\text{inv } F(y))q_n(y)$.

We begin with estimating $\text{inv } F(y)$ by the quantile function $Q_n(y)$. First we note that the law of iterated logarithm holds for the process $f(\text{inv } F(y))q_n(y)$ (cf. (5.3.1)), and hence we have

$$(5.5.24) \quad \sup_{0 < y < 1} |Q_n(y) - \text{inv } F(y)| \stackrel{a.s.}{=} O((\log \log n/n)^{1/2}),$$

provided $\inf_{0 < y < 1} f(\text{inv } F(y)) > 0$.

Next we estimate the density function $f(\cdot)$ by any of the empirical density functions $f_n(\cdot)$ of Chapter 6 for which the Glivenko–Cantelli theorem holds (cf. Theorem 6.2.1):

$$(5.5.25) \quad \sup_{-\infty < x < +\infty} |f_n(x) - f(x)| \stackrel{a.s.}{=} 0,$$

and prove

Theorem 5.5.2. Let $f_n(\cdot)$ be a sequence of empirical density functions satisfying the conditions of Theorem 6.2.1 and assume that $\inf_{a<x<b} f(x) > 0$ and $\sup_{a<x<b} |f'(x)| < \infty$ over $[a,b]$ of Theorem 4.5.6, assumed to be finite. There exists then a sequence of Brownian bridges $\{B_n(y); 0 \leq y \leq 1\}$ such that, as $n \to \infty$,

$$(5.5.26) \qquad \sup_{0 \leq y \leq 1} |f_n(Q_n(y)) q_n(y) - B_n(y)| \xrightarrow{P} 0.$$

Proof. By (4.5.22) it suffices to show that

$$\sup_{0 \leq y \leq 1} |f_n(Q_n(y)) q_n(y) - f(\operatorname{inv} F(y)) q_n(y)| \xrightarrow{P} 0.$$

This, in turn, is true because $q_n(y) \xrightarrow{\mathscr{D}} B(y)/f(\operatorname{inv} F(y))$ by (4.5.22), and $\sup_{0 \leq y \leq 1} |f_n(Q_n(y)) - f(\operatorname{inv} F(y))| \xrightarrow{\text{a.s.}} 0$. As to the latter statement, we write

$$\sup_{0 \leq y \leq 1} |f_n(Q_n(y)) - f(\operatorname{inv} F(y))|$$

$$\leq \sup_{0 \leq y \leq 1} |f_n(Q_n(y)) - f(Q_n(y))| + \sup_{0 \leq y \leq 1} |f(Q_n(y)) - f(\operatorname{inv} F(y))|$$

and use (5.5.24), (5.5.25) and continuity of $f(\cdot)$ to conclude that the right hand side of the above inequality goes to zero almost surely as $n \to \infty$.

As a consequence of (5.5.26) we have

$$(5.5.27) \qquad f_n(Q_n(y)) q_n(y) \xrightarrow{\mathscr{D}} B(y),$$

and, hence, distribution free confidence intervals for $\operatorname{inv} F(y)$ in terms of $Q_n(y)$ can be constructed under the conditions of Theorem 5.5.2.

In our Theorem 5.5.1 we investigated the Kolmogorov–Smirnov functionals of the standardized quantile process, i.e., that of the quantile process $q_n(y)$ with the weight function $f(\operatorname{inv} F(y))/\sqrt{y(1-y)}$. Now we are going to study a Cramér–von Mises functional of the quantile process $q_n(y)$ with the weight function $(f(\operatorname{inv} F(y)))^{1/2} (\operatorname{inv} F(y))^{(\lambda-1)/2}$, $\lambda = 1, 2, \ldots$. In order to describe the latter, we let

$$(5.5.28) \qquad q_n^0(y) = n^{1/2} f(\operatorname{inv} F(y))(Q_n^0(y) - \operatorname{inv} F(y))$$

where $Q_n^0(y) = X_k^{(n)}$ if $\dfrac{k-1}{n+1} < y \leq \dfrac{k}{n+1}$, $k = 1, 2, \ldots, n$, and define our

Cramér–von Mises type statistic as follows:

$$(5.5.29) \quad M_n^0(\lambda) = \sum_{k=1}^{n} \left\{ \left[q_n^0\left(\frac{k}{n+1}\right) \right]^2 \Big/ nf\left(\text{inv } F\left(\frac{k}{n+1}\right)\right) \right\} \left(\text{inv } F\left(\frac{k}{n+1}\right) \right)^{\lambda-1},$$
$$(\lambda = 1, 2, \ldots),$$

where it is clearly assumed that $f = F' \neq 0$.

We note that $Q_n^0(y)$ of (5.5.28) is slightly different from $Q_n(y)$ of Section 4.5. The former is introduced here for technical reasons. It is easy to see that our Theorem 4.5.7 remains true with $q_n^0(y)$ replacing $f(\text{inv } F(y))q_n(y) = n^{1/2} f(\text{inv } F(y))(Q_n(y) - \text{inv } F(y))$. Arguing heuristically with the latter theorem in mind, we should have

$$(5.5.30) \quad M_n^0(\lambda) \to M^0(\lambda) = \int_0^1 B^2(y) \lambda^{-1} d(\text{inv } F(y))^\lambda, \quad (\lambda = 1, 2, \ldots),$$

which coincides with the usual Cramér–von Mises limit if $F(y) \in U(0, 1)$ and $\lambda = 1$. If $F \notin U(0, 1)$, $M^0(\lambda)$ is seen to be a Cramér–von Mises type non-distribution free limit. It can be used to test the completely specified goodness-of-fit statistical hypothesis saying that a random sample is from a given F, provided we can also evaluate the distribution of $M^0(\lambda)$ for the underlying F.

Our aim now is to prove that (5.5.30) is indeed true. Towards this, we first prove

Lemma 5.5.5. *Assume that F is absolutely continuous with a density function f, strictly positive in the interval (a, b) of Theorem 4.5.6 and, in addition, we also have*

$$(5.5.31) \quad \lim_{y \to 0} y^{1/r} |\text{inv } F(y)| = \lim_{y \to 1} (1-y)^{1/r} \text{inv } F(y) = 0$$

for some $r > \lambda$. Then the integral $M^0(\lambda)$ of (5.5.30) exists, i.e.,

$$(5.5.32) \quad P\{|M^0(\lambda)| < +\infty\} = 1, \quad \lambda = 1, 2, \ldots.$$

Proof. Since $\{B(y); 0 \leq y \leq 1\} \overset{\mathscr{D}}{=} \{n^{-1/2} K(y, n); 0 \leq y \leq 1\}$, $n = 1, 2, \ldots$, by Corollary 1.15.2 we have

$$P\left\{ \sup_{0 < y < 1} \frac{B^2(y)}{y(1-y) \log\log(y(1-y))^{-1}} < +\infty \right\} = 1.$$

We have also

$$|M^0(\lambda)| \leq \sup_{0<y<1} \frac{B^2(y)}{y(1-y)\log\log(y(1-y))^{-1}} \cdot$$

$$\cdot \left| \int_0^1 y(1-y)\log\log(y(1-y))^{-1} d(\operatorname{inv} F(y))^\lambda \right|.$$

Since $B(y)$ is an almost surely continuous function, the statement of lemma follows from

$$\left| \int_0^1 y(1-y)\log\log(y(1-y))^{-1} d(\operatorname{inv} F(y))^\lambda \right|$$

$$\leq \int_0^m |\operatorname{inv} F(y)|^\lambda d\left(y \log\log \frac{1}{y}\right) + \int_m^1 |\operatorname{inv} F(y)|^\lambda d\left((1-y)\log\log\frac{1}{1-y}\right)$$

$$\leq \int_0^m y^{-\lambda/r} d\left(y \log\log \frac{1}{y}\right) + \int_m^1 (1-y)^{-\lambda/r} d\left((1-y)\log\log\frac{1}{1-y}\right) + \text{Const.}$$

$$< +\infty,$$

where m is such that $\operatorname{inv} F(m) = 0$ and the first inequality is by integration by parts combined with (5.5.31) and the second one is by the latter.

Next we prove

Theorem 5.5.3. *Let $X_1, X_2, \ldots, X_n$ be a random sample with a continuous distribution F which is also twice differentiable on (a,b) where a and b are as in Theorem 4.5.6, and $F' = f \neq 0$ on (a,b). Assume that F also satisfies conditions (4.5.10) and (4.5.12) of Theorem 4.5.6 and those of (5.5.31) with some $r > 2\lambda$. Then there exists a sequence of Brownian bridges $\{B_n\}$ such that*

(5.5.33)

$$\left| M_n^0(\lambda) - n^{-1} \sum_{k=1}^n \left\{ B_n^2\left(\frac{k}{n+1}\right) \bigg/ f\left(\operatorname{inv} F\left(\frac{k}{n+1}\right)\right) \right\} \left(\operatorname{inv} F\left(\frac{k}{n+1}\right)\right)^{\lambda-1} \right| \stackrel{a.s.}{=} o(1),$$

and

(5.5.34) $\quad \left| M_n^0(\lambda) - \int_0^1 B_n^2(y) \lambda^{-1} d(\operatorname{inv} F(y))^\lambda \right| \stackrel{P}{\to} 0, \quad \lambda = 1, 2, \ldots.$

Proof. We first note that applying condition (4.5.12) we get that the function $(\operatorname{inv} F(y))^{\lambda-1}/f(\operatorname{inv} F(y))$ is monotone on an interval to the right of a (to the left of b), and by condition (4.5.10) it is bounded away from infinity if y is bounded away from a and b. Let B_n be as in Theorem 4.5.7. Then the left-hand side of (5.5.33) is bounded above by

$$\sum_{k=1}^{n}\left\{\left|\left(q_n^0\left(\frac{k}{n+1}\right)\right)^2-B_n^2\left(\frac{k}{n+1}\right)\right|\Big/nf\left(\text{inv } F\left(\frac{k}{n+1}\right)\right)\right\}\left(\text{inv } F\left(\frac{k}{n+1}\right)\right)^{\lambda-1}$$

$$=\sum_{k=1}^{n}\left\{\left|q_n^0\left(\frac{k}{n+1}\right)-B_n\left(\frac{k}{n+1}\right)\right|\cdot\left|q_n^0\left(\frac{k}{n+1}\right)+B_n\left(\frac{k}{n+1}\right)\right|\Big/nf\left(\text{inv } F\left(\frac{k}{n+1}\right)\right)\right\}\cdot$$

$$\cdot\left(\text{inv } F\left(\frac{k}{n+1}\right)\right)^{\lambda-1}$$

$$\leq\left(\sup_{0<y<1}|q_n^0(y)-B_n(y)|\right)\left(\sup_{0<y<1}|B_n(y)-q_n^0(y)|+\sup_{0<y<1}|2q_n^0(y)|\right)$$

$$\cdot\left|\sum_{k=1}^{n}\left(\text{inv } F\left(\frac{k}{n+1}\right)\right)^{\lambda-1}\Big/nf\left(\text{inv } F\left(\frac{k}{n+1}\right)\right)\right|$$

$$\overset{\text{a.s.}}{=}O(r_i(n))(O(r_i(n))+O((\log\log n)^{1/2}))\left|\int_{(n+1)^{-1}}^{n(n+1)^{-1}}\lambda^{-1}d(\text{inv } F(y))^{\lambda}\right|$$

$$\overset{\text{a.s.}}{=}O(r_i(n))(\log\log n)^{1/2}\left|\left(\text{inv } F\left(1-\frac{1}{n+1}\right)\right)^{\lambda}-\left(\text{inv } F\left(\frac{1}{n+1}\right)\right)^{\lambda}\right|$$

$$\overset{\text{a.s.}}{=}o(1)\quad(i=1,2),$$

by (5.5.31), where the rates $r_i(n)$ $(i=1,2)$ are those of (4.5.24), and the first a.s. line above is by (4.5.24) applied twice and by the law of iterated logarithm for the process $q_n^0(y)$ (cf. Theorem 5.3.1). Hence (5.5.33) is proved, and (5.5.34) follows from (5.5.33) combined with Lemma 5.5.5.

A statistic, similar to $M_n^0(\lambda)$, was studied by DeWet and Venter (1972) in the special case of $F=\Phi$. Their statistic is

(5.5.35) $$L_n^0=\sum_{k=1}^{n}\left(X_k^{(n)}-\text{inv }\Phi\left(\frac{k}{n+1}\right)\right)^2-a_n^0$$

$$=\sum_{k=1}^{n}\frac{\left(q_n^0\left(\frac{k}{n+1}\right)\right)^2}{n\varphi^2\left(\text{inv }\Phi\left(\frac{k}{n+1}\right)\right)}-a_n^0,$$

where

$$a_n^0=n^{-1}\sum_{k=1}^{n}\left(\frac{k}{n+1}\right)\left(1-\frac{k}{n+1}\right)\Big/\varphi^2\left(\text{inv }\Phi\left(\frac{k}{n+1}\right)\right).$$

DeWet and Venter (1972) proved

(5.5.36) $$L_n^0 \xrightarrow{\mathscr{D}} \sum_{k=1}^{\infty} k^{-1}(Y_k^2 - 1),$$

where $Y_1, Y_2, \ldots$ are independent standard normal r.v.

Remark 5.5.2. Exactly the same way we proved Lemma 5.5.5 and Theorem 5.5.3, we can prove also with $\lambda = 1, 2, \ldots$ and $\delta = 0, 1$ that

(5.5.37) $\int_0^1 B(y)(\operatorname{inv} F(y))^\lambda dy$ exists with probability one,

(5.5.38) $$\left| n^{-1} \sum_{k=1}^{n} q_n^0\left(\frac{k}{n+1}\right) \left(\operatorname{inv} F\left(\frac{k}{n+1}\right)\right)^\lambda \right.$$
$$\left. - n^{-1} \sum_{k=1}^{n} B_n\left(\frac{k}{n+1}\right) \left(\operatorname{inv} F\left(\frac{k}{n+1}\right)\right)^\lambda \right| \xrightarrow{a.s.} 0,$$

(5.5.39) $$\left| n^{-1} \sum_{k=1}^{n} q_n^0\left(\frac{k}{n+1}\right) \left(\operatorname{inv} F\left(\frac{k}{n+1}\right)\right)^\lambda - \int_0^1 B_n(y)(\operatorname{inv} F(y))^\lambda dy \right| \xrightarrow{P} 0,$$

(5.5.40) $$\left| n^{-1} \sum_{k=1}^{n} \frac{q_n^0\left(\frac{k}{n+1}\right)}{f\left(\operatorname{inv} F\left(\frac{k}{n+1}\right)\right)} \left(\operatorname{inv} F\left(\frac{k}{n+1}\right)\right)^\delta \right.$$
$$\left. - \int_0^1 B_n(y) \frac{1}{\delta+1} d(\operatorname{inv} F(y))^{\delta+1} \right| \xrightarrow{P} 0,$$

provided all the conditions of Theorem 5.5.3 are assumed.

5.6. Asymptotic distribution results for some classical functionals of some k-sample empirical and quantile processes

Let X_{ji} $(1 \leq i \leq n(j))$ be k independent sequences $(k \geq 2)$ of i.i.d.r.v. with respective distribution functions $F_j(x)$ $(1 \leq j \leq k)$. A classical statistical problem is to test whether these k samples come from a common population with distribution function F, whose form might or might not be given to us. Thus we wish to test the following null hypotheses:

(5.6.1) $H_0: F_1 = F_2 = \ldots = F_k$ (homogenity),

(5.6.2) $H_0: F_1 = F_2 = \ldots = F_k = F$ (goodness-of-fit).

The k-sample based empirical processes which we are to study here are going to be constructed with the null assumptions (5.6.1) and (5.6.2) in mind. Towards this end let $F_{n(j)}^{(j)}(x) = F_{n(j)}(x)$ denote the empirical distribution functions based on the outcomes of the jth sample $X_{j1}, X_{j2}, \ldots, X_{jn(j)}$. For each vector $N = (n(1), n(2), \ldots, n(k))$ of positive integers, we define the k-sample empirical process $S_N(x)$ by

(5.6.3) $$S_N(x) = \sum_{j=1}^{k} c_j(N, x) \sqrt{n(j)} F_{n(j)}(x),$$

where the coefficients $c_j(N, x)$ satisfy

(5.6.4) $$\sum_{j=1}^{k} c_j(N, x) \sqrt{n(j)} = 0 \quad \text{for all } N \text{ and } x,$$

$$\sup_N \sup_x |c_j(N, x)| < \infty \quad \text{for all } j = 1, 2, \ldots, k.$$

Again for the sake of testing the hypotheses of (5.6.1) and (5.6.2), we define

(5.6.5) $$Z_N(y) = f(\operatorname{inv} F(y)) \sum_{j=1}^{k} c_j(N, y) \sqrt{n(j)} Q_{n(j)}(y), \quad 0 < y < 1,$$

where $Q_{n(j)} = Q_{n(j)}^{(j)}$ is the quantile function of the $X_{j1}, \ldots, X_{jn(j)}$. Then we have (cf. also Kiefer 1959; Burke, Csörgő 1976b):

Proposition 5.6.1. *Given any of the null hypotheses* (5.6.1), (5.6.2), *assuming that the true common distribution function F is continuous and that condition* (5.6.4) *holds, there exist k independent sequences $\{B_{n(j)}^{(j)}\} = \{B_{n(j)}\}$ of Brownian bridges and k independent Kiefer processes K_j such that for $S_N(x)$ of* (5.6.3) *we have*

(5.6.6) $$\sup_{-\infty < x < \infty} \left| S_N(x) - \sum_{j=1}^{k} c_j(N, x) B_{n(j)}(F(x)) \right|$$

$$\stackrel{\text{a.s.}}{=} O\left(\max_{1 \leq j \leq k} ((n(j))^{-1/2} \log n(j)) \right)$$

and

(5.6.7) $$\sup_{-\infty < x < \infty} \left| S_N(x) - \sum_{j=1}^{k} (n(j))^{-1/2} c_j(N, x) K_j(F(x), n(j)) \right|$$

$$\stackrel{\text{a.s.}}{=} O\left(\max_{1 \leq j \leq k} ((n(j))^{-1/2} \log^2 n(j)) \right),$$

as all $n(j) \to \infty$.

If we also assume (4.5.10) *and* (4.5.12) *of Theorem 4.5.6, then*

$$(5.6.8) \quad \sup_{0<y<1} \left| Z_N(y) - \sum_{j=1}^{k} c_j(N, y) B_{n(j)}(y) \right|$$

$$\stackrel{a.s.}{=} \begin{cases} O\left(\max_{1 \le j \le k} (n(j))^{-1/2}(\log n(j))\right) & \text{if } \gamma < 2 \\ O\left(\max_{1 \le j \le k} (n(j))^{-1/2}(\log\log n(j))^{\gamma}(\log n(j))^{(1+\varepsilon)(\gamma-1)}\right) & \text{if } \gamma \ge 2, \end{cases}$$

where γ is as in (4.5.10) *and $\varepsilon > 0$ is arbitrary; also*

$$(5.6.9) \quad \sup_{0<y<1} \left| Z_N(y) - \sum_{j=1}^{k} c_j(N, y)(n(j))^{-1/2} K_j(y, n(j)) \right|$$

$$\stackrel{a.s.}{=} O\left(\max_{1 \le j \le k} (n(j))^{-1/4}(\log\log n(j))^{1/4}(\log n(j))^{1/2}\right),$$

as all $n(j) \to \infty$.

Proof. Using (5.6.4), $S_N(x)$ can be written as $S_N(x) = \sum_{j=1}^{k} c_j(N, x)\sqrt{n(j)} \cdot (F_{n(j)}(x) - F(x))$. By Theorem 4.4.1 we can construct k sequences $\{B_{n(j)}(y); 0 \le y \le 1\}$ of Brownian bridges such that

$$\sup_x \left|\sqrt{n(j)}(F_{n(j)}(x) - F(x)) - B_{n(j)}(F(x))\right| \stackrel{a.s.}{=} O((n(j))^{-1/2} \log n(j)).$$

Since our k samples are assumed to be independent, the k sequences $\{B_{n(j)}\}$ can be constructed independently. Now our first assertion follows from having assumed (5.6.4). The proof of the rest of the statements goes along similar lines.

Corollary 5.6.1. *Suppose that the coefficients c_j depend only on x through F and write $c_j = c_j(N, F(x))$. Then, under the conditions of Proposition 5.6.1 leading up to* (5.6.6) *and* (5.6.7) *we have, when the $n(j) \to \infty$,*

$$(5.6.10) \quad \left| \int_{-\infty}^{\infty} S_N^2(x) dF(x) - \int_0^1 \left(\sum_{j=1}^{k} c_j(N, t)(n(j)^{-1/2} K_j(t, n(j))) \right)^2 dt \right|$$

$$\stackrel{a.s.}{=} O\left(\max_{1 \le j \le k} (n(j))^{-1/2} \log^2 n(j)(\log\log n(j))^{1/2}\right),$$

$$(5.6.11) \quad \left| \sup_{-\infty < x < +\infty} |S_N(x)| - \sup_{0 \le t \le 1} \left| \sum_{j=1}^{k} c_j(N, t)(n(j))^{-1/2} K_j(t, n(j)) \right| \right|$$

$$\stackrel{a.s.}{=} O\left(\max_{1 \le j \le k} ((n(j))^{-1/2} \log^2 n(j))\right),$$

and if we also assume (4.5.10) and (4.5.12) of Theorem 4.5.6, then

$$(5.6.12) \quad \left| \int_0^1 Z_N^2(y)\, dy - \int_0^1 \left(\sum_{j=1}^k c_j(N, y)(n(j))^{-1/2} K_j(y, n(j)) \right)^2 dy \right|$$

$$\stackrel{a.s.}{=} O\left(\max_{1 \le j \le k} (n(j))^{-1/4} (\log \log n(j))^{3/4} (\log n(j))^{1/2} \right),$$

$$(5.6.13) \quad \left| \sup_{0 < y < 1} |Z_N(y)| - \sup_{0 < y < 1} \left| \sum_{j=1}^k c_j(N, y)(n(j))^{-1/2} K_j(y, n(j)) \right| \right|$$

$$\stackrel{a.s.}{=} O\left(\max_{1 \le j \le k} (n(j))^{-1/4} (\log \log n(j))^{1/4} (\log n(j))^{1/2} \right).$$

Proof. The second statement is a direct consequence of Proposition 5.6.1 and of the transformation $t = F(x)$, since F is assumed to be continuous. The first statement is also based on Proposition 5.6.1 and it is proved like (4.4.25) of Corollary 4.4.1 upon observing that, under the conditions (5.6.4), the law of iterated logarithm holds for $S_N(x)$ (cf. Theorem 5.1.1) and also for $\sum_{j=1}^k c_j(N, t)(n(j))^{-1/2} K_j(t, n(j))$ (cf. Corollary 1.15.1). (5.6.12) and (5.6.13) are proved along similar lines.

We note here that the statements of Corollary 5.6.1 could be also stated in terms of Brownian bridges with $\log n(j)$ replacing $\log^2 n(j)$ in (5.6.10) and (5.6.11). Also, in (5.6.12) the new rate will be

$$O\left(\max_{1 \le j \le k} (n(j))^{-1/2} \log n(j) (\log \log n(j))^{1/2} \right),$$

while that of (5.6.13) will be that of (5.6.8).

Corollary 5.6.2. *Suppose that the coefficients c_j depend only on j and N and write $c_j = c_j(N)$. Assume that these $\{c_j(N)\}_{j=1}^k$ also satisfy, in addition to (5.6.4), the condition $\sum_{j=1}^k c_j^2(N) = 1$. Then*

$$B_N(t) = \sum_{j=1}^k c_j(N)(n(j))^{-1/2} K_j(y, n(j)) \stackrel{\mathscr{D}}{=} B(t),$$

a Brownian bridge for each N, and Corollary 5.6.1 holds accordingly.

On the basis of the above results the limiting distributions for the usual functionals can be written down immediately. We illustrate what we have in mind with the two sample situation, spelling out only a few examples. Applying Corollary 5.6.1 with

$$c_1 = \sqrt{\frac{n(2)}{n(1)+n(2)}}, \quad c_2 = -\sqrt{\frac{n(1)}{n(1)+n(2)}}, \quad \text{respectively with}$$

$$c_1 = \begin{cases} \sqrt{\dfrac{n(2)}{n(1)+n(2)}}\,\dfrac{1}{y}, & y = F(x) \geq \varepsilon > 0 \\ 0, & \text{otherwise,} \end{cases}$$

$$c_2 = \begin{cases} -\sqrt{\dfrac{n(1)}{n(1)+n(2)}}\,\dfrac{1}{y}, & y = F(x) \geq \varepsilon > 0 \\ 0, & \text{otherwise,} \end{cases}$$

we get

Corollary 5.6.3

$$(5.6.14) \quad \lim_{n(1),n(2)\to\infty} P\left\{ \sup_{-\infty<x<+\infty} \sqrt{\frac{n(1)n(2)}{n(1)+n(2)}}\,(F_{n(1)}(x) - F_{n(2)}(x)) \leq u \right\}$$

$$= \lim_{n(1),n(2)\to\infty} P\left\{ \sup_{0<y<1} f(\operatorname{inv} F(y))\sqrt{\frac{n(1)n(2)}{n(1)+n(2)}}\,(Q_{n(1)}(y) - Q_{n(2)}(y)) \leq u \right\}$$

$$= P\left\{ \sup_{0\leq y\leq 1} B(y) \leq u \right\} = 1 - e^{-2u^2}, \quad u \geq 0,$$

$$(5.6.15) \quad \lim_{n(1),n(2)\to\infty} P\left\{ \sup_{-\infty<x<+\infty} \sqrt{\frac{n(1)n(2)}{n(1)+n(2)}}\,|F_{n(1)}(x) - F_{n(2)}(x)| \leq u \right\}$$

$$= \lim_{n(1),n(2)\to\infty} P\left\{ \sup_{0<y<1} f(\operatorname{inv} F(y))\sqrt{\frac{n(1)n(2)}{n(1)+n(2)}}\,|Q_{n(1)}(y) - Q_{n(2)}(y)| \leq u \right\}$$

$$= P\left\{ \sup_{0\leq y\leq 1} |B(y)| \leq u \right\} = 1 - \sum_{k\neq 0} (-1)^{k+1} e^{-2k^2 u^2}, \quad u \geq 0,$$

$$(5.6.16) \quad \lim_{n(1),n(2)\to\infty} P\left\{ \sup_{\varepsilon \leq F(x)} \sqrt{\frac{n(1)n(2)}{n(1)+n(2)}}\,(F_{n(1)}(x) - F_{n(2)}(x))/F(x) \leq u \right\}$$

$$= \lim_{n(1),n(2)\to\infty} P\left\{ \sup_{\varepsilon \leq y} f(\operatorname{inv} F(y))\sqrt{\frac{n(1)n(2)}{n(1)+n(2)}}\,y^{-1}(Q_{n(1)}(y) - Q_{n(2)}(y)) \leq u \right\}$$

$$= P\left\{ \sup_{\varepsilon \leq y} B(y)/y \leq u \right\} = 2\Phi\left(u\left(\frac{\varepsilon}{1-\varepsilon}\right)^{1/2}\right) - 1, \quad u \geq 0,\ \varepsilon > 0,$$

$$(5.6.17) \quad \lim_{n(1),n(2)\to\infty} P\left\{ \frac{n(1)n(2)}{n(1)+n(2)} \int_{-\infty}^{+\infty} (F_{n(1)}(x) - F_{n(2)}(x))^2\, dF(x) \leq u \right\}$$

$$= \lim_{n(1),n(2)\to\infty} P\left\{ \frac{n(1)n(2)}{n(1)+n(2)} \int_0^1 (f(\operatorname{inv} F(y))(Q_{n(1)}(y) - Q_{n(2)}(y)))^2\, dy \leq u \right\}$$

$$= P\left\{ \int_0^1 B^2(y)\, dy \leq u \right\}, \quad u \geq 0,\ (\textit{cf. Theorem 1.5.2}).$$

Clearly, the above statements involving the empirical distributions hold true under the conditions of Proposition 5.6.1 leading up to (5.6.6). The statements in terms of the quantiles $Q_{n(1)}$ and $Q_{n(2)}$ are true if all the conditions of Proposition 5.6.1 are assumed. All the results (5.6.14)–(5.6.17) can be used to test the null hypothesis of (5.6.2). The statements of (5.6.14) and (5.6.15) concerning $F_{n(1)}$ and $F_{n(2)}$ are applicable to (5.6.1). Using the Glivenko–Cantelli Theorem, it can be easily shown that, in the statement of (5.6.16) involving $F_{n(1)}$ and $F_{n(2)}$, F can be replaced by any of $F_{n(1)}$, $F_{n(2)}$ and $(n(1)F_{n(1)}+n(2)F_{n(2)})/(n(1)+n(2))$. Thus modified, it can be used to test for H_0 of (5.6.1). Applying the method of Theorem 5.5.2, we can estimate $f(\text{inv } F(y))$ as there, and then all the statements of (5.6.14)–(5.6.17) which involve the two sample quantile process become applicable to (5.6.1). As to the application of (5.6.17) in case of the two sample empirical process to the problem of testing for (5.6.1), we prove

Corollary 5.6.4. *Given (5.6.1) and assuming that the common distribution function F of the two samples is continuous, we have*

$$(5.6.18) \quad \lim_{n(1), n(2) \to \infty} P\left\{ \frac{n(1)n(2)}{n(1)+n(2)} \int_{-\infty}^{+\infty} (F_{n(1)}(x) - F_{n(2)}(x))^2 \, dF_n(x) \right\}$$

$$= P\left\{ \int_0^1 B^2(y) \, dy \leq u \right\}, \quad u \geq 0, \text{ (cf. Theorem 1.5.2)}$$

where $n = n(1) + n(2)$ *and* $F_n(x) = \dfrac{n(1)F_{n(1)}(x) + n(2)F_{n(2)}(x)}{n}$.

Proof. Without loss of generality we can assume that $F(x) \in U(0, 1)$. The same way as we proved (5.6.10), we get

$$\left| \frac{n(1)n(2)}{n(1)+n(2)} \int_0^1 (F_{n(1)} - F_{n(2)})^2 \, dF_n - \int_0^1 B_n^2 \, dF_n \right| \xrightarrow{\text{a.s.}} 0.$$

Hence, in order to prove (5.6.18), it suffices to show that for any given $\varepsilon > 0$ and $0 < \delta < 1$ there exists an $n_0 = n_0(\varepsilon, \delta)$ such that

$$(5.6.19) \quad P\left\{ \left| \int_0^1 B_n^2(y) \, dF_n(y) - \int_0^1 B_n^2(y) \, dy \right| > \varepsilon \right\} < \delta \quad \text{whenever } n \geq n_0.$$

Now for any given integer m, $C>0$ and $\varepsilon>0$ we have

(5.6.20) $\quad P\left\{\left|\int_0^1 B_n^2(y)\,dF_n(y) - \int_0^1 B_n^2(y)\,dy\right| > \varepsilon\right\}$

$$\leq P\left\{\sum_{k=0}^{m-1}\left|\int_{k/m}^{(k+1)/m} B_n^2(y)\,d(F_n(y)-y)\right| > \varepsilon\right\}$$

$$\leq P\left\{\sum_{k=0}^{m-1}\left|\int_{k/m}^{(k+1)/m}\left(B_n^2(y)-B_n^2\left(\frac{k}{m}\right)\right)d(F_n(y)-y)\right|\right.$$

$$\left. +\sum_{k=0}^{m-1} B_n^2\left(\frac{k}{m}\right)\left|F_n\left(\frac{k+1}{m}\right)-F_n\left(\frac{k}{m}\right)-\frac{1}{m}\right| > \varepsilon\right\}$$

$$\leq P\left\{m\left(\max_{0\leq k\leq m-1}\sup_{0\leq s\leq\frac{1}{m}}\left|B_n^2\left(\frac{k}{m}+s\right)-B_n^2\left(\frac{k}{m}\right)\right|\right)\cdot\right.$$

$$\left. \cdot 2\left(\max_{0\leq k\leq m-1}\left|F_n\left(\frac{k}{m}\right)-\frac{k}{m}\right|+\frac{1}{m}\right) > \frac{\varepsilon}{2}\right\}$$

$$+P\left\{m\sup_{0\leq y\leq 1} B_n^2(y)\left(\max_{0\leq k\leq m-1}\left|F_n\left(\frac{k+1}{m}\right)-F_n\left(\frac{k}{m}\right)-\frac{1}{m}\right|\right) > \frac{\varepsilon}{2}\right\}$$

$$\leq P\left\{2\max_{0\leq k\leq m-1}\sup_{0\leq s\leq\frac{1}{m}}\left|B_n^2\left(\frac{k}{m}+s\right)-B_n^2\left(\frac{k}{m}\right)\right| > \frac{\varepsilon}{4}\right\}$$

$$+P\left\{2\max_{0\leq k\leq m-1}\sup_{0\leq s\leq\frac{1}{m}}\left|B_n^2\left(\frac{k}{m}+s\right)-B_n^2\left(\frac{k}{m}\right)\right| > \frac{\sqrt{\varepsilon}}{2}\right\}$$

$$+P\left\{m\left(\max_{0\leq k\leq m-1}\left|F_n\left(\frac{k}{m}\right)-\frac{k}{m}\right|\right) > \frac{\sqrt{\varepsilon}}{2}\right\}$$

$$+P\left\{\sup_{0\leq y\leq 1} B_n^2(y) > C\sqrt{\frac{\varepsilon}{2}}\right\}$$

$$+P\left\{m\left(\max_{0\leq k\leq m-1}\left|F_n\left(\frac{k+1}{m}\right)-F_n\left(\frac{k}{m}\right)-\frac{1}{m}\right|\right) > \frac{1}{C}\sqrt{\frac{\varepsilon}{2}}\right\}$$

$$= P_1 + P_2 + P_3 + P_4 + P_5.$$

Given $\varepsilon>0$ and $\delta>0$, we can choose m so big that $P_1 < \frac{\delta}{5}$ and $P_2 < \frac{\delta}{5}$.
Let $C>0$ be now so big that $P_4 < \frac{\delta}{5}$. For the already given m, $C>0$ and

$\varepsilon > 0$ we choose n so big that $P_3 < \frac{\delta}{5}$ and $P_5 < \frac{\delta}{5}$. This proves (5.6.19) and hence also (5.6.18).

Remark 5.6.1. In a similar way one can also show via (5.6.10) and an appropriate extension of (5.6.19) that

$$(5.6.21) \quad \int_{-\infty}^{+\infty} S_N^2(x) \, dF_n(x) \xrightarrow{\mathscr{D}} \int_0^1 \left(\sum_{j=1}^k c_j(N, t) K_j(t, 1) \right)^2 dt$$

where $F_n(x) = \sum_{j=1}^k n(j) F_{n(j)}(x)/n$, with $n = n(1) + \ldots + n(k)$.

5.7. Approximations of the empirical process when parameters are estimated

From a statistical point of view, Theorems 4.4.1 and 4.4.3 are useful to construct confidence intervals for an unknown distribution function F and also to construct goodness-of-fit tests for a completely specified F. Most goodness-of-fit problems arising in practice, however, do not usually specify F completely and, instead of one specific F, we are frequently given a whole parametric family of distribution functions $\{F(x; \theta); \theta \in \Theta \subseteq R^p\}$. From a goodness-of-fit point of view the unknown parameters θ are a nuisance (nuisance parameters), which render most goodness-of-fit null hypotheses to become composite ones. There are many possible ways of "getting rid of θ" so as to reduce composite goodness-of-fit null hypotheses to simple ones. As far as the empirical process is concerned, one natural way of doing this is to "estimate out θ" by using some kind of a "good estimator" sequence $\{\hat{\theta}_n\}$, based on random samples $X_1, X_2, \ldots, X_n$ ($n = 1, 2, \ldots$) on $F(x; \theta)$.

Concerning the classical Cramér–von Mises and Kolmogorov–Smirnov statistics, Darling (1955), and Kac, Kiefer and Wolfowitz (1955) investigated their asymptotic distributions when the unknown parameters of a *specified* distribution function were to be estimated first. Durbin (1973a) considered the more global question of weak convergence of the empirical process under a given sequence of alternative hypotheses when parameters of a continuous *unspecified* distribution function $F(x; \theta)$ are estimated from the data. The estimators themselves were to satisfy certain maximum likelihood-like conditions. Durbin (1973a) showed that, for such a general class of estimators, the estimated empirical process converged weakly to a Gaussian process, whose mean and covariance functions he also gave.

In this section we are going to use the strong approximation methodology of Chapter 4 to study the problem of obtaining asymptotic in-probability and almost sure representations, in terms of Gaussian processes, of the empirical process when parameters are estimated.

For an i.i.d. sequence $X_1, X_2, \ldots$ from a family of distribution functions $\{F(x; \theta); x \in R, \theta \in \Theta \subseteq R^p\}$, let $\{\hat{\theta}_n\} = \{(\hat{\theta}_{n1}, \ldots, \hat{\theta}_{np})\}$ be a sequence of estimators of the row vector θ based on the random sample $X_1, X_2, \ldots, X_n$. Consider the *estimated empirical process* defined by

$$(5.7.1) \qquad \hat{\beta}_n(x) = n^{1/2}[F_n(x) - F(x; \hat{\theta}_n)], \quad x \in R^1,$$

where F_n is the empirical distribution function of $X_1, \ldots, X_n$.

First we list the set of all conditions which will be used in our main theorem (cf. Theorem 5.7.1). We emphasize that only subsets of it will be used at appropriate places.

(5.7.2) (i) $\qquad n^{1/2}(\hat{\theta}_n - \theta_0) = n^{-1/2} \sum_{j=1}^{n} l(X_j, \theta_0) + \varepsilon_{1n},$

where $\theta_0 = (\theta_{01}, \ldots, \theta_{0p})$ is the true value of θ, $l(\cdot, \theta_0)$ is a measurable p-dimensional row vector valued function, and ε_{1n} converges to zero in a manner to be specified later on.

(ii) $El(X_j, \theta_0) = 0$.

(iii) $M(\theta_0) = E\{l(X_j, \theta_0)^t \cdot l(X_j, \theta_0)\}$ is a finite nonnegative definite matrix.

(iv) The vector $\nabla_\theta F(x; \theta)$ is uniformly continuous in x and $\theta \in \Lambda$ where Λ is the closure of a given neighbourhood of θ_0.

(v) Each component of the vector function $l(x, \theta_0)$ is of bounded variation on each finite interval.

(vi) The vector $\nabla_\theta F(x, \theta_0)$ is uniformly bounded in x, and the vector $\nabla_\theta^2 F(x; \theta)$ is uniformly bounded in x and $\theta \in \Lambda$, where Λ is as in (iv).

(vii) $\lim_{s \searrow 0} (s \log \log 1/s)^{1/2} \|l(\text{inv } F(s; \theta_0), \theta_0)\| = 0$

and

$\lim_{s \nearrow 1} ((1-s) \log \log 1/(1-s))^{1/2} \|l(\text{inv } F(s; \theta_0), \theta_0)\| = 0,$

where $\text{inv } F(s; \theta_0) = \inf\{x : F(x; \theta_0) \geq s\}$.

(viii) $s\|(\partial/\partial s)l(\text{inv } F(s;\theta_0),\theta_0)\| \leq C, \quad 0 < s < \tfrac{1}{2}$

and

$(1-s)\|(\partial/\partial s)l(\text{inv } F(x;\theta_0),\theta_0)\| \leq C, \quad \tfrac{1}{2} < s < 1$

for some positive constant C, where the vector of partial derivatives of the components of $l(\text{inv } F(x;\theta_0),\theta_0)$ with respect to s, $(\partial/\partial s)l(\text{inv } F(x;\theta_0),\theta_0)$, exists for all $s\in(0,1)$.

The estimated empirical process $\hat{\beta}_n(x)$ of (5.7.1) will be approximated by the two-parameter Gaussian process

$$(5.7.3) \quad G(x,n) = n^{-1/2}K(F(x;\theta_0),n)$$
$$- \left\{ \int l(x,\theta_0)d_x n^{-1/2} K(F(x;\theta_0),n) \right\} \nabla_\theta F(x;\theta_0)^t,$$

where K is the Kiefer process of Theorem 4.4.3 (cf. also Remark 4.4.3). G has mean function $EG(x,n)=0$ and covariance function

$$(5.7.4) \quad EG(x,n)G(y,m) = \min(n,m)\cdot(nm)^{-1/2}$$
$$\cdot \{F(\min(x,y);\theta_0) - F(x;\theta_0)F(y;\theta_0)$$
$$- J(x)\cdot \nabla_\theta F(y;\theta_0)^t - J(y)\cdot \nabla_\theta F(x;\theta_0)^t$$
$$+ \nabla_\theta F(x;\theta_0)\cdot M(\theta_0)\cdot \nabla_\theta F(y;\theta_0)^t\},$$

where $M(\theta_0)$ is defined by (5.7.2) (iii) and

$$J(x) = \int_{-\infty}^{x} l(z,\theta_0)\, d_z F(z;\theta_0).$$

(Here, of course, $F(\min(x,y);\theta_0) = \min(F(x;\theta_0), F(y;\theta_0))$.) Since $M(\theta_0)$ is nonnegative definite, there is a nonsingular matrix $D(\theta_0)$ such that

$$(5.7.5) \quad D(\theta_0)^t M(\theta_0) D(\theta_0) = \begin{pmatrix} I & 0 \\ 0 & 0 \end{pmatrix},$$

where I is the identity matrix and rank $I=$ rank $M(\theta_0)$. Hence $G(x,n)$ of (5.7.3) can be written as

$$(5.7.6) \quad G(x,n) = n^{-1/2}K(F(x;\theta_0),n) - n^{-1/2}W(n)\cdot D^{-1}(\theta_0)\cdot \nabla_\theta F(x;\theta_0)^t,$$

where $W(n) = \int l(x,\theta_0)d_x K(F(x;\theta_0),n)\cdot D(\theta_0)$ is a vector-valued Wiener process with covariance structure: $\min(n,m)$ multiplied by (5.7.5).

Clearly we have for each n that

$$(5.7.7) \quad G(x,n) \stackrel{\mathscr{D}}{=} D(x) = B(F(x;\theta_0))$$
$$- \left\{ \int l(x,\theta_0)d_x B(F(x;\theta_0)) \right\} \nabla_\theta F(x;\theta_0)^t,$$

where $\stackrel{\mathcal{D}}{=}$ stands for the equality of all finite-dimensional distributions and $B(x)$ is a Brownian bridge. Thus $ED(x)D(y)=\{\ \}$, where $\{\ \}$ is the right-hand side factor in 5.7.4.

Theorem 5.7.1 (Burke, Csörgő, Csörgő, Révész 1979). *Suppose that the sequence $\{\hat{\theta}_n\}$ satisfies (5.7.2) (i), (ii), (iii), and let*

$$\varepsilon_{2n} = \sup_{-\infty < x < \infty} |\hat{\beta}_n(x) - G(x, n)|.$$

Then

(a) $\varepsilon_{2n} \xrightarrow{P} 0$, *if conditions (5.7.2) (iv), (v) hold and* $\varepsilon_{1n} \xrightarrow{P} 0$;
(b) $\varepsilon_{2n} \xrightarrow{a.s.} 0$, *if conditions (5.7.2) (vi)–(viii) hold and* $\varepsilon_{1n} \xrightarrow{a.s.} 0$;
(c) $\varepsilon_{2n} \stackrel{a.s.}{=} O\{\max(h(n), n^{-\varepsilon})\}$ *for some* $\varepsilon > 0$, *if conditions (5.7.2) (vi)–(viii) hold and* $\varepsilon_{1n} \stackrel{a.s.}{=} O\{h(n)\}$, $h(n) > 0$, $h(n) \to 0$.

Remark 5.7.1. Durbin's (1973a) result (under his null hypotheses – Corollary 1 in Durbin (1973a)), i.e., $\hat{\beta}_n(\mathrm{inv}\, F(\cdot, \hat{\theta}_n)) \xrightarrow{\mathcal{D}} D(\mathrm{inv}\, F(\cdot; \theta_0))$, follows from part (a) of Theorem 5.7.1, because of (5.7.7). Here $\xrightarrow{\mathcal{D}}$ denotes weak convergence in the function space $D[0, 1]$. (This will also be the case under his sequences of alternatives (cf. Theorem 5.7.3 and Remark 5.7.4 concerning Durbin's original setup).) We should point out that Durbin used conditions (5.7.2) (i)–(iv), with $\varepsilon_{1n} \xrightarrow{P} 0$, to prove this weak convergence, but not (v). This slight regularity condition (5.7.2) (v) (satisfied, sure enough, in each practical situation) is the only price we pay for obtaining our in-probability representation of the limiting Gaussian process in both x and n. Nevertheless, if one still would like to get rid of this condition, then the use of Theorem 5.7.1 is still advantageous. As the proof of part (a) will show, we have (without (v))

(5.7.8) $$\sup_x |\hat{\beta}_n(x) - Y_n(F(x; \theta_0))| \xrightarrow{P} 0,$$

where

$$Y_n(s) = n^{-1/2} K(s, n) - \left\{n^{-1/2} \sum_{j=1}^{n} l(X_j, \theta_0)\right\} \nabla_\theta F(\mathrm{inv}\, F(s; \theta_0); \theta_0)^t.$$

In this way we could save a tightness-proof, since the tightness of $\{Y_n\}$ reduces to the a.s. continuity of the Kiefer process. But one still has to prove the convergence of the finite-dimensional distributions of Y_n to those of $D(\cdot)$ in (5.7.7), which is, at one hand, again easier than for $\hat{\beta}_n$, but, on the other hand, is essentially a repetition of the proof of Lemma 3 in Durbin (1973a). We should also note however, that, unlike in Durbin (1973a), the continuity of $F(x, \theta)$ in x is not used in Theorem 5.7.1. Conditions (5.7.2) (iv) and (vi) can be satisfied without the continuity of F (example: the binomial distribution).

Remark 5.7.2. Conditions (5.7.2) (vi)–(viii) are the extra ones used to obtain our a.s. representation (in case of part (c) with a rate sequence) of the limiting Gaussian process in both x and n. The thus gained results (cf. Theorems 5.7.1 and 5.7.3) are analogues of a Kiefer type approximation of the empirical process (cf. Theorem 4.4.3), while Durbin's result (cf. Remarks 5.7.1 and 5.7.4) is an analogue of Donsker's theorem (cf. Theorem 4.2.1). Commonly used distributions such as the normal and exponential and, in fact, all those density functions whose tail behaviour in the sense of the requirements (5.7.2) (vi)–(viii) is like that of the exponential density, satisfy these conditions when maximum likelihood estimators are employed.

Introduce the following

$$(5.7.9) \qquad \varepsilon_{3n}(s) = n^{1/2}[F_n(\text{inv } F(s; \theta_0)) - s] - n^{-1/2} K(s, n),$$

where K is the Kiefer process of Theorem 4.4.3. We have

$$\varepsilon_{3n}(F(x; \theta_0)) = n^{1/2}[F_n(x) - F(x; \theta_0)] - n^{-1/2} K(F(x; \theta_0), n).$$

Our proof of Theorem 5.7.1 hinges on the following two lemmas.

Lemma 5.7.1 *Suppose that the vector function $l(x, \theta_0)$ satisfies conditions (5.7.2) (iii) and (v). Then, as $n \to \infty$,*

$$L_n = \int l(x, \theta_0) \, d_x \varepsilon_{3n}(F(x; \theta_0)) \xrightarrow{P} 0.$$

Proof. Let $T_j(x)$ denote the total variation of the jth component $l_j(\cdot, \theta_0)$ of $l(\cdot, \theta_0)$ on the interval $[-x, x]$, $j = 1, \ldots, p$, and let $T(x) = (T_1(x), \ldots, T_p(x))$. Clearly we can choose a sequence of positive numbers u_n tending so slowly to infinity that $\|T(u_n)\| n^{-1/2} \log^2 n \to 0$. (If $\|T(n)\|$ is bounded, then any $u_n \to \infty$ sequence will suffice, while if $\|T(n)\| \nearrow \infty$, then we take $u_n = \text{inv } T(v_n)$, where $v_n \nearrow \infty$ so that $v_n = o\{n^{1/2}/\log^2 n\}$, and $\text{inv } T(y) = \inf\{x : \|T(x)\| \geq y\}$). With this u_n then, consider

$$L_n = \int_{|x| > u_n} l(x, \theta_0) \, d_x n^{1/2}[F_n(x) - F(x; \theta_0)]$$

$$- \int_{|x| > u_n} l(x, \theta_0) \, d_x n^{-1/2} K(F(x; \theta_0), n)$$

$$+ \int_{|x| \leq u_n} l(x, \theta_0) \, d_x \varepsilon_{3n}(F(x; \theta_0)) = L_{1n} - L_{2n} + L_{3n}.$$

Integrating by parts and using Theorem 4.4.3 one obtains

$$\|L_{3n}\| \leq \left\| \int_{-u_n}^{u_n} \varepsilon_{3n}(F(x;\theta_0))\, dl(x,\theta_0) \right\| + \left\| [\varepsilon_{3n}(F(x;\theta_0))l(x,\theta_0)]_{x=-u_n}^{u_n} \right\|$$

$$\stackrel{\text{a.s.}}{=} O\{n^{-1/2}\log^2 n\}\|T(u_n)\| \to 0.$$

If the components in L_{1n} and L_{2n} are denoted respectively by $L_{1n}^{(j)}$ and $L_{2n}^{(j)}$, $j=1,\ldots,p$, then we have $EL_{1n}^{(j)} = EL_{2n}^{(j)} = 0$ and

(5.7.10) $$E(L_{1n}^{(j)})^2 = E(L_{2n}^{(j)})^2 = \int_{|x|>u_n} l_j^2(x,\theta_0)\, dF(x;\theta_0)$$

$$- \Bigl(\int_{x \leq -u_n} l_j(x,\theta_0)\, dF(x;\theta_0) \Bigr)^2 - \Bigl(\int_{x \leq u_n} l_j(x,\theta_0)\, dF(x;\theta_0) \Bigr)^2.$$

Whence, by the Chebishev inequality with $\varepsilon > 0$,

$$P\{\|L_{1n}\| + \|L_{2n}\| > 2\varepsilon\} \leq \frac{2}{\varepsilon^2} \sum_{j=1}^{p} \int_{|x|>u_n} l_j^2(x,\theta_0)\, dF(x;\theta_0),$$

and this latter bound tends to zero by condition (5.7.2) (iii), since $u_n \to \infty$.

Lemma 5.7.2. *Suppose that the vector function $l(\text{inv } F(x;\theta_0), \theta_0)$ satisfies conditions (5.7.2) (vii) and (viii). Then, as $n \to \infty$,*

$$L_n = \int_0^1 l(\text{inv } F(s;\theta_0), \theta_0)\, d\varepsilon_{3n}(s) \stackrel{\text{a.s.}}{=} O\{n^{-\varepsilon}\},$$

for some $\varepsilon > 0$, where $\varepsilon_{3n}(s)$ is again that of (5.7.9).

Proof. We have

$$L_n = \int_0^1 \varepsilon_{3n}(s)(\partial/\partial s)\, l(\text{inv } F(s;\theta_0), \theta_0)\, ds.$$

This latter equality is correct provided the function $\varepsilon_{3n}(s)l(\text{inv } F(s;\theta_0), \theta_0)$ at $s=0$ and $s=1$ is almost surely the zero vector. This, in turn, is true by condition (5.7.2) (vii) and by the fact that the Kiefer process $K(s,n)$ (cf. Theorem 1.4.1) and the empirical process $n^{1/2}[F_n(\text{inv } F(s;\theta_0)) - s]$ behave like $(s \log \log 1/s)^{1/2}$ and $((1-s) \log \log 1/(1-s))^{1/2}$ as $s \searrow 0$ and $s \nearrow 1$, respectively.

Consider now

$$L_n = \int_0^{n^{-1/3}} + \int_{n^{-1/3}}^{1/2} + \int_{1/2}^{1-n^{-1/3}} + \int_{1-n^{-1/3}}^{1} = L_{1n}^* + L_{2n}^* + L_{3n}^* + L_{4n}^*.$$

By Theorem 4.4.3 and the first part of (5.7.2) (viii) we have almost surely

$$\|L_{2n}^*\| \leq O\{n^{-1/2}\log^2 n\} \int_{n^{-1/3}}^{1/2} \|(\partial/\partial s)l(\text{inv } F(s; \theta_0), \theta_0)\| \, ds$$

$$\leq O\{n^{-1/2}\log^2 n\} \int_{n^{-1/3}}^{1/2} s^{-1} \, ds$$

$$= O\{n^{-1/2}\log^3 n\}.$$

Also,

$$\|L_{1n}^*\| \leq \int_0^{n^{-1/3}} |\beta_n(\text{inv } F(s; \theta_0))| \cdot \|(\partial/\partial s)l(\text{inv } F(s; \theta_0), \theta_0)\| \, ds$$

$$+ \int_0^{n^{-1/3}} |n^{-1/2}K(s,n)| \cdot \|(\partial/\partial s)l(\text{inv } F(s; \theta_0), \theta_0)\| \, ds.$$

Since by (S.5.1.4)

(5.7.11) $\quad \sup_{0 < s < n^{-1/3}} |\beta_n(\text{inv } F(s; \theta_0))(s(1-s))^{-1/2}| \stackrel{\text{a.s.}}{=} O\{\log n\},$

and by (1.15.1)

(5.7.12) $\quad \overline{\lim}_{n \to \infty} \sup_{0 < s < 1} |K(s,n)[4ns(1-s)\log\log(n/(s(1-s)))]^{-1/2}| \stackrel{\text{a.s.}}{=} 1,$

we have by the first part of (5.7.2) (viii)

$$\|L_{1n}^*\| \leq O\{\log n\} \int_0^{n^{-1/3}} (\log\log(n/s))^{1/2} s^{-1/2} \, ds, \quad \text{a.s.},$$

$$\stackrel{\text{a.s.}}{=} O\{n^{-1/7}\}.$$

The terms L_{3n}^* and L_{4n}^* are estimated similarly and hence the lemma.

Proof of Theorem 5.7.1. Using the one-term Taylor expansion of F with respect to θ_0 we obtain

(5.7.13) $\quad \hat{\beta}_n(x) = n^{1/2}[F_n(x) - F(x; \theta_0)] - n^{1/2}[F(x; \hat{\theta}_n) - F(x; \theta_0)]$

$$= n^{-1/2}K(F(x; \theta_0), n) - n^{1/2}(\hat{\theta}_n - \theta_0)\nabla_\theta F(x; \theta_n^*)^t + \varepsilon_{3n}(F(x; \theta_0))$$

$$= n^{-1/2}K(F(x; \theta_0), n) - n^{-1/2}(\hat{\theta}_n - \theta_0)\nabla_\theta F(x; \theta_0)^t + \varepsilon_{3n}(F(x; \theta_0)) + \varepsilon_{4n}(x),$$

where ε_{3n} is defined by (5.7.9), and by Theorem 4.4.3

(5.7.14) $\quad \sup_x |\varepsilon_{3n}(F(x; \theta_0))| \stackrel{\text{a.s.}}{=} O\{n^{-1/2}\log^2 n\},$

while $\|\theta_n^* - \theta_0\| \leq \|\hat{\theta}_n - \theta_0\|$ and

$$\varepsilon_{4n}(x) = n^{1/2}(\hat{\theta}_n - \theta_0)(\nabla_\theta F(x; \theta_0) - \nabla_\theta F(x; \theta_n^*))^t.$$

It follows from (5.7.2) (i), (ii) and (iii) that $n^{1/2}(\hat{\theta}_n-\theta_0)$ is asymptotically a normal vector, and thus $\|\hat{\theta}_n-\theta_0\| \xrightarrow{P} 0$. Hence, using also (5.7.2) (iv), we have

$$\sup_x \|\varepsilon_{4n}(x)\| \xrightarrow{P} 0. \tag{5.7.15}$$

Also, by conditions (5.7.2) (i) and (ii)

$$\begin{aligned}
n^{1/2}(\hat{\theta}_n-\theta_0) &= n^{-1/2} \sum_{j=1}^{n} l(X_j, \theta_0) + \varepsilon_{1n} \\
&= \int l(x, \theta_0) \, d_x n^{1/2} F_n(x) + \varepsilon_{1n} \\
&= \int l(x, \theta_0) \, d_x n^{1/2} [F_n(x) - F(x; \theta_0)] + \varepsilon_{1n} \\
&= \int l(x, \theta_0) \, d_x n^{-1/2} K(F(x; \theta_0), n) + L_n + \varepsilon_{1n},
\end{aligned} \tag{5.7.16}$$

where L_n is of Lemma 5.7.1. Since the vector $\nabla_\theta F(x; \theta_0)$ is uniformly bounded in x by (5.7.2) (iv), part (a) of the theorem follows from (5.7.14), (5.7.15) and (5.7.16).

To prove parts (b) and (c) we use the two-term Taylor expansion of F with respect to θ_0 in the second term of the first row in (5.7.13). Applying also (5.7.16), we obtain

$$\begin{aligned}
\hat{\beta}_n(x) &= n^{-1/2} K(F(x; \theta_0), n) - n^{1/2}(\hat{\theta}_n-\theta_0) \nabla_\theta F(x; \theta_0)^t \\
&\quad - \tfrac{1}{2} n^{1/2}(\hat{\theta}_n-\theta_0)^2 \nabla_\theta^2 F(x, \theta_n^*)^t + \varepsilon_{3n}(F(x; \theta_0)) \\
&= G(x, n) + (L_n + \varepsilon_{1n}) \nabla_\theta F(x; \theta_0)^t \\
&\quad - \tfrac{1}{2} n^{1/2}(\hat{\theta}_n-\theta_0)^2 \nabla_\theta^2 F(x, \theta_n^*)^t + \varepsilon_{3n}(F(x; \theta_0)),
\end{aligned}$$

where $\|\theta_n^*-\theta_0\| \leq \|\hat{\theta}_n-\theta_0\|$. If $\varepsilon_{1n} \xrightarrow{a.s.} 0$, then it follows from (5.7.2) (i) that $\theta_n^* \xrightarrow{a.s.} \theta_0$. Hence the vector $\nabla_\theta^2 F(x; \theta_n^*)$ is almost surely uniformly bounded in x and n by (5.7.2) (vi). Because of (5.7.2) (iii) the law of iterated logarithm can be applied componentwise to the partial sum sequence in (5.7.2) (i). Whence we get $\|n^{1/2}(\hat{\theta}_n-\theta_0)^2\| \stackrel{a.s}{=} O\{n^{-1/2} \log \log n\}$, that is

$$\sup_x |n^{1/2}(\hat{\theta}_n-\theta_0)^2 \nabla_\theta^2 F(x; \theta_n^*)^t| \stackrel{a.s.}{=} O\{n^{-1/2} \log \log n\}.$$

Thus, if (5.7.2) (i), (ii), (iii) and (vi) hold and $\varepsilon_{1n} \stackrel{a.s.}{=} 0$, then

$$\hat{\beta}_n(x) - G(x, n) \stackrel{a.s.}{=} \varepsilon_{5n}(x) + O\{n^{-1/2} \log^2 n\},$$

where $\varepsilon_{5n}(x) = (L_n + \varepsilon_{1n})\nabla_\theta F(x; \theta_0)^t$. If (5.7.2) (vii), (viii) hold, then by Lemma 5.7.2 $L_n \xrightarrow{\text{a.s.}} O\{n^{-\varepsilon}\}$. Whence by (5.7.2) (vi)

$$\sup_x |\varepsilon_{5n}(x)| \xrightarrow{\text{a.s.}} 0,$$

and if, in addition, $\varepsilon_{1n} \xrightarrow{\text{a.s.}} O\{h(n)\}$, $h(n) > 0$, $h(n) \to 0$, then

$$\sup_x |\varepsilon_{5n}(x)| \xrightarrow{\text{a.s.}} O\{\max(h(n), n^{-\varepsilon})\}.$$

The last sentence also completes the proof of parts (b) and (c) of Theorem 5.7.1.

The limiting Gaussian process G of Theorem 5.7.1 depends, in general, not only on F but also on θ_0, the true value of θ. Thus, in general, Theorem 5.7.1 cannot be used to test the composite hypothesis

$$H_0: F \in \{F(x; \theta): \theta \in \Theta \subseteq R^p\}.$$

In order to give an asymptotic theoretical solution to the latter problem, we define the process $\hat{G}(x, n)$ by

(5.7.17) $\quad \hat{G}(x, n) = n^{-1/2} K(F(x; \hat{\theta}_n), n) - n^{-1/2} W(n) \cdot D^{-1}(\hat{\theta}_n) \cdot \nabla_\theta F(x; \hat{\theta}_n)^t,$

where $M(\theta)$ (cf. (5.7.2) (iii)) is assumed to exist and is nonnegative definite for $\theta \in \Lambda$, and $D(\theta)^t M(\theta) D(\theta)$ is assumed to satisfy (5.7.5). For $W(n)$ see (5.7.6). We have

Theorem 5.7.2. Suppose that the column vectors $q_j(\theta) = (\partial/\partial \theta_j) D^{-1}(\theta) \cdot \nabla_\theta F(x; \theta)^t$, $1 \leq j \leq p$, of partial derivatives exist and are uniformly bounded on $R \times \Lambda$. Then, under the conditions (5.7.2) (i), (ii), (iii) and (vi)

(5.7.18) $\quad\quad\quad \sup_{-\infty < x < \infty} |\hat{G}(x, n) - G(x, n)| = \varepsilon_{6n},$

where $\varepsilon_{6n} \xrightarrow{P} 0$ if $\varepsilon_{1n} \xrightarrow{P} 0$, and $\varepsilon_{6n} \xrightarrow{\text{a.s.}} O(n^{-\delta})$ for some $\delta > 0$, if $\varepsilon_{1n} \xrightarrow{\text{a.s.}} 0$ as $n \to \infty$. Consequently,

(5.7.19) $\quad\quad\quad \sup_{-\infty < x < \infty} |\hat{\beta}(x) - \hat{G}(x, n)| = \varepsilon_{2n}^*,$

where ε_{2n}^* converges to zero like ε_{2n} of (a) or (b) or (c) in Theorem 5.7.1, $\hat{G}$ is defined by (5.7.17) and ε_{1n} by (5.7.2) (i).

Proof. Assume $\varepsilon_{1n} \xrightarrow{\text{a.s.}} 0$, for it will be clear from the proof where to make the obvious changes to arrive at the conclusion of Theorem 5.7.2

in the case when $\varepsilon_{1n} \xrightarrow{P} 0$. We have

$$\hat{G}(x, n) - G(x, n) = n^{-1/2}\{K(F(x; \hat{\theta}_n), n) - K(F(x; \theta_0), n)\}$$
$$- n^{-1/2}W(n)\{D^{-1}(\hat{\theta}_n)\nabla_\theta F(x; \hat{\theta}_n)^t - D^{-1}(\theta_0)\nabla_\theta F(x; \theta_0)^t\}.$$

On letting $h_n = n^{-1/2}(\log \log n)^{1/2}$ in Theorem 1.15.2, we have $\gamma_n = \{2(n \log \log n)^{1/2} \log [n^{1/2}(\log \log n)^{-1/2}]\}^{-1/2} \approx \{2(n \log \log n)^{1/2} \log n^{1/2}\}^{-1/2}$ and hence, using Taylor's theorem,

(5.7.20) $\quad n^{-1/2} \sup_x |K[F(x; \hat{\theta}_n), n] - K[F(x; \theta_0), n]|$

$$= n^{-1/2} \sup_x |K[F(x; \theta_0) + (\hat{\theta}_n - \theta_0)\nabla_\theta F(x; \theta_0)^t$$
$$+ \tfrac{1}{2}(\hat{\theta}_n - \theta_0)^2 \nabla_\theta^2 F(x; \theta_n^*)^t, n] - K[F(x; \theta_0), n]|$$
$$\stackrel{\text{a.s.}}{=} O\{n^{-1/4}(\log \log n)^{1/4}(\log n)^{1/2}\},$$

where $\|\theta_n^* - \theta_0\| \leq \|\hat{\theta}_n - \theta_0\|$. The latter equality of (5.7.20) holds by condition (5.7.2) (vi) and the fact that $\|\hat{\theta}_n - \theta_0\| \stackrel{\text{a.s.}}{=} O\{n^{-1/2}(\log \log n)^{1/2}\}$ if $\varepsilon_{1n} \xrightarrow{\text{a.s.}} 0$.

Let $Q(x, \theta)$ be the $p \times p$ matrix whose jth column is the vector $q_j(\theta)$. Then we have

$$n^{-1/2}W(n) \cdot [D^{-1}(\hat{\theta}_n) \cdot \nabla_\theta F(x; \hat{\theta}_n)^t - D^{-1}(\theta_0) \cdot \nabla_\theta F(x; \theta_0)^t]$$
$$= n^{-1/2}W(n)[(\hat{\theta}_n - \theta_0) \cdot Q(x, \theta_n^*)]^t$$
$$\stackrel{\text{a.s.}}{=} O\{n^{-1/2} \log \log n\},$$

by the law of the iterated logarithm for the Wiener process $W(n)$ and for the partial sum sequence of (5.7.2) (i), and the uniform boundedness of Q on $R \times \Lambda$, where $\|\theta_n^* - \theta_0\| \leq \|\hat{\theta}_n - \theta_0\|$. This, together with (5.7.20), implies $\varepsilon_{6n} \stackrel{\text{a.s.}}{=} O(n^{-\delta})$ for some $\delta > 0$ if $\varepsilon_{1n} \xrightarrow{\text{a.s.}} 0$, and hence the theorem.

Remark 5.7.3. (5.7.9) says that $\hat{G}(x, n)$ is just as good an approximation of $\hat{\beta}_n$ as $G(x, n)$ was. Now let $m(\theta, \alpha)$ ($0 < \alpha < 1$; $\theta \in \Theta$) be the number for which

$$P\left\{\sup_{-\infty < x < +\infty} |G(x, n; \theta)| > m(\theta, \alpha)\right\} = \alpha.$$

Clearly the function $m(\theta, \alpha)$ is uniquely determined in this way and it is continuous in θ and α. This implies

$$\alpha \stackrel{\text{a.s.}}{=} \lim_{n \to \infty} P\left\{\sup_{-\infty < x < +\infty} |G(x, n; \hat{\theta}_n)| > m(\hat{\theta}_n, \alpha)\right\}$$
$$\stackrel{\text{a.s.}}{=} \lim_{n \to \infty} P\left\{\sup_{-\infty < x < +\infty} |\hat{\beta}_n(x)| > m(\hat{\theta}_n, \alpha)\right\},$$

provided ε_{1n} of (5.7.2) (i) goes to zero almost surely. Theoretically, one can therefore propose the following test of level α: reject the composite hypothesis $H_0: F \in \{F(x, \theta); \theta \in \Theta \subseteq R^p\}$ if $\sup_{-\infty < x < +\infty} |\hat{\beta}_n(x)| > m(\theta_1, \alpha)$, where θ_1 is the numerical value of $\hat{\theta}_n$ in a given experiment. We should note, of course, that the evaluation of $m(\theta, \alpha)$ itself appears to be quite difficult for any given F.

Our method can be also applied to give an analogue of Theorem 5.7.1 under a sequence of alternatives. Suppose that the continuous distribution function of the i.i.d. sequence is $F(x; \lambda, \theta)$, where λ is a p_1-dimensional vector of parameters which is assumed to be known, and θ is a p_2-dimensional vector of unknown parameters which is estimated by $\{\hat{\theta}_n\}$, based on $X_1, X_2, \ldots, X_n$. Consider the null hypothesis

(5.7.21) $$H_0: (\lambda, \theta) = (\lambda_0, \theta_0),$$

where θ_0 stands for the true value of θ. Let

(5.7.22) $$\hat{\beta}_n(x) = n^{1/2}[F_n(x) - F(x; \lambda_0, \hat{\theta}_n)], \quad x \in R^1,$$

where F_n is the empirical distribution function. In addition to H_0, we also wish to study $\hat{\beta}_n$ under a sequence of alternatives $\{H_n\}$ defined as follows:

Let $\{\lambda_n\}$ be a sequence of p_1-dimensional (nonrandom) vectors satisfying the condition

(5.7.23) $$\lambda_n = \lambda_0 + \gamma \cdot n^{-1/2},$$

where γ is a given constant vector. Let Λ_1 denote the closure of a given neighbourhood of λ_0 and let $m = \min \{k; \lambda_n \in \Lambda_1, \text{ for all } n \geq k > 2\}$. Then, consider

(5.7.24) $$H_n: (\lambda, \theta) = (\lambda_0, \theta_0),$$

for $n = m, m+1, \ldots$ where λ_n satisfies (5.7.23). If we choose $\lambda_n = \lambda_0$ for all n, i.e., $\gamma = 0$, then H_n and H_0 are identical.

First we list all the conditions whose appropriate subcollections will be used in Theorem 5.7.3. These conditions are, of course, parallel to those of (5.7.2).

(5.7.25) (i) Under H_n:

$$n^{1/2}(\hat{\theta}_n - \theta_0) = n^{-1/2} \sum_{j=1}^n l(X_j, \lambda_0, \theta_0) + A\gamma^t + \varepsilon_{7n},$$

where A is a given finite matrix of order $p_2 \times p_1$, l is a measurable p_2-dimensional vector valued function, and ε_{7n} converges to zero in a manner to be specified.

(ii) $E\{l(X_j, \lambda_0, \theta_0)|H_n\} = 0$ for $n=0$ and $n \geq m$.

(iii) $E\{l(X_j, \lambda_0, \theta_0)^t l(X_j, \lambda_0, \theta_0)|H_n\} = M(\lambda_n, \theta_0)$, a finite non-negative definite matrix for each $n \geq m$ which converges to a finite non-negative matrix $M = M(\lambda_0, \theta_0)$ as $n \to \infty$.

(iv) The vector $\nabla_\lambda F(x; \lambda, \theta_0)$ is uniformly continuous in x and $\lambda \in \Lambda_1$, and the vector $\nabla_\theta F(x; \lambda_0, \theta)$ is uniformly continuous in x and $\theta \in \Lambda_2$, where Λ_2 is the closure of a given neighbourhood of θ_0.

(v) Each component of $l(x, \lambda_0, \theta_0)$ is of bounded variation on each finite interval.

(vi) The vectors $\nabla_\lambda F(x; \lambda_0, \theta_0)$, $\nabla_\theta F(x; \lambda_0, \theta_0)$ are uniformly bounded in x, while the vector $\nabla_\lambda^2 F(x; \lambda, \theta_0)$ is uniformly bounded in x and $\lambda \in \Lambda_1$, and the vector $\nabla_\theta^2 F(x; \lambda_0, \theta)$ is uniformly bounded in x and $\theta \in \Lambda_2$.

(vii) Condition (5.7.2) (vii) holds for the vector
$l(\text{inv } F(s; \lambda_0, \theta_0), \lambda_0, \theta_0)$, where $\text{inv } F(s; \lambda, \theta)$
$= \inf \{x : F(x; \lambda, \theta) \geq s\}$.

(viii) Condition (5.7.2) (viii) holds for the vector
$l(\text{inv } F(s; \lambda_0, \theta_0), \lambda_0, \theta_0)$.

The estimated empirical process $\hat{\beta}_n(x)$ of (5.7.22), under the sequence of alternatives $\{H_n\}$ of (5.7.24), will be estimated by the two-parameter Gaussian process

(5.7.26) $\quad Z(x, n) = G(x, n) - A\gamma^t \nabla_\theta F(x; \lambda_0, \theta_0)^t + \gamma \nabla_\lambda F(x; \lambda_0, \theta_0)^t$,

with

$$G(x, n) = n^{-1/2} K(F(x; \lambda_0, \theta_0), n)$$
$$- \left\{ \int l(x, \lambda_0, \theta_0) d_x n^{-1/2} K(F(x; \lambda_0, \theta_0), n) \right\} \nabla_\theta F(x; \lambda_0, \theta_0)^t.$$

This process $G(x, n)$ is the same process as defined by (5.7.3). The mean of Z is

$$EZ(x, n) = -A\gamma^t \nabla_\theta F(x; \lambda_0, \theta_0)^t + \gamma \nabla_\lambda F(x; \lambda_0, \theta_0)^t,$$

and its covariance is given by (5.7.4), with the obvious changes in notation.

On letting

$$\hat{Z}(x, n) = \hat{G}(x, n) - A\gamma^t \nabla_\theta F(x; \lambda_0, \hat{\theta}_n)^t + \gamma \nabla_\lambda F(x; \lambda_0, \hat{\theta}_n)^t,$$

where $\hat{G}$ is defined by (5.7.17) (with the notation suitably modified), the results corresponding to Theorem 5.7.2 continue to hold under the sequence of alternatives $\{H_n\}$.

Theorem 5.7.3. *Suppose that conditions* (5.7.25) (i)–(iii) *hold, and let*

$$\varepsilon_{8n} = \sup_{x \in R^1} |\hat{\beta}_n(x) - Z(x, n)|.$$

Then, under the sequence of alternatives $\{H_n\}$,

(a) $\varepsilon_{8n} \xrightarrow{P} 0$, *if conditions* (5.7.25) (iv), (v) *hold and* $\varepsilon_{7n} \xrightarrow{P} 0$;

(b) $\varepsilon_{8n} \xrightarrow{a.s.} 0$, *if conditions* (5.7.25) (vi)–(viii) *hold and* $\varepsilon_{7n} \xrightarrow{a.s.} 0$;

(c) $\varepsilon_{8n} \stackrel{a.s.}{=} O\{\max(h(n), n^{-\varepsilon})\}$ *for some* $\varepsilon > 0$, *if conditions* (5.7.25) (vi)–(viii) *hold and* $\varepsilon_{7n} \stackrel{a.s.}{=} O\{h(n)\}$, $h(n) > 0$, $h(n) \to 0$.

Remark 5.7.4. Here the whole content (modified to the present situation) of Remarks 5.7.1 and 5.7.2 can be repeated. Specifically, Durbin (1973a) proved the weak convergence of $\hat{\beta}_n(\text{inv } F(\cdot\,; \lambda_0, \hat{\theta}_n))$, under $\{H_n\}$, to a process that can be represented by putting a Brownian bridge $B(\cdot)$ into the definition (5.7.26) of Z in place of $n^{-1/2}K(\cdot, n)$. He used conditions (5.7.25) (i)–(iv) (with $\varepsilon_{7n} \xrightarrow{P} 0$) to prove this (he requires (5.7.25) (iv) in a slightly stronger form than ours), and the extra condition that $F(x; \lambda, \theta)$ is continuous in x for all (λ, θ) in some neighbourhood of (λ_0, θ_0). If we want to prove this weak convergence (but not the two-parameter representation in (a)) without condition (5.7.25) (v), then the method proposed in Remark 5.7.1 works again with

$$Y_n(s) = n^{-1/2} K(s, n)$$

$$- \left\{ n^{-1/2} \sum_{j=1}^n l(X_j, \lambda_0, \theta_0) + A\gamma^t \right\} \nabla_\theta F(\text{inv } F(s; \lambda_0, \theta_0); \lambda_0, \theta_0)^t$$

$$+ \gamma \nabla_\lambda F(\text{inv } F(s; \lambda_0, \theta_0); \lambda_0, \theta_0)^t.$$

Remark 5.7.5. Durbin (1973a) proves the weak convergence of

$$\hat{\delta}_n(s) = n^{1/2}[\hat{F}_n(s) - s], \quad 0 \le s \le 1,$$

where $\hat{F}_n(s)$ is the proportion of $F(X_1; \lambda_0, \hat{\theta}_n), \ldots, F(X_n; \lambda_0, \hat{\theta}_n)$ which satisfy $F(X_j; \lambda_0, \hat{\theta}_n) \le s$. The processes $\hat{\delta}_n$ and $\hat{\beta}_n$ are asymptotically equivalent. In order to see this we first note that $\hat{\delta}_n(s) = \hat{\beta}_n(\text{inv } F(s; \lambda_0, \hat{\theta}_n))$, where now $\text{inv } F(s; \lambda_0, \theta) = \sup\{x : F(x; \lambda_0, \theta) \le s\}$. Secondly, if we assume that we have already carried out the program of the last sentence of Remark 5.7.4 concerning the finite dimensional distributions of Y_n, we prove also easily that

$$\sup_x |\hat{\beta}_n(x) - Y_n(F(x; \lambda_0, \hat{\theta}_n))| \xrightarrow{P} 0, \quad \text{as} \quad n \to \infty.$$

Hence on putting $x = \text{inv } F(s; \lambda_0, \hat{\theta}_n)$, we have

$$\sup_{0 \leq s \leq 1} |\hat{\delta}_n(s) - Y_n(s)| \xrightarrow{P} 0, \quad \text{as} \quad n \to \infty,$$

which establishes asymptotic equivalence.

The proof of Theorem 5.7.3 is similar to that of Theorem 5.7.1. For details we refer to Burke, Csörgő, Csörgő, Révész (1979).

Remark 5.7.6. If we assume that the function l possesses not only a finite second (cf. (5.7.25) (iii)) but a finite absolute moment of order r, $r > 2$, then we can proceed the following way. Let $D(\lambda_n, \theta_0)$ be the nonsingular matrix for which $D(\lambda_n, \theta_0)^t M(\lambda_n, \theta_0) D(\lambda_n, \theta_0)$ satisfies (5.7.5). Then, by Theorem 2.6.3,

$$\left\| \sum_{j=1}^n l(X_j, \lambda_0, \theta_0) D(\lambda_n, \theta_0)^{-1} - W(n) \right\| \stackrel{\text{a.s.}}{=} o(n^{1/r}),$$

where $W(n)$ is a vector-valued Wiener process (cf. (5.7.6)). If the underlying probability space is still richer (if necessary), then there exists a Kiefer process $\tilde{K}$ such that $W(n) = \int l(x, \lambda_0, \theta_0) d_x \tilde{K}(F(x; \lambda_0, \theta_0), n) D(\lambda_n, \theta_0)$. Let $\varepsilon_{9n} = \sup_x |\hat{\beta}_n(x) - \tilde{Z}(x, n)|$, where $\tilde{Z}(x, n)$ has the same form as $Z(x, n)$ in (5.7.26) with $\tilde{G}(x, n)$ in place of $G(x, n)$, where

(5.7.27) $\quad \tilde{G}(x, n) = n^{-1/2} K(F(x; \lambda_0, \theta_0), n)$

$\quad - \left\{ \int l(x; \lambda_0, \theta_0) d_x n^{-1/2} \tilde{K}(F(x; \lambda_0, \theta_0), n) \right\} \nabla_\theta F(x; \lambda_0, \theta_0)^t.$

The above proof shows the following. Under the rth moment condition and (only) (5.7.25) (i)–(iv) we have $\varepsilon_{9n} \xrightarrow{P} 0$, if $\varepsilon_{7n} \xrightarrow{P} 0$, while under the rth moment condition and (only) (5.7.25) (i)–(iii), (vi) we have $\varepsilon_{9n} \xrightarrow{\text{a.s.}} 0$, if $\varepsilon_{7n} \xrightarrow{\text{a.s.}} 0$. Moreover, if in the latter case $\varepsilon_{7n} = O\{h(n)\}$, then $\varepsilon_{9n} \stackrel{\text{a.s.}}{=} O\{\max(h(n), n^{-\tau})\}$, for some $\tau > 0$ (cf. Remark 8 in Burke, Csörgő, Csörgő, Révész (1979)). Naturally, the same type of "results" hold in the simpler setting of Theorem 5.7.1. These "results" are entirely useless at the present stage, since we do not know anything about the joint distribution of K and $\tilde{K}$ in (5.7.27). Since $\varepsilon_{9n} \to 0$, it follows from Durbin's weak convergence theorem that $\tilde{G}(\cdot, n)$ converges weakly to $D(\cdot)$ of (5.7.7.). The problem is how to replace $\tilde{K}$ by K in (5.7.27), so that we should not have to fall back to Durbin's weak convergence theorem in order to make sense out of $\tilde{G}(x, n)$. This was achieved in Lemmas 5.7.1 and 5.7.2 by imposing (very mild) extra restrictions on l. However, we conjecture that this should be possible without the latter restrictions. It appears that the proof of this conjecture (under the rth moment condition and only (5.7.25) (i)–(iii)) would require an extension of the proof of Theorem 4.4.3 to simultaneously approximating β_n and $\sum_{j=1}^{n} l(X_j)$.

5.8. Asymptotic quadratic quantile tests for composite goodness-of-fit

In the previous section we investigated the problem of testing for the composite goodness-of-fit hypothesis

$$H_0: F \in \mathscr{F} = \{F(x; \theta): \theta \in \Theta \subseteq R^p\}$$

via the estimated empirical process and proposed a Kolmogorov–Smirnov type statistic (cf. Theorem 5.7.2) for the latter H_0. It is clear that applying the method of that section we could have also talked about the estimated quantile process. Other statistics, like, for example, the Cramér–von Mises types, could have been also proposed in terms of the estimated empirical as well as the estimated quantile process. In this section we continue to study this problem, using the estimated quantile process, and will propose a Cramér–von Mises type statistic. However, we will restrict ourselves to the case when we have only scale and shift as nuisance parameters. A description of this problem now follows.

Let F be a continuous distribution function with unknown location and scale parameters $-\infty < \mu < +\infty$ and $\sigma > 0$ respectively, and assume that F is of the form $F(x; \mu, \sigma) = F_0\left(\frac{x-\mu}{\sigma}\right)$, $x \in R^1$, where F_0 is a known distribution function with mean zero and variance one. Let $\mathscr{F}$ be the class of all continuous distribution functions of this latter form, i.e.,

(5.8.1) $\quad \mathscr{F} = \left\{F(x; \mu, \sigma): F(x; \mu, \sigma) = F_0\left(\frac{x-\mu}{\sigma}\right), -\infty < \mu < +\infty, \sigma > 0\right\}$

where F_0 is a known distribution function with

$$\int_{-\infty}^{+\infty} x\,dF_0 = 0, \quad \int_{-\infty}^{+\infty} x^2\,dF_0 = 1, \quad \int_{-\infty}^{+\infty} x^4\,dF_0 < \infty.$$

Further let $X_1, X_2, \ldots, X_n$ be a random sample from a distribution F. Our task is to test the composite null hypothesis

(5.8.2) $$H_0: F \in \mathscr{F}.$$

In case of the normal family, i.e., when F_0 of (5.8.1) is equal to Φ, the literature is enormous. We intend to give here only those steps of development concerning this specific problem of testing for normality, which led us to our approach to the more general problem of (5.8.2). We begin with the Shapiro–Wilk (1965) approach. In order to describe the latter, we first note that our random sample is then of the form

(5.8.3) $$X_i = \sigma Z_i + \mu \quad (i = 1, 2, \ldots, n),$$

where the Z_i are i.i.d.r.v. with distribution function $F_0 = \Phi$. Let the elements of the ordered random sample be denoted by $Z_i^{(n)}$ and $X_i^{(n)}$ ($i = 1, 2, \ldots, n$), the expectation of $Z_i^{(n)}$ by m_i, the column vectors with coordinates $Z_i^{(n)}, X_i^{(n)}, m_i$ by Z, X, m, and the covariance matrix of Z by V. The minimum variance linear estimate of σ, based on X, is

(5.8.4) $$\hat{\sigma}_n = \sum_{i=1}^n c_i X_i^{(n)} \equiv \frac{m^t V^{-1} X}{m^t V^{-1} m}, \quad \text{where } (c_1, \ldots, c_n) = \frac{m^t V^{-1}}{m^t V^{-1} m}.$$

We note that the latter is the same as the minimum variance linear estimator of σ when the mean μ is known. The Shapiro–Wilk (1965) statistic for testing H_0 of (5.8.2) with $F_0 = \Phi$ is

(5.8.5) $$W_n = \frac{\hat{\sigma}_n^2}{\sum_{i=1}^n (X_i - \bar{X}_n)^2}, \quad \text{with } \bar{X}_n = \frac{1}{n}\sum_{i=1}^n X_i.$$

In order to evaluate W_n, one needs to know the elements of V, which are evaluated only for sample sizes up to 20 (cf. Sarahan, Greenberg 1956). In their quoted paper Shapiro and Wilk developed approximations for calculating the elements of V up to sample sizes 50. They also tabulated the critical values of the distribution of W_n for $n = 1, 2, \ldots, 50$. Prompted by these tabulation difficulties and arguing heuristically that for large n V should be close to the $n \times n$ identity matrix I, Shapiro and Francia (1972) proposed to substitute V^{-1} in (5.8.4) by I for large samples and tabulated the critical values of their new statistic

$$(5.8.6) \qquad W'_n = \frac{\left(\sum_{i=1}^{n} b_i X_i^{(n)}\right)^2}{\sum_{i=1}^{n} (X_i - \overline{X}_n)^2}, \quad \text{with} \quad (b_1, \ldots, b_n) = \frac{m^t}{m^t m},$$

for $50 \leq n \leq 100$. (For a justification of replacing V^{-1} by I we refer to Ali and Chan (1964).) Now only the values of m remain to be known and they are given by Harter (1961) for sample sizes $2 \leq n \leq 100$ and for some specific n up to 400. W'_n is closely related to the statistic

$$(5.8.7) \qquad L'_n = \sum_{i=1}^{n} \left(\frac{X_i^{(n)} - \overline{X}_n}{S_n} - m_i\right)^2,$$

where, as before, $m_i = E(Z_i^{(n)})$, and $S_n^2 = \frac{1}{n} \sum_{i=1}^{n} (X_i - \overline{X}_n)^2$. In fact, the following relationship can be seen by elementary calculations:

$$(5.8.8) \quad L'_n = 2n^{1/2}(\sum m_i^2)^{1/2}(1 - (n \sum m_i^2)^{-1/2}(W'_n)^{1/2}) + (n^{1/2} - (\sum m_i^2)^{1/2})^2.$$

Now using the statistic L'_n, it is natural to reject the hypothesis H_0 of (5.8.2) with $F_0 = \Phi$ for large values of L'_n which is equivalent to rejecting the same H_0 for small values of W'_n resp. for those of W_n (cf. (5.8.8)).

DeWet and Venter (1972) proposed to substitute m_i of (5.8.7) by inv $\Phi\left(\frac{i}{n+1}\right)$, the approximate expectation of $Z_i^{(n)}$ under H_0, and introduced the statistic

$$(5.8.9) \qquad L_n = \sum_{i=1}^{n} \left(\frac{X_i^{(n)} - \overline{X}_n}{S_n} - \text{inv } \Phi\left(\frac{i}{n+1}\right)\right)^2 - a_n,$$

where a_n is a sequence of norming factors, defined in terms of $\varphi(\text{inv } \Phi(\cdot))$, whose approximate values they also tabulated. They also derived the asymptotic distribution of L_n with tables provided. The mentioned asymptotic distribution of L_n can be viewed as the first large sample theory for the Shapiro–Wilk W_n and the Shapiro–Francia W'_n tests for normality.

Considerations, similar to those which led us to introduce the statistic $M_n^0(\lambda)$ of (5.5.29) suggest that, for the family $\mathscr{F}$ of (5.8.1), we should consider the *estimated quantile process*

$$\text{(5.8.10)} \quad \hat{q}_n(y) = \hat{q}_n(y; \overline{X}_n, S_n)$$
$$= n^{1/2} f_0(\text{inv } F_0(y))((Q_n^0(y) - \overline{X}_n)/S_n - \text{inv } F_0(y)), \quad 0 < y < 1,$$

and the statistic

$$\text{(5.8.11)}$$
$$M_n(\lambda) = \sum_{k=1}^{n} \left\{ n^{-1} \left(\hat{q}_n\left(\frac{k}{n+1}\right) \right)^2 \bigg/ f_0\left(\text{inv } F_0\left(\frac{k}{n+1}\right)\right) \right\} \left(\text{inv } F_0\left(\frac{k}{n+1}\right) \right)^{\lambda-1}$$
$$= \sum_{k=1}^{n} f_0\left(\text{inv } F_0\left(\frac{k}{n+1}\right)\right) \left(\frac{X_k^{(n)} - \overline{X}_n}{S_n} - \text{inv } F_0\left(\frac{k}{n+1}\right) \right)^2 \left(\text{inv } F_0\left(\frac{k}{n+1}\right) \right)^{\lambda-1},$$
$$\lambda = 1, 2, \ldots,$$

for testing the composite goodness-of-fit null hypothesis H_0 of (5.8.2). An advantage of looking at $M_n(\lambda)$ instead of L_n is in that no normalizing factor like a_n is needed for the asymptotic distribution of $M_n(\lambda)$ to exist. Hence, the calculation of $M_n(\lambda)$ is easier than that of L_n. We also note that, in the definition of $M_n(\lambda)$, F_0 is not assumed to be the unit normal distribution.

The main result of this section is

Theorem 5.8.1 (Csörgő, Révész 1980). *Let $X_1, X_2, \ldots, X_n$ be a random sample with a distribution function $F \in \mathscr{F}$ of (5.8.1). Assume that F_0 also satisfies conditions (4.5.10) and (4.5.12) of Theorem 4.5.6. Further assume (5.5.31) with $r > 2(\lambda+1)$ and*

$$\text{(5.8.12)} \quad \inf_{0 < y < 1} \frac{|f_0'(\text{inv } F_0(y))|}{y^\delta} > 0, \quad \inf_{0 < y < 1} \frac{|f_0'(\text{inv } F_0(y))|}{(1-y)^\delta} > 0$$

for some $1 \leq \delta < 3/2$,

$$\text{(5.8.13)} \quad \int_{-\infty}^{+\infty} f_0^3(x)\, dx < +\infty.$$

Then, there exists a sequence of Brownian bridges $\{B_n\}$ such that for $\lambda = 1, 2, \ldots$ we have

$$\text{(5.8.14)} \quad |M_n(\lambda) - G_n(\lambda)| \xrightarrow{P} 0,$$

where

(5.8.15) $G_n(\lambda) = \int_0^1 B_n^2(y) \lambda^{-1} d(\text{inv } F_0(y))^\lambda + (\mathcal{T}_n^0)^2 J^{(\lambda-1)} + (\mathcal{T}_n^1)^2 J^{(\lambda+1)}$

$$- 2(\mathcal{T}_n^0 \mathcal{R}_n^{(\lambda-1)} + \mathcal{T}_n^1 \mathcal{R}_n^{(\lambda)} - \mathcal{T}_n^0 \mathcal{T}_n^1 J^{(\lambda)}),$$

and where

(5.8.16) $\mathcal{T}_n^0 = \int_0^1 B_n(y) \, d(\text{inv } F_0(y)), \quad \mathcal{T}_n^1 = \int_0^1 B_n(y) 2^{-1} d(\text{inv } F_0(y))^2,$

(5.8.17) $J^{(\alpha)} = \int_0^1 f_0(\text{inv } F_0(y))(\text{inv } F_0(y))^\alpha \, dy = \int_{-\infty}^{+\infty} f_0^2(x) x^\alpha \, dx,$
$$\alpha = \lambda-1, \lambda, \lambda+1,$$

(5.8.18) $\mathcal{R}_n^{(\alpha)} = \int_0^1 B_n(y)(\text{inv } F_0(y))^\alpha \, dy, \quad \alpha = \lambda-1, \lambda.$

Remark 5.8.1. We note that by Lemma 5.5.5 and (5.5.37), all the random integrals occurring in the definition of $G_n(\lambda)$ exist on assuming (5.5.31) with $r > 2(\lambda+1)$ and (5.8.13) implies the existence of the non-random ones.

The proof of this theorem is based on a number of preliminary results.

For the sake of further reference we repeat here our zero mean, variance one initial conditions on F_0 of $\mathcal{F}$ of (5.8.1) as:

(5.8.19) $\int_0^1 \text{inv } F_0(y) \, dy = \int_{-\infty}^{+\infty} x \, dF_0(x) = 0,$

(5.8.20) $\int_0^1 (\text{inv } F_0(y))^2 \, dy = \int_{-\infty}^{+\infty} x^2 \, dF_0(x) = 1.$

Lemma 5.8.1. *Given (5.8.19) and (5.5.31) with $r > 2\lambda$, we have*

(5.8.21) $\left| n^{-1} \sum_{k=1}^n \text{inv } F_0\left(\frac{k}{n+1}\right) \right| = o(n^{-1/2}).$

Proof.

$$\left| n^{-1} \sum_{k=1}^n \text{inv } F_0\left(\frac{k}{n+1}\right) \right| = \left| n^{-1} \sum_{k=1}^n \text{inv } F_0\left(\frac{k}{n+1}\right) - \int_0^1 \text{inv } F_0(y) \, dy \right|$$

$$\leq \int_0^{1/n} |\text{inv } F_0(y)| \, dy + \int_{1-1/n}^1 \text{inv } F_0(y) \, dy = o(n^{-1/2}).$$

Corollary 5.8.1. If, in addition to (5.8.19) and (5.8.20), $F_0'=f_0\neq 0$ on (a, b), where a and b are as in Theorem 4.5.6, then

$$(5.8.22) \quad \frac{\bar{X}_n-\mu}{\sigma} = n^{-1}\sum_{k=1}^{n}\frac{X_k^{(n)}-\mu}{\sigma}$$

$$= n^{-1}\sum_{k=1}^{n}\left(\frac{X_k^{(n)}-\mu}{\sigma} - \text{inv } F_0\left(\frac{k}{n+1}\right)\right) + o(n^{-1/2})$$

$$= n^{-1}\sum_{k=1}^{n}\frac{q_n^0\left(\frac{k}{n+1}\right)}{n^{1/2}f_0\left(\text{inv } F_0\left(\frac{k}{n+1}\right)\right)} + o(n^{-1/2})$$

$$= n^{-1/2}I_n^0 + o(n^{-1/2}),$$

where $I_n^0 = n^{-1}\sum_{k=1}^{n}\dfrac{q_n^0\left(\frac{k}{n+1}\right)}{f_0\left(\text{inv } F_0\left(\frac{k}{n+1}\right)\right)}$, and q_n^0 is as in (5.5.28) with $F=F_0$.

Lemma 5.8.2. Given (5.8.20) and (5.5.31) with $r>2(\lambda+1)$, we have

$$(5.8.23) \quad \left|n^{-1}\sum_{k=1}^{n}\left(\text{inv } F_0\left(\frac{k}{n+1}\right)\right)^2 - 1\right| = o(n^{-1/2}).$$

Proof. By (5.8.20) and an argument similar to that of (5.8.21).

Lemma 5.8.3. Assume that F_0 of (5.8.1) satisfies condition (4.5.10) of Theorem 4.5.6. Assume also (5.8.12). Then

$$(5.8.24) \quad n^{-1/2}\sum_{k=1}^{n}\frac{\left(q_n^0\left(\frac{k}{n+1}\right)\right)^2}{nf_0^2\left(\text{inv } F_0\left(\frac{k}{n+1}\right)\right)} \xrightarrow{P} 0.$$

Proof. By Markov's inequality it suffices to show that

$$(5.8.25) \quad En^{-1/2}\sum_{k=1}^{n}\frac{\left(q_n^0\left(\frac{k}{n+1}\right)\right)^2}{nf_0^2\left(\text{inv } F_0\left(\frac{k}{n+1}\right)\right)} \approx n^{-1/2}\sum_{k=1}^{n}\frac{\frac{k}{n+1}\left(1-\frac{k}{n+1}\right)}{nf_0^2\left(\text{inv } F_0\left(\frac{k}{n+1}\right)\right)}$$

$$\approx n^{-1/2}\int_{1/n}^{1-1/n}\frac{y(1-y)}{f_0^2(\text{inv } F_0(y))}\,dy = o(1).$$

Now by condition (4.5.10) and (5.8.12) we have

$$(5.8.26) \quad \int_{1/n}^{1-1/n} \frac{y(1-y)}{f_0^2(\text{inv } F_0(y))} dy \leq \gamma \int_{1/n}^{1-1/n} \frac{1}{|f_0'(\text{inv } F_0(y))|} dy$$

$$\leq \gamma \int_{1/n}^{1-1/n} \frac{1}{(1-y)^\delta} dy + \gamma \int_{1/n}^{1-1/n} \frac{1}{y^\delta} dy$$

$$= \begin{cases} O(\log n) & \text{if } \delta = 1 \\ O(n^{\delta-1}) & \text{if } 1 < \delta < 3/2, \end{cases}$$

which, in turn, implies (5.8.25), and Lemma 5.8.3 is proved.

Lemma 5.8.4. *If, in addition to* (5.8.19), (5.8.20) *and* (5.5.31) *with* $r > 2(\lambda+1)$, F_0 *also satisfies the conditions of Lemma 5.8.3, then*

$$(5.8.27) \quad \frac{S_n - \sigma}{\sigma} = n^{-1/2} I_n^1 + o_P(n^{-1/2}),$$

where

$$I_n^1 = n^{-1} \sum_{k=1}^{n} \frac{q_n^0\left(\frac{k}{n+1}\right)}{f_0\left(\text{inv } F_0\left(\frac{k}{n+1}\right)\right)} \left(\text{inv } F_0\left(\frac{k}{n+1}\right)\right).$$

Proof. By Corollary 5.8.1, (5.8.23) and (5.8.24), we have

$$\frac{S_n - \sigma}{\sigma} = \left\{\frac{1}{n} \sum_{k=1}^{n} \left(\frac{X_k^{(n)} - \mu}{\sigma}\right)^2 - \left(\frac{\overline{X}_n - \mu}{\sigma}\right)^2\right\}^{1/2} - 1$$

$$= \left\{\frac{1}{n} \sum_{k=1}^{n} \left(\frac{q_n^0\left(\frac{k}{n+1}\right)}{n^{1/2} f_0\left(\text{inv } F_0\left(\frac{k}{n+1}\right)\right)} + \text{inv } F_0\left(\frac{k}{n+1}\right)\right)^2 \right.$$

$$\left. - (n^{-1/2} I_n^0 + o(n^{-1/2}))^2\right\}^{1/2} - 1$$

$$= \left\{\frac{1}{n^2} \sum_{k=1}^{n} \frac{\left(q_n^0\left(\frac{k}{n+1}\right)\right)^2}{f_0^2\left(\text{inv } F_0\left(\frac{k}{n+1}\right)\right)} + 1 + o(n^{-1/2}) \right.$$

$$\left. + \frac{2}{n^{3/2}} \sum_{k=1}^{n} \frac{q_n^0\left(\frac{k}{n+1}\right)\left(\text{inv } F_0\left(\frac{k}{n+1}\right)\right)}{f_0\left(\text{inv } F_0\left(\frac{k}{n+1}\right)\right)} - \frac{(I_n^0)^2}{n} + o_P(n^{-1/2})\right\}^{1/2} - 1$$

$$= (1 + 2n^{-1/2} I_n^1 + o_P(n^{-1/2}))^{1/2} - 1$$

$$= n^{-1/2} I_n^1 + o_P(n^{-1/2}),$$

where $o_p(n^{-1/2})$ of the third equality above results from observing that, by Corollary 5.8.1, $I_n^0 n^{-1} = o_P(n^{-1/2})$.

Corollary 5.8.2. *If the conditions of Lemma 5.8.4 are satisfied, then the estimated quantile process $\hat{q}_n\left(\frac{k}{n+1}\right)$ (cf. (5.8.10)) can be written in terms of the process $q_n^0\left(\frac{k}{n+1}\right)$ as follows:*

$$(5.8.28) \quad \hat{q}_n\left(\frac{k}{n+1}\right) = \hat{q}_n\left(\frac{k}{n+1}; \bar{X}_n, S_n\right)$$

$$= \frac{1}{1+o_P(1)}\left\{q_n^0\left(\frac{k}{n+1}\right) - I_n^0 f_0\left(\operatorname{inv} F_0\left(\frac{k}{n+1}\right)\right)\right.$$

$$\left. - I_n^1 f_0\left(\operatorname{inv} F_0\left(\frac{k}{n+1}\right)\right)\left(\operatorname{inv} F_0\left(\frac{k}{n+1}\right)\right)\right\} + o_P(1) f_0\left(\operatorname{inv} F_0\left(\frac{k}{n+1}\right)\right)$$

$$+ o_P(1) f_0\left(\operatorname{inv} F_0\left(\frac{k}{n+1}\right)\right)\left(\operatorname{inv} F_0\left(\frac{k}{n+1}\right)\right).$$

Proof. By Corollary 5.8.1 and Lemma 5.8.4 we have

$$\hat{q}_n\left(\frac{k}{n+1}\right) = n^{1/2} f_0\left(\operatorname{inv} F_0\left(\frac{k}{n+1}\right)\right)\left\{\frac{X_k^{(n)} - \bar{X}_n}{S_n} - \operatorname{inv} F_0\left(\frac{k}{n+1}\right)\right\}$$

$$= n^{1/2} f_0\left(\operatorname{inv} F_0\left(\frac{k}{n+1}\right)\right) \frac{\sigma}{S_n}\left\{\frac{X_k^{(n)} - \mu}{\sigma} - \operatorname{inv} F_0\left(\frac{k}{n+1}\right) - \frac{\bar{X}_n - \mu}{\sigma}\right.$$

$$\left. - \left(\frac{S_n}{\sigma} - 1\right)\left(\operatorname{inv} F_0\left(\frac{k}{n+1}\right)\right)\right\}$$

$$= \frac{1}{1 + n^{-1/2} I_n^1 + o_P(n^{-1/2})}\left\{q_n^0\left(\frac{k}{n+1}\right) - (I_n^0 + o_P(1)) f_0\left(\operatorname{inv} F_0\left(\frac{k}{n+1}\right)\right)\right.$$

$$\left. - (I_n^1 + o_P(1)) f_0\left(\operatorname{inv} F_0\left(\frac{k}{n+1}\right)\right)\left(\operatorname{inv} F_0\left(\frac{k}{n+1}\right)\right)\right\},$$

and the statement of (5.8.28) is proved upon observing also that, by Corollary 5.8.1, $n^{-1/2} I_n^1 = o_P(1)$.

Proof of Theorem 5.8.1. In the light of Corollary 5.8.2 $M_n(\lambda)$ can be written as

(5.8.29)
$$M_n(\lambda) = \sum_{k=1}^{n}\left\{n^{-1}\left(\hat{q}_n\left(\frac{k}{n+1}\right)\right)^2 \Big/ f_0\left(\operatorname{inv} F_0\left(\frac{k}{n+1}\right)\right)\right\}\left(\operatorname{inv} F_0\left(\frac{k}{n+1}\right)\right)^{\lambda-1}$$

$$= \frac{1}{1+o_P(1)}\{M_n^0(\lambda) + (I_n^0)^2 J_n^{(\lambda-1)} + (I_n^1)^2 J_n^{(\lambda+1)}$$
$$- 2(I_n^0 R_n^{(\lambda-1)} + I_n^1 R_n^{(\lambda)} - I_n^0 I_n^1 J_n^{(\lambda)})\}$$
$$+ o_P(1)\{J_n^{(\lambda-1)} + J_n^{(\lambda+1)}\} + 2o_P(1)\{R_n^{(\lambda-1)} - I_n^0 J_n^{(\lambda-1)}$$
$$- I_n^1 J_n^{(\lambda)} + R_n^{(\lambda)} - I_n^0 J_n^{(\lambda)} - I_n^1 J_n^{(\lambda+1)}\}, \quad (\lambda = 1, 2, \ldots),$$

where $M_n^0(\lambda)$, I_n^0 and I_n^1 are respectively defined in (5.5.29), Corollary 5.8.1 and Lemma 5.8.4, and

(5.8.30)
$$J_n^{(\alpha)} = n^{-1}\sum_{k=1}^{n} f_0\left(\operatorname{inv} F_0\left(\frac{k}{n+1}\right)\right)\left(\operatorname{inv} F_0\left(\frac{k}{n+1}\right)\right)^{\alpha}, \quad \alpha = \lambda-1, \lambda, \lambda+1,$$

(5.8.31) $\quad R_n^{(\alpha)} = n^{-1}\sum_{k=1}^{n} q_n^0\left(\frac{k}{n+1}\right)\left(\operatorname{inv} F_0\left(\frac{k}{n+1}\right)\right)^{\alpha}, \quad \alpha = \lambda-1, \lambda.$

We have already remarked that the integral $J^{(\alpha)}$ of (5.8.17) exists, and hence

(5.8.32) $\quad J_n^{(\alpha)} \to J^{(\alpha)}$, as $n \to \infty$, $(\alpha = \lambda-1, \lambda, \lambda+1)$.

By (5.5.39) and (5.8.18) we also have

(5.8.33) $\quad |R_n^{(\alpha)} - \mathcal{R}_n^{(\alpha)}| \xrightarrow{P} 0$, as $n \to \infty$, $(\alpha = \lambda-1, \lambda)$.

Hence, in (5.8.29) the terms $R_n^{(\alpha)}$ ($\alpha = \lambda-1, \lambda$) have limit distributions and the terms $J_n^{(\alpha)}$ ($\alpha = \lambda-1, \lambda, \lambda+1$) have finite limits. Also, by Corollary 5.8.1, resp. by Lemma 5.8.4, I_n^0 resp. I_n^1 converges in distribution to a $\mathcal{N}(0, 1)$ r.v. resp. to a $\mathcal{N}\left(0, \frac{1}{4}EX_1^4 - \frac{1}{4}\right)$ r.v. These facts together imply that the term

$$o_P(1)\{J_n^{(\lambda-1)} + J_n^{(\lambda+1)}\} +$$
$$+ 2o_P(1)\{R_n^{(\lambda-1)} - I_n^0 J_n^{(\lambda-1)} - I_n^1 J_n^{(\lambda)} + R_n^{(\lambda)} - I_n^0 J_n^{(\lambda)} - I_n^1 J_n^{(\lambda+1)}\}$$

of (5.8.29) goes to zero in probability for $\lambda = 1, 2, \ldots$. Thus, by Theorem 5.5.3, (5.5.40) with $\delta = 0$ and 1, and by (5.8.32) and (5.8.33) we get our statement (5.8.14).

Theorem 5.8.1 now, in principle, gives the asymptotic distribution of $M_n(\lambda)$. Let

(5.8.34) $\quad G(\lambda) = \int_0^1 B^2(y)\lambda^{-1} d(\operatorname{inv} F_0(y))^\lambda + (\mathcal{T}^0)^2 J^{(\lambda-1)} + (\mathcal{T}^1)^2 J^{(\lambda+1)}$
$$- 2(\mathcal{T}^0 \mathcal{R}^{(\lambda-1)} + \mathcal{T}^1 \mathcal{R}^{(\lambda)} - \mathcal{T}^0 \mathcal{T}^1 J^{(\lambda)})$$

where $\lambda = 1, 2, \ldots$ and

(5.8.35) $\quad \mathcal{T}^0 = \int_0^1 B(y) \, d(\operatorname{inv} F_0(y)), \quad \mathcal{T}^1 = \int_0^1 B(y) 2^{-1} d(\operatorname{inv} F_0(y))^2,$

(5.8.36) $\quad \mathcal{R}^{(\alpha)} = \int_0^1 B(y)(\operatorname{inv} F_0(y))^\alpha \, dy, \quad \alpha = \lambda - 1, \lambda.$

Since, for every n, $G_n(\lambda) \xrightarrow{\mathcal{D}} G(\lambda)$, ($\lambda = 1, 2, \ldots$), we have

Corollary 5.8.3. *The conditions of Theorem 5.8.1 imply*

(5.8.37) $\quad\quad\quad\quad M_n(\lambda) \xrightarrow{\mathcal{D}} G(\lambda), \quad \lambda = 1, 2, \ldots.$

In the case when the parameters μ and σ are known then, naturally, we use the statistic $M_n^0(\lambda)$ of Theorem 5.5.3 instead of $M_n(\lambda)$ of Theorem 5.8.1, and the proof of the latter reduces to that of the former. In fact in the μ and σ specified case, $\mathcal{T}_n^0$ and $\mathcal{T}_n^1$ resp. $\mathcal{T}^0$ and $\mathcal{T}^1$ should be simply replaced by zero in (5.8.15) resp. in (5.8.34).

We note that the distribution of $G(\lambda)$ does not depend on the unknown parameters μ and σ of $\mathcal{F}$, i.e., the nuisance parameters of H_0 of (5.8.2) are now eliminated. This means that via Corollary 5.8.3 we have a possibility of testing for H_0, provided we can evaluate the distribution of $G(\lambda)$ for a given F_0. However this task will not be simple in general. We observe, however, that the distribution of $G(\lambda)$ is somewhat simpler when F_0 is symmetric around zero. In the latter case $J^{(\lambda)} = 0$ if λ is an odd integer, and $J^{(\lambda-1)} = J^{(\lambda+1)} = 0$ if λ is an even integer; i.e.,

(5.8.38) $\quad G(\lambda) = \int_0^1 B^2(y)\lambda^{-1} d(\operatorname{inv} F_0(y))^\lambda + (\mathcal{T}^0)^2 J^{(\lambda-1)} + (\mathcal{T}^1)^2 J^{(\lambda+1)}$
$$- 2(\mathcal{T}^0 \mathcal{R}^{(\lambda-1)} + \mathcal{T}^1 \mathcal{R}^{(\lambda)}),$$

if λ is an odd integer and F_0 is symmetric around zero, while

(5.8.39)
$$G(\lambda) = \int_0^1 B^2(y)\lambda^{-1} d(\operatorname{inv} F_0(y))^\lambda - 2(\mathcal{T}^0 \mathcal{R}^{(\lambda-1)} + \mathcal{T}^1 \mathcal{R}^{(\lambda)} - \mathcal{T}^0 \mathcal{T}^1 J^{(\lambda)}),$$

if λ is an even integer and F_0 is symmetric around zero.

Further simplifications are gained if one of the parameters μ and σ are assumed to be known. If σ is assumed to be known, then $\mathcal{T}^1$ in (5.8.34) should be replaced by zero and then (5.8.37) reads

$$(5.8.40) \quad M_n(\lambda) \xrightarrow{\mathcal{D}} \int_0^1 B^2(y)\lambda^{-1} d(\operatorname{inv} F_0(y))^\lambda + (\mathcal{T}^0)^2 J^{(\lambda-1)} - 2\mathcal{T}^0 \mathcal{R}^{(\lambda-1)},$$

while, if μ is assumed to be known, then $\mathcal{T}^0$ should be taken to be zero, and then (5.8.37) results in

$$(5.8.41) \quad M_n(\lambda) \xrightarrow{\mathcal{D}} \int_0^1 B^2(y)\lambda^{-1} d(\operatorname{inv} F_0(y))^\lambda + (\mathcal{T}^1)^2 J^{(\lambda+1)} - 2\mathcal{T}^1 \mathcal{R}^{(\lambda)}.$$

Combining now (5.8.39) with (5.8.40) and (5.8.41), we get

$$(5.8.42) \quad M_n(\lambda) \xrightarrow{\mathcal{D}} \int_0^1 B^2(y)\lambda^{-1} d(\operatorname{inv} F_0(y))^\lambda - 2\mathcal{T}^0 \mathcal{R}^{(\lambda-1)},$$

if λ is an even integer, F_0 is symmetric around zero and σ is assumed to be known, and

$$(5.8.43) \quad M_n(\lambda) \xrightarrow{\mathcal{D}} \int_0^1 B^2(y)\lambda^{-1} d(\operatorname{inv} F_0(y))^\lambda - 2\mathcal{T}^1 \mathcal{R}^{(\lambda)},$$

if λ is an even integer, F_0 is symmetric around zero and μ is assumed to be known.

However, it is of no use to combine (5.8.38) with (5.8.40) and (5.8.41), for then we get again (5.8.40) and (5.8.41) only.

Going back to the definition of $M_n(\lambda)$ (cf. (5.8.11)), we observe that it is a positive r.v. when λ is an odd integer and it is a real valued r.v. when λ is an even integer (the same observation holds for $M_n^0(\lambda)$ of (5.5.29)). Hence, if λ is an odd integer, we should reject the null hypothesis H_0 of (5.8.2) when $M_n(\lambda)$ is too large and, if λ is an even integer, we should reject H_0 when $M_n(\lambda)$ is too large or too small. We also observe that, if λ is an odd integer, and if H_0 is not true then $M_n(\lambda) \xrightarrow{\text{a.s.}} \infty$ as $n \to \infty$. This means that the proposed test is *consistent* against any alternative hypothesis when λ is an odd integer. On the other hand, the proposed test is not necessarily consistent against all alternatives when λ is an even integer. For example, if F_0 is symmetric around zero then the proposed test is definitely not consistent against symmetric alternatives in the case of λ even.

5.9. On testing for exponentiality

In the definition of the estimated quantile process $\hat{q}_n$ of (5.8.10) the sample mean $\bar{X}_n$ and the sample standard deviation S_n were used to estimate μ and σ respectively. Any other method of estimation for μ and σ, like for example maximum likelihood estimators, could have been used. Indeed, it is desirable to work out an analogue of Theorem 5.8.1 in a more general context, like that of Section 5.7, for example. Instead of taking this general route, however, in this section we are going to concentrate only on testing for exponentiality.

We consider the family of exponential density functions

$$(5.9.1) \quad \mathrm{Exp}\,(A, B) = \{f(x; A, B): f(x; A, B) = B^{-1} f_0\left(\frac{x-A}{B}\right)$$
$$= B^{-1} \exp\,(-(x-A)/B),\ x > A \in R^1,\ B > 0\}.$$

First we observe that $f_0(x) = e^{-x}$, $x > 0$ is not like F_0 of (5.8.1) in the sense that here we have $\int_0^\infty x f_0(x)\,dx = \int_0^1 (x-1)^2 f_0(x)\,dx = 1$ instead of mean zero and variance one. However, the methodology of our preceeding section is applicable.

Let $X_1, X_2, \ldots, X_n$ ($n \geq 2$) be i.i.d.r.v. with a density function f and, on the basis of this random sample, we wish to test the following composite null hypotheses

$$(5.9.2) \quad H_0^{(1)}: f \in \mathrm{Exp}\,(0, B),\ B > 0,$$

and

$$(5.9.3) \quad H_0^{(2)}: f \in \mathrm{Exp}\,(A, B),\ A \in R^1,\ B > 0.$$

First we deal with $H_0^{(1)}$ of (5.9.2). In this case $X_1, X_2, \ldots, X_n$ are assumed to be independent positive random variables and, given $H_0^{(1)}$, $\bar{X}_n$ is the maximum likelihood estimator of B of $\mathrm{Exp}\,(0, B)$. It is natural then to define the *estimated quantile process of the family* $\mathrm{Exp}\,(0, B)$ by

$$(5.9.4)$$
$$\hat{q}_n\left(\frac{k}{n+1}\right) = n^{1/2} \left(\frac{X_k^{(n)}}{\bar{X}_n} - \mathrm{inv}\,F_0\left(\frac{k}{n+1}\right)\right) f_0\left(\mathrm{inv}\,F_0\left(\frac{k}{n+1}\right)\right),\ k = 1, 2, \ldots, n,$$

where, given $H_0^{(1)}$ of (5.9.2), $X_k^{(n)}$ ($1 \leq k \leq n$) are the order statistics of the independent $\mathrm{Exp}\,(0, B)$ r.v. X_k ($0 \leq k \leq n$).

We note that the above estimated quantile process of (5.9.4) is different from that of (5.8.10). Also $F_0(x)$ of (5.9.2) is $F_0(x) = 1 - e^{-x}$, $x \geq 0$, and hence $\operatorname{inv} F_0\left(\frac{k}{n+1}\right) = \log\left[1 \big/ \left(1 - \frac{k}{n+1}\right)\right]$ and $f_0\left(\operatorname{inv} F_0\left(\frac{k}{n+1}\right)\right) = 1 - \frac{k}{n+1}$ $(1 \leq k \leq n)$.

Now the process q_n^0 of the family $\operatorname{Exp}(0, B)$ is

(5.9.5)
$$q_n^0\left(\frac{k}{n+1}\right) = n^{1/2} f_0\left(\operatorname{inv} F_0\left(\frac{k}{n+1}\right)\right)\left(\frac{X_k^{(n)}}{B} - \operatorname{inv} F_0\left(\frac{k}{n+1}\right)\right), \quad 1 \leq k \leq n,$$

and our Theorem 4.5.7 holds for the latter with $\gamma = 1$ in (4.5.10). Also, since $\int_0^\infty x\, dF_0(x) = 1$, instead of Lemma 5.8.1 we now have

(5.9.6) $\left| n^{-1} \sum_{k=1}^n \operatorname{inv} F_0\left(\frac{k}{n+1}\right) - 1 \right| = o(n^{-1/2}),$

and hence

(5.9.7)
$$\frac{\bar{X}_n}{B} = n^{-1} \sum_{k=1}^n \frac{X_k^{(n)}}{B} = n^{-1} \sum_{k=1}^n \left(\frac{X_k^{(n)}}{B} - \operatorname{inv} F_0\left(\frac{k}{n+1}\right)\right) + n^{-1} \sum_{k=1}^n \operatorname{inv} F_0\left(\frac{k}{n+1}\right)$$
$$= n^{-1/2} I_n^0 + 1 + o(n^{-1/2}),$$

where I_n^0 is as in (5.8.22), but now defined with q_n^0 of (5.9.5).

Consequently we have (cf. (5.9.4), (5.9.5) and (5.9.7))

(5.9.8) $\hat{q}_n\left(\frac{k}{n+1}\right) = n^{1/2} \frac{B}{\bar{X}_n} \left\{ \frac{X_k^{(n)}}{B} - \operatorname{inv} F_0\left(\frac{k}{n+1}\right) - \left(\frac{\bar{X}_n}{B} - 1\right) \cdot \right.$
$$\left. \cdot \left(\operatorname{inv} F_0\left(\frac{k}{n+1}\right)\right)\right\} f_0\left(\operatorname{inv} F_0\left(\frac{k}{n+1}\right)\right)$$
$$= \frac{1}{1 + o_P(1)} \left\{ q_n^0\left(\frac{k}{n+1}\right) - I_n^0 f_0\left(\operatorname{inv} F_0\left(\frac{k}{n+1}\right)\right)\left(\operatorname{inv} F_0\left(\frac{k}{n+1}\right)\right)\right\}$$
$$+ o_P(1) f_0\left(\operatorname{inv} F_0\left(\frac{k}{n+1}\right)\right)\left(\operatorname{inv} F_0\left(\frac{k}{n+1}\right)\right),$$

where $\frac{B}{\bar{X}_n} = \frac{1}{1 + o_P(1)}$ via (5.9.7).

As to the appropriate analogue of the statistic $M_n(\lambda)$ of (5.8.11) with $\hat{q}_n$ as in (5.9.8), for $\lambda = 1, 2, \ldots$ we have

$$(5.9.9) \quad M_n(\lambda) = \sum_{k=1}^{n} \left(\frac{X_k^{(n)}}{\bar{X}_n} - \log \frac{1}{1 - \frac{k}{n+1}} \right)^2 \left(1 - \frac{k}{n+1}\right) \left(\log \frac{1}{1 - \frac{k}{n+1}} \right)^{\lambda - 1}$$

$$= \frac{1}{1 + o_P(1)} \{ M_n^0(\lambda) + (I_n^0)^2 J_n^{(\lambda+1)} - 2I_n^0 R_n^{(\lambda)} \} + o_P(1) J_n^{(\lambda+1)} + 2 o_P(1) \cdot$$
$$\cdot (R_n^{(\lambda)} - I_n^0 J_n^{(\lambda+1)}),$$

where $M_n^0(\lambda)$ is as in (5.5.29), but now defined in terms of q_n^0 of (5.9.5), $J_n^{(\lambda+1)}$ is as in (5.8.30) and $R_n^{(\lambda)}$ is as in (5.8.31) with q_n^0 of (5.9.5).

Consequently, for $M_n(\lambda)$ of (5.9.9) we have the following (5.8.14) type statement

(5.9.10)
$$\left| M_n(\lambda) - \left\{ \int_0^1 B_n^2(y) \lambda^{-1} d \left(\text{inv } F_0(y) \right)^{\lambda} + (\mathcal{T}_n^0)^2 J^{(\lambda+1)} - 2 \mathcal{T}_n^0 \mathcal{R}_n^{(\lambda)} \right\} \right| = o_P(1),$$
$$\lambda = 1, 2, \ldots,$$

where $\mathcal{T}_n^0$ is as in (5.8.16), $J^{(\lambda+1)}$ is as in (5.8.17) and $\mathcal{R}_n^{(\lambda)}$ is as in (5.8.18), all of them in terms of $F_0(x) = 1 - e^{-x}$, $x \geq 0$.

Whence we have

Theorem 5.9.1. *Given $H_0^{(1)}$ of (5.9.2), and $M_n(\lambda)$ as in (5.9.9), $\mathcal{T}^0$ resp. $\mathcal{R}^{(\lambda)}$ as in (5.8.35) resp. in (5.8.36),*

$$(5.9.10) \quad M_n(\lambda) \xrightarrow{\mathscr{D}} \left\{ \int_0^1 B^2(y) \lambda^{-1} d \left(\text{inv } F_0(y) \right)^{\lambda} + (\mathcal{T}^0)^2 J^{(\lambda+1)} - 2 \mathcal{T}^0 \mathcal{R}^{(\lambda)} \right.$$

$$= \int_0^1 \frac{B^2(y)}{1-y} \left(\log \frac{1}{1-y} \right)^{\lambda-1} dy + \left(\int_0^1 \frac{B(y)}{1-y} dy \right)^2 \int_0^1 (1-y) \left(\log \frac{1}{1-y} \right)^{\lambda+1} dy$$

$$- 2 \int_0^1 \frac{B(y)}{1-y} dy \int_0^1 B(y) \left(\log \frac{1}{1-y} \right)^{\lambda} dy, \quad \lambda = 1, 2, \ldots.$$

If $\lambda = 1$, then (5.9.10) implies

$$(5.9.11) \quad M_n(1) \xrightarrow{\mathscr{D}} \int_0^1 \frac{B^2(y)}{1-y} dy + \frac{1}{4} \left(\int_0^1 \frac{B(y)}{1-y} dy \right)^2$$

$$- 2 \int_0^1 \frac{B(y)}{1-y} dy \int_0^1 B(y) \left(\log \frac{1}{1-y} \right) dy.$$

As to testing for $H_0^{(1)}$ of (5.9.2), we should reject $H_0^{(1)}$ when $M_n(\lambda)$ is too large, whatever the value of λ might be, since inv $F_0(y) = \log\frac{1}{1-y} > 0$ ($0 < y < 1$).

If we are to test $H_0^{(2)}$ of (5.9.3), we can make the following transformation: $X_1 - X_1^{(n)}$, $X_2 - X_1^{(n)}$, ..., $X_n - X_1^{(n)}$ and delete from it the term which equals to zero (with probability one there is only one such term). Let us denote the resulting variables by $Y_1, Y_2, ..., Y_{n-1}$ and let $Y_1^{(n-1)} < Y_2^{(n-1)} < ... < Y_{n-1}^{(n-1)}$ be their order statistics. Given $H_0^{(2)}$ of (5.9.3), $Y_1, Y_2, ..., Y_{n-1}$ are independent Exp $(0, B)$ r.v., and the above procedure thus modified, can now be used also to test for $H_0^{(2)}$ of (5.9.3).

We note in passing, that the very first test for exponentiality, which was proposed by Fisher (1929), can be viewed as a predecessor of our test statistic $M_n(\lambda)$ of (5.9.9). In order to test for $H_0^{(1)}$ of (5.9.2), Fisher (1929) proposed the statistic $X_n^{(n)}/n\bar{X}_n$, and tabulated its critical values.

Supplementary remarks

Section 5.1. An analogue of Theorem S.1.15.1 for the uniform empirical process is immediate. Define $\varkappa((y_1, y_2]) = n^{1/2}(\alpha_n(y_2) - \alpha_n(y_1))$. We have

Theorem S. 5.1.1. *Let $0 < \varepsilon_N < \frac{1}{2}$, $0 < a_N \leq N$, where a_N is a sequence of integers, such that ε_N and a_N/N are non-increasing and a_N is non-decreasing in N. Assume also that*

(S.5.1.1) $$a_N \varepsilon_N (\log N)^{-4} \geq C > 0.$$

Then

(S.5.1.2) $$\varlimsup_{N \to \infty} \sup_{1 \leq n \leq N - a_N} \sup_{0 \leq y \leq 1 - \varepsilon_N} \sup_{0 \leq s \leq \varepsilon_N} \beta_N |\varkappa((y, y+s], n+a_N) - \varkappa((y, y+s], n)|$$
$$= \varlimsup_{N \to \infty} \sup_{1 \leq n \leq N - a_N} \sup_{0 \leq y \leq 1 - \varepsilon_N} \beta_N |\varkappa((y, y+\varepsilon_N], n+a_N) - \varkappa((y, y+\varepsilon_N], n)| \stackrel{a.s.}{=} 1,$$

where $$\beta_N = \left(2 a_N \varepsilon_N (1 - \varepsilon_N)\left(\log\frac{N}{\varepsilon_N a_N} + \log\log N\right)\right)^{-1/2}.$$

If we also have

(S.5.1.3) $$\lim_{N \to \infty} \frac{\log\frac{1}{\varepsilon_N}\frac{N}{a_N}}{\log \log N} = \infty,$$

then $\varlimsup_{N \to \infty}$ in (S.5.1.2) can be replaced by $\lim_{N \to \infty}$.

This result can be viewed as an analogue to Theorem 3.1.1 and the Erdős–Rényi law (cf. Theorem 2.4.3). Clearly, one can formulate several other analogues of these Erdős–Rényi-type laws. We do not study this question here, and for further insight we refer to the paper of S. Csörgő (1979).

The following question, which can be considered as an analogue of Theorem 5.1.4, was studied in several papers: if $g(y)$ is a function on $[0, 1]$, then what are the limit properties of the sequence

$$\sup_{-\infty < x < +\infty} \frac{|F_n(x) - F(x)|}{g(F(x))} \text{ as } n \to \infty \text{ (cf., e.g., James 1975).}$$

Studying the process β_n over the whole real line Csáki (1975) proved also

(S.5.1.4) $$\varlimsup_{n \to \infty} \left(\sup_{0 < F(x) < 1} \frac{|\beta_n(x)|}{\sqrt{F(x)(1-F(x))}} \right)^{\frac{1}{\log \log n}} \stackrel{a.s.}{=} \sqrt{e},$$

provided F is continuous.

Section 5.3. The analogue of Theorem S.5.1.1 for quantile processes is straightforward.

Section 5.4. The herewith given proof of Corollary 5.4.3 differs slightly from that in the paper of S. Csörgő (1976). The present version is also due to him.

Section 5.5. Theorem 5.5.2 proposes a two-step estimation of $f(\operatorname{inv} F(y))$ via estimating f and $\operatorname{inv} F$ separately. In a recent paper (Csörgő, Révész 1980b) we proposed a direct, one-step estimation of $\dfrac{1}{f(\operatorname{inv} F(y))}$ for the sake of producing confidence band for $\operatorname{inv} F(y)$.

Section 5.7. The special case of Theorems 5.7.1 and 5.7.2 when $\theta \in \Theta \subseteq R^1$ was studied in Csörgő, Komlós, Major, Révész, Tusnády (1977). For some further elaboration on this problem under a sequence of alternatives we refer to Burke, Csörgő (1976).

Maximum likelihood estimators often satisfy (5.7.2) (i) with $\varepsilon_{1n} \xrightarrow{a.s.} 0$ or $\varepsilon_{1n} \xrightarrow{P} 0$ (cf., e.g., Ibragimov, Has'minskiĭ 1972, 1973a) and

$$l(x, \theta_0) = \nabla_\theta \log f(x; \theta_0) \cdot I^{-1}(\theta_0),$$

where f is the density function of F and $I^{-1}(\theta_0)$ is the inverse of Fisher's information matrix:

$$I(\theta_0) = E\left(\nabla_\theta \log f(x; \theta_0)\right)^t \cdot \left(\nabla_\theta \log f(x; \theta_0)\right).$$

For illuminating comments re. this matter (the familiar Cramér-type conditions) we refer to Section 4 in Durbin (1973a). In particular, we find ourselves in agreement with his suggestion that for any particular problem a maximum likelihood or other putative efficient estimator should be first constructed and then the validity of (5.7.2) (i) should be checked directly. It can be shown, however, that maximum likelihood estimators, under certain regularity conditions, have a sum representation with $\varepsilon_{1n} \stackrel{a.s.}{=} O(n^{-\varepsilon})$, for some $\varepsilon > 0$, in (5.7.2) (i), via extending the technique for θ one-dimensional of Ibragimov and Has'minskiĭ (1973b) to the θ multi-dimensional case (cf. Burke, Csörgő, Csörgő, Révész (1979)).

Durbin (1973b) points out that if θ is estimated by a sequence of maximum likelihood estimators based on a randomly chosen half of $X_1, X_2, ..., X_n$, then the resultant empirical process converges weakly to a Brownian bridge (cf. also Durbin 1976). This line of thinking was initiated by K. C. Rao (1972). Let

$$\bar{\beta}_n(x) = n^{1/2}[F_n(x) - F(x; \bar{\theta}_n)],$$

where $\{\bar{\theta}_n\}$ is a sequence of maximum likelihood estimators based on a randomly chosen half of $X_1, X_2, ..., X_n$ and F_n is based on the full sample. Then assuming that ε_{1n} of (5.7.2) (i) goes to zero in probability, resp. a.s., $\bar{\beta}_n(x)$ can be approximated in probability, resp. almost surely, by a Kiefer process $n^{-1/2} \bar{K}(F(x; \theta_0), n)$, uniformly in $x \in R^1$. Thus, $\bar{\beta}_n$ behaves asymptotically as if θ were completely specified. For a proof of this result, we refer to Csörgő, Komlós, Major, Révész, Tusnády (1977) and to Burke and Csörgő (1976). (Cf. also Theorem 4.2 in Burke, Csörgő, Csörgő, Révész 1979.)

6. A Study of Further Empirical Processes with the Help of Strong Approximation Methods

6.0. Introduction

Let $X_1, X_2, \ldots$ be a sequence of i.i.d.r.v. with distribution function $F(x)$, density function $f(x)$ and characteristic function $c(t)$. We saw in Chapters 4 and 5 that the empirical distribution function $F_n(x)$ of the sample $X_1, X_2, \ldots, X_n$ is a natural estimation of $F(x)$ and also that the quantile function $Q_n(y)$ can play a similarly natural role when estimating inv $F(y)$. In this chapter we intend to study some related problems concerning density and characteristic functions. A similar problem arises when a regression function $r(x) = E(Y_1 | X_1 = x)$ is to be estimated from a sequence $(X_1, Y_1), (X_2, Y_2), \ldots, (X_n, Y_n)$ of i.i.d.r.v.

In the above formulated three estimation problems a fundamental question is to find the natural estimators. This chapter is divided into four sections. The first two are on empirical densities, Section 6.3 is concerned with empirical regression, while Section 6.4 is on empirical characteristic functions.

6.1. Strong invariance principles and limit distributions for empirical densities

Let $X_1, X_2, \ldots, X_n$ be a sequence of i.i.d.r.v. with density function $f(x)$. Let $K_n(a, b)$ be the number of elements of the sample $X_1, X_2, \ldots, X_n$ lying in the interval (a, b). Then the probability $P(a < X_1 < b) = \int_a^b f(t)\, dt$ can be estimated by the relative frequency $n^{-1} K_n(a, b)$ (if n is big enough) and the value of a continuous $f(x)$ $(a < x < b)$ can be estimated by $(b-a)^{-1} \int_a^b f(t)\, dt$ (if the interval (a, b) is short enough), i.e.,

$$f(x) \sim (b-a)^{-1} \int_a^b f(t)\, dt \sim (n(b-a))^{-1} K_n(a, b) \quad (a < x < b).$$

(Here the sign $\sim$ does not stand for any precise mathematical statement; it only indicates an intuitive near equality.)

Hence an empirical density function $f_n(x)$, in the interval (a, b), can be defined as
$$f_n(x) = (n(b-a))^{-1} K_n(a, b) \quad (a < x < b).$$

Roughly speaking this is the idea behind most of the definitions of an empirical density function. When attempting a definition of this type one makes two errors. The first one occurs when the function $f(x)$ is estimated by its integral mean $(b-a)^{-1} \int_a^b f(t) dt$, and the second error, in turn, is made when the probability $\int_a^b f(t) dt$ is estimated by the relative frequency $n^{-1} K_n(a, b)$. The first error is small if the interval is short while the second one is small if the interval (a, b) contains a large enough number of elements of the sample, that is when (a, b) is not too short. Hence one of the main problems is how to find a good compromise between these two opposing tendencies.

Now we turn to some exact definitions of the empirical density function. In these definitions and also in the sequel of this chapter, $\{h_n\}$ is a decreasing sequence of real numbers tending to 0 and $\{l_n\}$ is an increasing sequence of integers tending to $+\infty$.

Definition 1 (Hystogram). Let
$$\ldots < x_{-1}(n) < x_0(n) < x_1(n) \ldots$$
be a partition of the real line such that
$$x_{i+1}(n) - x_i(n) = h_n \quad (i = 0, \pm 1, \pm 2, \ldots)$$
and define
$$f_n(x) = f_n^{(1)} = (nh_n)^{-1} K_n(x_i(n), x_{i+1}(n)) \quad \text{if} \quad x_i(n) < x < x_{i+1}(n).$$

The basic idea of the next definition is to collect those elements of the sample which are near to a fixed x.

Definition 2. Let
$$f_n(x) = f_n^{(2)}(x) = (nh_n)^{-1} K_n\left(x - \frac{h_n}{2}, x + \frac{h_n}{2}\right) \quad \text{for any } x.$$

This definition can be reformulated as follows: set
$$e(x) = \begin{cases} 1 & \text{if } -\tfrac{1}{2} < x < \tfrac{1}{2}, \\ 0 & \text{otherwise.} \end{cases}$$

Then clearly we have

(6.1.1) $$f_n^{(2)} = (nh_n)^{-1} \sum_{i=1}^{n} e((x-X_i)h_n^{-1}).$$

Rosenblatt (1956) (cf. also Parzen (1962)) proposed a generalization of (6.1.1) so that $e(x)$ should be replaced by an arbitrary density function λ. Whence the following:

Definition 3. Let $\lambda(x)$ be an arbitrary density function and define

$$f_n(x) = f_n^{(\lambda)} = (nh_n)^{-1} \sum_{i=1}^{n} \lambda((x-X_i)h_n^{-1}) = h_n^{-1} \int_{-\infty}^{+\infty} \lambda((x-y)h_n^{-1}) \, dF_n(y),$$

where F_n is the empirical distribution function based on the sample $X_1, X_2, ..., X_n$. For an optimal choice of λ we refer to Epanechnikov (1969).

Suppose that the density function f is vanishing outside an interval $-\infty \leq A < B \leq +\infty$ and is square-integrable inside. Let $\varphi = \{\varphi_k(x)\}_{k=1}^{\infty}$ be a complete orthonormal sequence defined on (A, B). Consider the Fourier expansion of f:

$$\sum_{k=1}^{\infty} c_k \varphi_k(x), \quad \left(c_k = \int_A^B \varphi_k(x) f(x) \, dx \right).$$

The next definition of an empirical density function is based on an estimation of the Fourier coefficients c_k. Since $c_k = \int_A^B \varphi_k(x) f(x) \, dx = E\varphi_k(X_1)$, we give our

Definition 4 (Čencov 1962, van Ryzin 1966, Schwartz 1967). Let

$$f_n(x) = f_n^{(\varphi)} = \sum_{k=1}^{l_n} \hat{c}_k \varphi_k(x) \quad (A < x < B)$$

where

$$\hat{c}_k = n^{-1} \sum_{j=1}^{n} \varphi_k(X_j).$$

Finally we give a very general definition which contains the previous ones as special cases.

Definition 5 (Földes, Révész 1974). Suppose that $f(x)$ is vanishing outside the interval $-\infty \leq C < D \leq +\infty$ and let $\psi = \{\psi_k(x, y)\}$ be a sequence of Borel-measurable functions defined on the square $(A, B)^2$ where

$(C, D) \subseteq (A, B)$. Then an empirical density function can be defined as follows:

$$f_n(x) = f_n^{(\psi)} = n^{-1} \sum_{k=1}^{n} \psi_n(x, X_k) = \int_A^B \psi_n(x, y) \, dF_n(y).$$

Choosing now

(6.1.2) $$\psi_n(x, y) = \sum_{j=1}^{l_n} \varphi_j(x) \varphi_j(y)$$

resp.

(6.1.3) $$\psi_n(x, y) = h_n^{-1} \lambda((x-y) h_n^{-1})$$

we get our Definition 4 resp. Definition 3. In order to get Definition 1 as a special case, simply let

(6.1.4) $$\psi_n(x, y) = h_n^{-1} \sum_{j=-\infty}^{+\infty} \alpha_j^{(n)}(x) \alpha_j^{(n)}(y),$$

where

$$\alpha_j^{(n)}(x) = \begin{cases} 1 & \text{if } x_{j-1}(n) \leq x < x_j(n), \\ 0 & \text{otherwise} \end{cases}$$

and $x_{j+1}(n) - x_j(n) = h_n$.

Now we say that the sequence $\psi = \{\psi_k(x, y)\}$ (defined on $(A, B)^2$) produces an asymptotically unbiased estimation of f if for $f_n = f_n^{(\psi)}$

(6.1.5) $$Ef_n = E \int_A^B \psi_n(x, y) \, dF_n(y) = \int_A^B \psi_n(x, y) \, dF(y) \to f(x)$$

for every $x \in (A, B)$. The estimator $f_n^{(\psi)}$ will be called uniformly asymptotically unbiased if the convergence in (6.1.5) is uniform in x. $f_n^{(\psi)}$ will be called a $\pi = \pi(n)$ estimation of f if

(6.1.6) $$\sup_{A \leq x \leq B} |Ef_n(x) - f(x)| = \pi(n) = \pi(n; A, B).$$

Suppose that $f_n^{(\psi)}$ is a π estimator of f. Then, applying Theorem 4.4.1 resp. Theorem 4.4.3 and assuming that the functions $\psi_k(x, y)$ have uniformly bounded variations, say, $\sup_x \text{var}_y \psi_k(x, y) = V_k$, and

(6.1.7) $$\lim_{y \searrow A} |\psi_n(x, y)| \sqrt{F(y) \log \log \frac{1}{F(y)}}$$

$$= \lim_{y \nearrow B} |\psi_n(x, y)| \sqrt{(1-F(y)) \log \log \frac{1}{1-F(y)}} = 0, \quad x \in (A, B),$$

we get
$$f_n(x)-f(x) = (f_n - Ef_n) + (Ef_n - f)$$
$$= n^{-1/2} \int_A^B \psi_n(x,y) \, dn^{1/2}(F_n(y)-F(y)) + Ef_n - f$$
$$= n^{-1/2} \int_A^B n^{1/2}(F_n(y)-F(y)) \, d_y \psi_n(x,y) + Ef_n - f$$
$$\overset{a.s.}{=} n^{-1/2} \int_A^B B_n(F(y)) \, d_y \psi_n(x,y) + O(n^{-1} V_n \log n) + Ef_n - f$$
$$\overset{a.s.}{=} n^{-1/2} \int_0^1 B_n(y) \, d_y \psi_n(x, \operatorname{inv} F(y)) + O(n^{-1} V_n \log n + \pi(n)),$$

resp.
$$n^{1/2}(f_n - f) \overset{a.s.}{=} n^{-1/2} \int_A^B K(F(y), n) \, d_y \psi_n(x, y) + O(n^{-1/2} V_n \log^2 n + n^{1/2} \pi(n))$$
$$\overset{a.s.}{=} n^{-1/2} \int_0^1 K(y, n) \, d_y \psi_n(x, \operatorname{inv} F(y)) + O(n^{-1/2} V_n \log^2 n + n^{1/2} \pi(n)),$$

where $\{B_n\}$ is a suitable sequence of Brownian bridges and K is a suitable Kiefer process.

Consider the following Gaussian processes:

(6.1.8) $\quad \Gamma_n(x) = \int_0^1 \psi_n(x, \operatorname{inv} F(y)) \, dB_n(y) = \int_0^1 B_n(y) \, d_y \psi_n(x, \operatorname{inv} F(y)),$

and

(6.1.9)
$$\bar{\Gamma}(x,n) = \int_0^1 \psi_n(x, \operatorname{inv} F(y)) \, dK(y,n) = \int_0^1 K(y,n) \, d_y \psi_n(x, \operatorname{inv} F(y)),$$

where integration by parts is justified by (6.1.7). Then the above results can be summarized as follows:

Theorem 6.1.1. *Let $f(x)$ be a density function vanishing outside an interval $-\infty \leq C < D \leq +\infty$. Further let $\psi = \{\psi_n(x,y)\}$ be a sequence of functions defined on $(A,B)^2$ with $-\infty \leq A \leq C < D \leq B \leq +\infty$ and*

(6.1.10) $\qquad \sup_x \operatorname{var}_y \psi_n(x,y) = V_n = V_n(A,B).$

Suppose $f_n^{(\psi)}$ is a π estimator of f. Then

(6.1.11) $\quad \sup_x |n^{1/2}(f_n-f)-\Gamma_n| = O(n^{-1/2}V_n \log n + n^{1/2}\pi(n))$

and

(6.1.12) $\quad \sup_x |n(f_n-f)-\bar\Gamma(x,n)| = O(V_n \log^2 n + n\pi(n))$.

In order to get concrete results we have to evaluate the value of V_n and $\pi(n)$ in specific cases. This can be achieved by standard methods of classical analysis and the required results are known for several ψ. We list a number of these in the sequel.

Lemma 6.1.1 (Révész, 1972). *Let ψ be defined by* (6.1.4) *and assume that $h_n \to 0$ as $n \to \infty$. Then*
$$Ef_n(x) \to f(x) \quad (n \to \infty),$$
if x is a continuity point of f. If we assume also that $f(x)$ is uniformly continuous on an interval $-\infty \leq A < B \leq +\infty$, then for any $\varepsilon > 0$ we also have
$$\sup_{A+\varepsilon \leq x \leq B-\varepsilon} |Ef_n(x)-f(x)| \to 0.$$
If $f(x)$ has a bounded derivative on an interval $-\infty \leq A < B \leq +\infty$, then for any $\varepsilon > 0$ we have
$$\sup_{A+\varepsilon \leq x \leq B-\varepsilon} |Ef_n(x)-f(x)| = O(h_n).$$
Further we have
$$\sup_x \mathrm{var}_y \, \psi_n(x,y) = V_n = h_n^{-1}.$$

(Clearly we mean $-\infty + \varepsilon = -\infty$, $+\infty - \varepsilon = +\infty$.)

Lemma 6.1.2 (Bochner 1955, Parzen 1962, Révész 1972). *Let ψ be defined by* (6.1.3). *Assume that $\lambda(x)$ is a bounded density function for which*
$$\lim_{|x| \to \infty} x\lambda(x) = 0.$$
Then
$$Ef_n(x) \to f(x) \quad (n \to \infty),$$
for any continuity point of f.
If we also assume that:
(i) $f(x)$ *has a bounded second derivative on an interval* $-\infty \leq A < B \leq +\infty$,

(ii) $\quad \lim_{|x| \to \infty} x^4 \lambda(x) = 0$

and

(iii) $$\int_{-\infty}^{+\infty} x\lambda(x)\,dx = 0,$$

then, for any $\varepsilon > 0$, we have

$$\sup_{A+\varepsilon \leq x \leq B-\varepsilon} |Ef_n(x) - f(x)| = O(h_n^2).$$

Further if λ is a function of bounded variation, then we have

$$\sup_x \operatorname{var}_y \psi_n(x, y) = V_n = O(h_n^{-1}).$$

Lemma 6.1.3 (Zygmund 1968, p. 64, Theorem 10G). *Let ψ be defined by (6.1.2) where $(A, B) = (0, \pi)$ and $\varphi_k(x) = \sqrt{2/\pi} \cos kx$. Suppose that $f(x)$ is differentiable on $(0, \pi)$ and its derivative has a bounded variation. Then, for any $\varepsilon > 0$*

$$\sup_{\varepsilon \leq x \leq \pi - \varepsilon} |Ef_n - f| = O(l_n^{-1})$$

and

$$\int_0^\pi (Ef_n - f)^2\,dx = O(l_n^{-3}).$$

Further

$$\sup_x \operatorname{var}_y \psi_n(x, y) = V_n = O(l_n).$$

Theorem 6.1.1 and Lemmas 6.1.1, 6.1.2 and 6.1.3 enable us to study the Gaussian processes $\Gamma_n(x)$ and $\bar{\Gamma}_n(x)$ of (6.1.8) and (6.1.9) instead of directly studying the process $n^{1/2}(f_n(x) - f(x))$, which is usually more difficult to handle. In order to do so, the following simple theorem for general Gaussian processes will be useful in the sequel.

Theorem 6.1.2. *Let $G_n(x)$ ($A < x < B$; $n = 1, 2, \ldots$) be a sequence of Gaussian processes with*

$$EG_n(x) \equiv 0, \quad R_n(u, v) = EG_n(u)G_n(v)$$

and

$$E\int_A^B G_n^2(x)\,dx = \int_A^B EG_n^2(x)\,dx = \int_A^B R_n(x, x)\,dx = m_n < +\infty.$$

Assume that $R_n(u, v)$ $((u, v) \in (A, B)^2)$ is continuous at any point (u, u) ($A < u < B$), square integrable,

(6.1.13) $$\Delta_n^2 = \operatorname{Var} \int_A^B G_n^2(x)\,dx = 2\int_A^B \int_A^B R_n^2(u, v)\,du\,dv \to \infty \quad (n \to \infty)$$

and

(6.1.14) $$\frac{\int_A^B \left(\int_A^B R_n(u,v) f(v)\, dv\right)^2 du}{\int_A^B \int_A^B R_n^2(u,v)\, du\, dv} \to 0 \quad (n \to \infty)$$

for any $f \in L^2(A, B)$.
Then

(6.1.15) $$\Delta_n^{-1} \left(\int_A^B G_n^2(x)\, dx - m_n \right) \xrightarrow{\mathscr{D}} \mathcal{N}(0, 1).$$

Proof. Define the integral operator $\mathscr{R}_n \in L^2(A, B) \to L^2(A, B)$ by

$$\mathscr{R}_n f = \int_A^B R_n(u, v) f(v)\, dv$$

and let $\varphi_k^{(n)}$ resp. $\lambda_k^{(n)}$ be the sequence of eigenfunctions resp. eigenvalues of $\mathscr{R}_n$. Then the Karhunen–Loève expansion of G_n (cf., e.g., Yeh 1973, p. 283, Theorem 19.4) is

(6.1.16) $$G_n(x) = \sum_{i=1}^{\infty} (\lambda_i^{(n)})^{1/2} \varphi_i^{(n)}(x) N_i^{(n)},$$

where, for any n, $\{N_i^{(n)}\}_{i=1}^{\infty}$ is a sequence of independent $\mathcal{N}(0, 1)$ r.v., and by (6.1.16) we mean

$$\int_A^B \left(G_n(x) - \sum_{i=1}^{\infty} (\lambda_i^{(n)})^{1/2} \varphi_i^{(n)}(x) N_i^{(n)} \right)^2 dx \xrightarrow{\text{a.s.}} 0 \quad \text{as} \quad n \to \infty.$$

Now (6.1.16) implies

(6.1.17) $$\int_A^B G_n^2(x)\, dx = \sum_{i=1}^{\infty} \lambda_i^{(n)} (N_i^{(n)})^2$$

and

(6.1.18) $$m_n = \sum_{i=1}^{\infty} \lambda_i^{(n)}, \quad \Delta_n^2 = 2 \sum_{i=1}^{\infty} (\lambda_i^{(n)})^2.$$

Since (6.1.14) implies

(6.1.19) $$\frac{\max_k \lambda_k^{(n)}}{\left(\sum_{k=1}^{\infty} (\lambda_k^{(n)})^2 \right)^{1/2}} \to 0 \quad (n \to \infty),$$

we have (6.1.15) by (6.1.16) and (6.1.18) and the central limit theorem.

Now we are in the position to study the Gaussian process Γ_n of (6.1.8). First we show that for some ψ the conditions of Theorem 6.1.2 are satisfied, so we can conclude that (6.1.15) holds in such cases.

As an initial step, instead of Γ_n, we investigate the Gaussian processes

$$(6.1.20) \quad \Gamma_n^*(x) = \int_0^1 \psi_n(x, \operatorname{inv} F(y))\, dW(y) \quad (A < x < B)$$

where W is a Wiener process. Clearly, the covariance function of the process Γ_n^* is

(6.1.21)
$$R_n^*(u,v) = E\Gamma_n^*(u)\Gamma_n^*(v) = \int_0^1 \psi_n(u, \operatorname{inv} F(y))\psi_n(v, \operatorname{inv} F(y))\, dy$$
$$= \int_A^B \psi_n(u, y)\psi_n(v, y) f(y)\, dy.$$

The next lemma can be obtained by elementary computations.

Lemma 6.1.4. Let $R_n^*(u,v)$ be defined by (6.1.21). Then for the m_n and Δ_n defined by (6.1.18), and when R_n is replaced by R_n^*, we have

$$(6.1.22) \quad m_n = \begin{cases} h_n^{-1} & \text{if } \psi \text{ is defined by (6.1.4)}, \\ h_n^{-1} \int_A^B \lambda^2(u)\, du & \text{if } \psi \text{ is defined by (6.1.3)} \\ & \text{and } \lambda \text{ is square integrable,} \\ \int_A^B f(y) \sum_{j=1}^{l_n} \varphi_j^2(y)\, dy & \text{if } \psi \text{ is defined by (6.1.2)}, \end{cases}$$

$$(6.1.23) \quad \Delta_n^2 = \begin{cases} h_n^{-1} \int_A^B f^2(u)\, du & \text{if } \psi \text{ is defined by (6.1.4)}, \\ (1+o(1)) h_n^{-1} \int_A^B f^2(u)\, du \int_A^B \left(\int_A^B \lambda(x+y)\lambda(x)\, dx\right)^2 dy \\ & \text{if } \psi \text{ is defined by (6.1.3)}, \\ \int_A^B f^2(u) \sum_{j=1}^{l_n} \varphi_j^2(u)\, du & \text{if } \psi \text{ is defined by (6.1.2)}, \end{cases}$$

provided $f \in L^2(A,B)$.

Let $g \in L^2(A,B)$. Then

$$(6.1.24) \quad \int_A^B \left(\int_A^B R_n^*(u,v) g(v)\, dv\right)^2 du = O(1) \quad \text{if } \psi \text{ is defined by (6.1.4)},$$
$$(6.1.3) \text{ or } (6.1.2).$$

This lemma and Theorem 6.1.2 together imply

Theorem 6.1.3. *Let* Γ_n^* *be defined by* (6.1.20). *Then*

$$\Delta_n^{-1}\left(\int_A^B (\Gamma_n^*(x))^2\,dx - m_n\right) \xrightarrow{\mathcal{D}} \mathcal{N}(0,1),$$

where m_n resp. Δ_n is defined by (6.1.18) resp. (6.1.13).

Having our Theorem 6.1.3 for Γ_n^* and taking into consideration that $B(x) = W(x) - xW(1)$ and that by (6.1.13) $\Delta_n \to \infty$, by (6.1.24) we get

Theorem 6.1.4. *Let* Γ_n *be defined by* (6.1.8) *and suppose that* ψ *is any one of the functions defined by* (6.1.2), (6.1.3) *and* (6.1.4). *Then*

$$\Delta_n^{-1}\left(\int_A^B \Gamma_n^2(x)\,dx - m_n\right) \xrightarrow{\mathcal{D}} \mathcal{N}(0,1),$$

where m_n resp. Δ_n is defined by (6.1.18) resp. (6.1.13).

The relationship (6.1.11) stated that $n^{1/2}(f_n - f)$ can be uniformly estimated by the Gaussian process Γ_n if ψ satisfies some regularity conditions, that is to say, if V_n and $\pi(n)$ defined by (6.1.10) resp. (6.1.6) are small enough. Up to now we have evaluated $\pi(n) = \pi(n; A+\varepsilon, B-\varepsilon)$ and V_n for some concrete ψ. Now, applying Theorem 6.1.4, the statement of (6.1.11), Lemmas 6.1.1, 6.1.2 and 6.1.3 for V_n and $\pi(n)$ we get:

Theorem 6.1.5. *Suppose that* f *is vanishing outside a finite interval* $-\infty < C < D < +\infty$.
(a) *Let* ψ *be defined by* (6.1.4) *and assume that* f *has a bounded derivative on* (C, D). *Then*

$$\left(h_n \int_C^D f^2\,dx\right)^{-1/2}\left[nh_n \int_C^D (f_n - f)^2\,dx - 1\right] \xrightarrow{\mathcal{D}} \mathcal{N}(0,1),$$

provided that $n^{-1}h_n^{-3/2}\log^2 n \to 0$ and $h_n^{5/2} n \to 0$. For example if $h_n = n^{-\alpha}$ then it is assumed that $2/5 < \alpha < 2/3$.
(b) *Let* ψ *be defined by* (6.1.3) *and assume that*
 (i) λ *is vanishing outside a finite interval* (A, B) *with* $\operatorname*{var}_{x \in (A,B)} \lambda(x) \leq C$ *and* $\int_A^B x\lambda(x)\,dx = 0$,
 (ii) f *has a bounded second derivative on* (C, D). *Then, with*

$$\sigma^2 = \int_C^D f^2(u)\,du \int_A^B \left(\int_A^B \lambda(x+y)\lambda(x)\,dx\right)^2 dy,$$

we have

$$h_n^{-1/2}\sigma^{-1}n\left[h_n\int_A^B (f_n-f)^2\,dx - \int_A^B \lambda^2(x)\,dx\right] \xrightarrow{\mathscr{D}} \mathcal{N}(0,1),$$

provided that $(\log^2 n)n^{-1}h_n^{-3/2}\to 0$ and $nh_n^{9/2}\to 0$. For example if $h_n=n^{-\alpha}$ then it is assumed that $2/9<\alpha<2/3$.

(c) Let ψ be defined by (6.1.2) where $\varphi_k(x)=\sqrt{2/\pi}\cos kx$ ($0\leq x\leq \pi$). Suppose that f is vanishing outside the interval $[0,\pi]$ and absolutely continuous inside. Suppose also that $f'(x)$ ($0<x<\pi$) has a bounded variation. Then

$$l_n^{1/2}\left(\int_0^\pi f^2(u)\,du\right)^{-1}\left[nl_n^{-1}\int_0^\pi (f_n-f)^2\,du - 1\right] \xrightarrow{\mathscr{D}} \mathcal{N}(0,1),$$

provided that $(\log^2 n)n^{-1}l_n^{3/2}\to 0$ and $nl_n^{-7/2}\to 0$. For example if $l_n=n^\beta$, then it is assumed that $2/7<\beta<2/3$.

We note that the proof of (c) also hinges on the following simple facts

$$\frac{1}{n}\sum_{k=1}^n \int_0^\pi f(u)\varphi_k^2(u)\,du \to 1 \quad \text{and} \quad \frac{1}{n}\sum_{k=1}^n \int_0^\pi f^2(u)\varphi_k^2(u)\,du \to \int_0^\pi f^2(u)\,du.$$

Remark 6.1.1. In Theorem 6.1.5 we had a great freedom to choose the sequences $\{h_n\}$ and $\{l_n\}$, but we did not say anything about their optimal choice, if any. One possible criterion of optimality could be to choose them so that $E\int_C^D (f_n-f)^2\,dx$ should be minimum. In this remark it will be assumed that f is vanishing outside of a finite interval (C,D).

Since $E\int (f_n-f)^2 = E\int (f_n-Ef_n)^2 + \int (Ef_n-f)^2$, one can estimate the two terms of the right-hand side separately. By Lemmas 6.1.1, 6.1.2, and 6.1.3 we have:

$$\int (Ef_n-f)^2 = \begin{cases} O(h_n^2) & \text{if } \psi \text{ is defined by (6.1.4) and } f \text{ has a bounded} \\ & \text{derivative in } (C,D), \\ O(h_n^4) & \text{if } \psi \text{ is defined by (6.1.3) and } f \text{ and } \lambda \text{ satisfy} \\ & \text{the conditions of Lemma 6.1.2,} \\ O(l_n^{-3}) & \text{if } \psi \text{ is defined by (6.1.2), where } \varphi_j=\sqrt{2/\pi}\cos jx \\ & \text{and } f \text{ has a derivative of bounded variation in} \\ & (0,\pi) \text{ and vanishing outside the latter interval,} \end{cases}$$

and, given the above conditions for estimating $\int (Ef_n-f)^2$, elementary calculations also yield (cf. Révész 1972)

$$E\int (f_n-Ef_n)^2 = \begin{cases} O(h_n^{-1}n^{-1}) & \text{if } \psi \text{ is defined by (6.1.4)}, \\ O(h_n^{-1}n^{-1}) & \text{if } \psi \text{ is defined by (6.1.3)}, \\ O(l_n n^{-1}) & \text{if } \psi \text{ is defined by (6.1.2)}. \end{cases}$$

On equating their respective error terms, the above two formulae imply that the expectation $E\int (f_n-f)^2$ will be the smallest possible if
 (i) $h_n^2 = O(h_n^{-1}n^{-1})$, i.e., $h_n = O(n^{-1/3})$ in the case of (6.1.4),
 (ii) $h_n^4 = O(h_n^{-1}n^{-1})$, i.e., $h_n = O(n^{-1/5})$ in the case of (6.1.3),
 (iii) $l_n^{-3} = O(l_n n^{-1})$, i.e., $l_n = O(n^{1/4})$ in the case of (6.1.2).

Comparing these optimal choices with the respective bounds in Theorem 6.1.5, one can see that, unfortunately, the results of the latter cannot be applied when $\{h_n\}$ resp. $\{l_n\}$ are chosen in the best possible way. Indeed, there are no Theorem 6.1.5 type results available with $\{h_n\}$ and $\{l_n\}$ chosen optimally.

6.2. Strong theorems for empirical densities

Under very weak restrictions on ψ one can prove that Definition 5 of Section 1 produces a uniformly strongly consistent estimator of f; that is to say $\sup_x |f_n(x)-f(x)| \xrightarrow{\text{a.s.}} 0$. A simple result of this type is:

Theorem 6.2.1. Let $f_n^{(\psi)}$ be a π estimator of f (cf. (6.1.6)) and assume that (6.1.7) holds, and

(6.2.1) $$\sup_x \operatorname{var}_y \psi_n(x,y) = V_n = o(n^{1/2}(\log \log n)^{-1/2})$$

and

(6.2.2) $$\pi(n) = o(1).$$

Then

$$\limsup_{n \to \infty} \sup_x |f_n(x)-f(x)| \xstackrel{\text{a.s.}}{=} 0.$$

Proof. By (6.1.12) it is enough to prove that

$$\limsup_{n \to \infty} \sup_x n^{-1}|\bar{\Gamma}(x,n)| \xstackrel{\text{a.s.}}{=} 0.$$

By (6.1.7) and (6.1.9) we have

$$\sup_x n^{-1}|\bar{\Gamma}(x,n)| \leq \sup_x n^{-1} \int_0^1 |K(y,n)|\, d_y \psi_n(x, \mathrm{inv}\, F(y))$$

$$\leq n^{-1} \sup_{0 \leq y \leq 1} |K(y,n)| \cdot V_n,$$

and our theorem now follows by (6.2.1) and the law of iterated logarithm for a Kiefer process (cf. Corollary 1.15.1).

It is of some interest to check for the conditions (6.2.1) and (6.2.2) in some special cases. It will be seen that the latter will hold only over some restricted intervals (cf. (6.2.3) and (6.2.4)).

Corollary 6.2.1. *Let* ψ_n *be defined by* (6.1.4) *and assume that* f *is uniformly continuous on an interval* $-\infty \leq A < B \leq +\infty$,

$$h_n^{-1} = o(n^{1/2} (\log \log n)^{-1/2}) \quad \text{and} \quad h_n \to 0.$$

Then, for any $\varepsilon > 0$

(6.2.3) $$\lim_{n \to \infty} \sup_{A+\varepsilon < x < B-\varepsilon} |f_n(x) - f(x)| \stackrel{\text{a.s.}}{=} 0.$$

Corollary 6.2.2. *Let* ψ_n *be defined by* (6.1.3) *and assume that the conditions* (i), (ii), (iii) *of Lemma* 6.1.2 *are satisfied*, $h_n^{-1} = o(n^{1/2}(\log \log n)^{-1/2})$ *and* $h_n \to 0$. *Then* 6.2.3 *holds true.*

Corollary 6.2.3. *Let* ψ_n *be defined by* (6.1.2) *and assume that the conditions of Lemma* 6.1.3 *are satisfied*, $l_n = o(n^{1/2}(\log \log n)^{-1/2})$ *and* $l_n \to \infty$. *Then for any* $\varepsilon > 0$ *we have*

(6.2.4) $$\lim_{n \to \infty} \sup_{\varepsilon < x < \pi - \varepsilon} |f_n(x) - f(x)| = 0 \quad \text{a.s.}$$

Theorem 6.2.1, as well as Corollaries 6.2.1, 6.2.2 and 6.2.3, are far from being best possible. Especially the restrictions on $\{h_n\}$ and $\{l_n\}$ are too strong (cf. Révész 1972). A much sharper result will be presented when Definition 3 of Section 6.1 is used. However, further conditions will have to be assumed on f for the sake of this sharper result. In order to handle this case, first we present some further theorems for Gaussian process which are similar to $\Gamma_n(x)$ and $\bar{\Gamma}(x, n)$.

From now on it will be assumed that

1.a. f is vanishing outside the interval $[0, 1]$,
1.b. f is twice differentiable over $(0, 1)$ and $|f''| \leq C$,
1.c. f is strictly positive on $(0, 1)$, say $f \geq \alpha > 0$.

Let λ be a density function for which

2.a. $\lambda \leq C$,
2.b. $\lambda(-x) = \lambda(x)$,
2.c. $\lim_{x \to +\infty} x^4 \lambda(x) = 0$,
2.d. λ is twice differentiable on an interval $-\infty \leq -a < +a \leq +\infty$.

Define the Gaussian process (cf. (6.1.3) and (6.1.7))

$$G = G_h(x) = (f(x))^{-1/2} \int_0^1 W(F(y)) \, d_y \lambda((x-y)h^{-1})$$

$$= (f(x))^{-1/2} \int_0^1 \lambda((x-y)h^{-1}) \, d_y W(F(y))$$

$$= (f(x))^{-1/2} \int_0^1 \lambda((x - \text{inv } F(y))h^{-1}) \, dW(y).$$

Then we have

Theorem 6.2.2. *For any $0 < \varepsilon < \frac{1}{2}$ there exists a $C = C(\varepsilon) > 0$ and an $h_0 = h_0(\varepsilon) > 0$ such that*

(6.2.5) $\qquad P\{\sup_{0 < s < h} \sup_{\varepsilon < x < 1-\varepsilon} |G_s(x)| \geq \Lambda h^{1/2} z\} \leq Ch^{-1} e^{-\frac{z^2}{2+\varepsilon}}$

for any $0 < h < h_0$ and $z > 0$ where $\Lambda^2 = \int_{-\infty}^{+\infty} \lambda^2(x) \, dx$.

The proof of this theorem is based on the following two lemmas.

Lemma 6.2.1. *Using the conditions and notations of Theorem 6.2.2 we have*

(6.2.6) $\qquad P\{\sup_{\varepsilon < x < 1-\varepsilon} |G_h(x)| \geq \Lambda h^{1/2} z\} \leq Ch^{-1} e^{-\frac{z^2}{2+\varepsilon}}.$

Proof. For any $\varepsilon < x < 1 - \varepsilon$ we have

$$EG_h^2(x) = (f(x))^{-1} \int_0^1 \lambda^2((x - \text{inv } F(y))h^{-1}) \, dy$$

$$= (f(x))^{-1} \int_0^1 \lambda^2((x-y)h^{-1}) f(y) \, dy = \Lambda^2 h + O(h^3)$$

and
$$E(G_h(x+\Delta x)-G_h(x))^2 = \text{Const.}\ \Delta x$$
if $\varepsilon < x < 1-\varepsilon$ and $0 < \Delta x < h \leq h_0$.

Let T be a positive integer. Then

(6.2.7) $\quad P\{\sup\limits_{\varepsilon \leq iT^{-1} \leq 1-\varepsilon} |G(iT^{-1})| \geq \Lambda h^{1/2} z\} \leq CTe^{-\frac{z^2}{2+\varepsilon}}$

and

(6.2.8) $\quad P\{\sup\limits_{\varepsilon \leq i2^{-k}T^{-1} \leq 1-\varepsilon} |G((i+1)2^{-k}T^{-1})-G(i2^{-k}T^{-1})| \geq Cx_k(2^kT)^{-1/2}\}$

$$\leq C2^k Te^{-\frac{x_k^2}{2+\varepsilon}}.$$

Since
$$\sup\limits_{\varepsilon \leq x \leq 1-\varepsilon} |G(x)| \leq \sup\limits_{\varepsilon \leq iT^{-1} \leq 1-\varepsilon} |G(iT^{-1})|$$
$$+ \sum_{k=1}^{\infty} \sup\limits_{\varepsilon \leq i2^{-k}T^{-1} \leq 1-\varepsilon} |G((i+1)2^{-k}T^{-1})-G(i2^{-k}T^{-1})|,$$

(6.2.7) and (6.2.8) together imply

$$P\left\{\sup\limits_{\varepsilon \leq x \leq 1-\varepsilon} |G(x)| \geq \Lambda h^{1/2} z + T^{-1/2} C \sum_{k=1}^{\infty} x_k 2^{-k/2}\right\}$$
$$\leq CTe^{-\frac{z^2}{2+\varepsilon}} + CT \sum_{k=1}^{\infty} 2^k e^{-\frac{x_k^2}{2+\varepsilon}}.$$

Let $T = Mh^{-1}\Lambda^{-2}$ and $x_k = (4k+z^2)^{1/2}$, where M is a positive constant, to be specified later on. Then

$$\Lambda h^{1/2} z + CT^{-1/2} \sum_{k=1}^{\infty} x_k 2^{-k/2} \leq \Lambda h^{1/2} z \left(1 + CM^{-1/2} \sum_{k=1}^{\infty} 2^{-k/2}\right)$$
$$+ \Lambda Ch^{1/2} M^{-1/2} \sum_{k=1}^{\infty} 2k^{1/2} 2^{-k/2} \leq \Lambda h^{1/2}[z(1+\varepsilon)+\varepsilon]$$

if M is large enough. Further

$$CTe^{-\frac{z^2}{2+\varepsilon}} + CT \sum_{k=1}^{\infty} 2^k e^{-\frac{x_k^2}{2+\varepsilon}} \leq CMh^{-1}\Lambda^{-2} e^{-\frac{z^2}{2+\varepsilon}} \left(1 + \sum_{k=1}^{\infty} 2^k e^{-\frac{4k}{2+\varepsilon}}\right).$$

By the choice $z(1+\varepsilon)+\varepsilon = u$, we get (6.2.6).

The next lemma can be proved the same way as the latter one, hence the details will be omitted.

Lemma 6.2.2. Let $0 < \Delta h < h < h_0$. Then

(6.2.9) $\quad P\{\sup_{\varepsilon \leq x \leq 1-\varepsilon} |G_{h+\Delta h}(x) - G_h(x)| \geq C(\Delta h)^{1/2} z\} \leq C(\Delta h)^{-1} e^{-\frac{z^2}{2+\varepsilon}}.$

Proof of Theorem 6.2.2. Let T be a positive integer. Then by Lemmas 6.2.1 and 6.2.2 we have

(6.2.10) $\quad P\{\sup_{0 \leq i \leq T} \sup_{\varepsilon \leq x \leq 1-\varepsilon} |G_{ihT^{-1}}(x)| \geq \Lambda h^{1/2} z\}$

$\leq P\{\sup_{0 \leq i \leq T} \sup_{\varepsilon \leq x \leq 1-\varepsilon} |G_{ihT^{-1}}(x)| \geq \Lambda h^{1/2} T^{-1/2} z\} \leq CTh^{-1} e^{-\frac{z^2}{2+\varepsilon}},$

and

(6.2.11) $\quad P\{\sup_{0 < i \leq 2^k T} \sup_{\varepsilon < x < 1-\varepsilon} |G_{(2i+1)h2^{-k}T^{-1}}(x) - G_{2ih2^{-k}T^{-1}}(x)|$

$\geq C(h2^{-k}T^{-1})^{1/2} x_k\} \leq C \cdot 2^{2k} T^2 h^{-1} e^{-\frac{x_k^2}{2+\varepsilon}}.$

Since

$\sup_{0 < \psi < h} \sup_{\varepsilon \leq x \leq 1-\varepsilon} |G_\psi(x)| \leq \sup_{0 \leq i \leq T} \sup_{\varepsilon \leq x \leq 1-\varepsilon} |G_{ihT^{-1}}(x)|$

$+ \sum_{k=1}^{\infty} \sup_{0 < i \leq 2^k T} \sup_{\varepsilon \leq x \leq 1-\varepsilon} |G_{(2i+1)h2^{-k}T^{-1}}(x) - G_{2ih2^{-k}T^{-1}}(x)|,$

(6.2.10) and (6.2.11) together imply

$P\left\{\sup_{0 < \psi \leq h} \sup_{\varepsilon \leq x \leq 1-\varepsilon} |G_\psi(x)| \geq \Lambda h^{1/2} z + C \sum_{k=1}^{\infty} (h2^{-k} T^{-1})^{1/2} x_k\right\}$

$\leq CTh^{-1} e^{-\frac{z^2}{2+\varepsilon}} + C \sum_{k=1}^{\infty} 2^{2k} T^2 h^{-1} e^{-\frac{x_k^2}{2+\varepsilon}}.$

Let $x_k = (5k + z^2)^{1/2}$. Then

$\Lambda h^{1/2} z + C \sum_{k=1}^{\infty} (h2^{-k} T^{-1})^{1/2} x_k$

$\leq \Lambda h^{1/2} z \left(1 + C\Lambda^{-1} T^{-1/2} \sum_{k=1}^{\infty} 2^{-k/2}\right) + Ch^{1/2} T^{-1/2} \sum_{k=1}^{\infty} (5k)^{1/2} 2^{-k/2}$

$\leq \Lambda h^{1/2} z (1+\varepsilon) + h^{1/2} \varepsilon$

if T is big enough. Further

$CTh^{-1} e^{-\frac{z^2}{2+\varepsilon}} + C \sum_{k=1}^{\infty} 2^{2k} T^2 h^{-1} e^{-\frac{x_k^2}{2+\varepsilon}} \leq CTh^{-1} e^{-\frac{z^2}{2+\varepsilon}} \left(1 + \sum_{k=1}^{\infty} 2^{2k} Te^{-\frac{5k}{2+\varepsilon}}\right).$

By the choice $\Lambda z(1+\varepsilon) + \varepsilon = u\Lambda$, we get (6.2.5).

Define the Gaussian process

(6.2.12) $\quad \Gamma_h(x, y) = (f(x))^{-1/2} \int_0^1 \lambda((x-u)h^{-1}) \, d_u W(F(u), y).$

Now we prove:

Theorem 6.2.3. *We have*

(6.2.13) $\quad P\left\{ \sup_{0 < y \leq Y} \sup_{0 < s \leq h} \sup_{\varepsilon \leq x \leq 1-\varepsilon} |\Gamma_s(x, y)| \geq \Lambda h^{1/2} Y^{1/2} z \right\}$

$$\leq Ch^{-1} e^{-\frac{z^2}{2+\varepsilon}}.$$

Proof. First we observe that

$$M_y = \sup_{0 < s \leq h} \sup_{\varepsilon \leq x \leq 1-\varepsilon} |\Gamma_s(x, y)|$$

is a sub-martingale (cf. Lemma 1.14.1), i.e.,

$$E(M_y | M_t, \, t \leq x) \geq M_x \quad \text{a.s.} \quad (y > x).$$

Consequently $e^{tM_y^2}$ is also a sub-martingale. Now by the sub-martingale inequality we have

$$P\left\{ \sup_{y \leq Y} M_y \geq (hY)^{1/2} \Lambda z \right\} = P\left\{ \sup_{y \leq Y} \exp\left(\frac{tM_y^2}{\Lambda^2 hY}\right) \geq \exp tz^2 \right\}$$

$$\leq e^{-tz^2} E \exp\left(\frac{tM_Y^2}{\Lambda^2 hY}\right).$$

By Theorem 6.2.2 $E \exp\left(\dfrac{tM_y^2}{\Lambda^2 hy}\right)$ is bounded above whenever $t < \tfrac{1}{2}$, and hence we have our statement.

With the help of the inequality (6.2.13) it is going to be easy now to prove a strong theorem for the process $\Gamma_h(x, y)$. Let $\{h_n\}$ be a sequence of positive numbers for which the following two conditions hold

3.a. $\qquad\qquad\qquad h_n \searrow 0, \quad nh_n \nearrow \infty,$

3.b. $\qquad\qquad \dfrac{\log^4 n}{nh_n \log h_n^{-1}} \to 0, \quad \dfrac{nh_n^5}{\log h_n^{-1}} \to 0$

and define the Gaussian process (cf. (6.2.12))

$$\Gamma(x, n) = (nh_n)^{-1/2} \Gamma_{h_n}(x, n) = (nh_n f(x))^{-1/2} \int_0^1 \lambda((x-u)h_n^{-1}) \, d_u W(F(u), n).$$

Then we have

Theorem 6.2.4. *For any $\varepsilon > 0$ we have*

(6.2.14) $$\lim_{n \to \infty} (2\Lambda^2 \log h_n^{-1})^{-1/2} \sup_{\varepsilon \leq x \leq 1-\varepsilon} |\Gamma(x,n)| \stackrel{a.s.}{=} 1.$$

Proof. Step 1. Let $\theta > 1$ and put

$$A_n = \sup_{[\theta^n] \leq k < [\theta^{n+1}]} (2\Lambda^2 \log h_k^{-1})^{-1/2} \sup_{\varepsilon \leq x \leq 1-\varepsilon} |\Gamma(x,k)|.$$

At first we prove

(6.2.15) $$\varlimsup_{n \to \infty} A_n \leq 1 \quad \text{a.s.}$$

By Theorem 6.2.3 we have

$$\sum_{n=1}^{\infty} P\{A_n \geq 1+\varepsilon\} \leq \sum_{n=1}^{\infty} h_{[\theta^n]}^\delta$$

for a suitable $\delta = \delta(\varepsilon) > 0$. Our conditions on $\{h_n\}$ imply the convergence of the series $\sum_{n=1}^{\infty} h_{[\theta^n]}^\delta$. Hence we have (6.2.15) by the Borel–Cantelli lemma.

Step 2. For any $\varepsilon > 0$ one can find a positive number $Q = Q_\varepsilon$ and a density function $\lambda_\varepsilon(x)$ satisfying Conditions 2.a–2.d of this section such that

$$\int_{-\infty}^{+\infty} (\lambda(x) - \lambda_\varepsilon(x))^2 dx \leq \varepsilon$$

and $\lambda_\varepsilon(x) = 0$ if $|x| \geq Q$. Put

$$\Gamma_\varepsilon(x,n) = (nh_n f(x))^{-1/2} \int_0^1 \lambda_\varepsilon((x-u)h_n^{-1}) d_u W(F(u),n),$$

and

$$B_n = \max_{0 < k \leq [(2Qh_n)^{-1}]} (2\Lambda^2 \log h_n^{-1})^{-1/2} \Gamma_\varepsilon((2k-1)(4Qh_n)^{-1}, n).$$

Now we prove that

(6.2.16) $$\varliminf_{n \to \infty} B_n \geq 1.$$

Clearly we have

$$\sum_{n=1}^{\infty} P(B_n \leq 1-\varepsilon) \leq \sum_{n=1}^{\infty} \left\{ 1 - \frac{h_n^{1-\varepsilon}}{(\log h_n^{-1})^{1/2}} \right\}^{\frac{1}{2Qh_n}}$$

and this series is convergent by our conditions. Hence we have (6.2.16), and our theorem follows from (6.2.15) and (6.2.16).

Theorem 6.2.4 and the trivial relation

$$(n \log h_n^{-1})^{-1/2} h_n^{1/2} W(1, n) \xrightarrow{\text{a.s.}} 0$$

clearly imply:

Corollary 6.2.4. *Let*

$$\tilde{\Gamma}(x, n) = (nh_n f(x))^{-1/2} \int_0^1 \lambda((x-u)h_n^{-1}) \, d_y K(F(y), n).$$

Then for any $\varepsilon > 0$ we have

(6.2.17) $\qquad \lim_{n \to \infty} (2\Lambda^2 \log h_n^{-1})^{1/2} \sup_{\varepsilon < x < 1-\varepsilon} |\tilde{\Gamma}(x, n)| \overset{\text{a.s.}}{=} 1.$

Now we can present a much sharper result than that of Theorem 6.2.1, but only in the case when Definition 3 is used. Our result is summarized by

Theorem 6.2.5. *Suppose that Conditions 1.a–1.c, 2.a–2.d and 3.a, 3.b of this section are satisfied. Then for any $\varepsilon > 0$ we have*

(6.2.18) $\qquad \lim_{n \to \infty} \left(\frac{nh_n}{2\Lambda^2 \log h_n^{-1}} \right)^{1/2} \sup_{\varepsilon < x < 1-\varepsilon} \left| \frac{f_n(x) - f(x)}{f^{1/2}(x)} \right| \overset{\text{a.s.}}{=} 1$

where $\Lambda^2 = \int_{-\infty}^{+\infty} \lambda^2(x) \, dx$.

Proof. This theorem simply follows from Corollary 6.2.4, Theorem 6.1.1 and Lemma 6.1.2.

6.3. Empirical regression

Let $(X, Y), (X_1, Y_1), (X_2, Y_2), \ldots$ be a sequence of i.i.d.r.v. with

(6.3.1) $\qquad 0 \leq X \leq 1, \quad Y \in R^1,$

(6.3.2) $\qquad E(Y|X = x) = r(x) \quad (0 \leq x \leq 1),$

where $r(x)$ is differentiable with $|r'(x)| \leq C < \infty$,

(6.3.3) $\qquad P(X \leq t) = F(t) = \int_0^t f(x) \, dx \quad \text{where} \quad f(x) \geq \varepsilon > 0,$

(6.3.4) $\qquad P(Y - r(X) \leq t | X = x) = G(t) \quad (-\infty < t < +\infty, \, 0 \leq x \leq 1),$

(6.3.5) $\qquad E((Y - r(X))^2 | X = x) = \sigma^2 > 0 \quad (0 \leq x \leq 1),$

(6.3.6) $\qquad \int_{-\infty}^{+\infty} e^{xt} \, dG(x) < \infty \quad \text{on some interval } |t| < t_0.$

Here we are interested in introducing and investigating two non-parametric estimators of the regression function $r(x)$ based on the sample $(X_1, Y_1), (X_2, Y_2), \ldots, (X_n, Y_n)$.

Our first definition is based on the k_n-nearest neighbour method. Let $0 < X_1^{(n)} < X_2^{(n)} < \ldots < X_n^{(n)}$ be the ordered sample based on the sample $X_1, X_2, \ldots, X_n$, and let $Y_i^{(n)}$ be that Y_j which corresponds to $X_i^{(n)}$, that is to say

$$Y_i^{(n)} = Y_j \quad \text{if} \quad X_i^{(n)} = X_j.$$

Further let k_n be an increasing sequence of even integers for which

(6.3.7) $\quad k_n n^{-2/3} \log n \to 0, \quad k_n^{-1} (\log n)^3 \to 0 \quad (n \to \infty).$

Now we give the following definition of an empirical regression function

$$r_n(x) = \begin{cases} k_n^{-1} \sum_{j=1}^{k_n} Y_j^{(n)} & \text{if } x < X_{k_n}^{(n)}, \\ k_n^{-1} \sum_{j=i-k_n/2}^{i+k_n/2-1} Y_j^{(n)} & \text{if } X_i^{(n)} \leq x < X_{i+1}^{(n)} \\ & (i = k_n/2, k_n/2+1, \ldots, (n+1)-k_n/2), \\ k_n^{-1} \sum_{j=n-k_n+1}^{n} Y_j^{(n)} & \text{if } x \geq X_{n+2-k_n}^{(n)}. \end{cases}$$

Making use of this definition of an empirical regression function we can get:

Theorem 6.3.1 (Révész 1979). *Assume that conditions* (6.3.1)–(6.3.7) *are fulfilled. Then*

(6.3.8) $\quad \lim_{n\to\infty} P\{k_n^{1/2} \sigma^{-1} \sup_{0 \leq x \leq 1} |r_n(x) - r(x)| \leq a(n/k_n, y)\}$

$$= \exp(-2e^{-y}) \quad (-\infty < y < +\infty),$$

where

$$a(u, v) = (2 \log u + \tfrac{1}{2} \log \log u - \tfrac{1}{2} \log \pi + v)(2 \log u)^{-1/2}$$

$$(u > e, -\infty < v < +\infty).$$

In any possible application of this theorem the major difficulty is coming from the presence of σ in (6.3.8), for σ is unknown in most practical cases. This difficulty can be overcome via introducing the estimator

$$\sigma_n = \left(n^{-1} \sum_{i=1}^{n} (Y_i - r_n(X_i))^2 \right)^{1/2}$$

and proving:

Theorem 6.3.2 (Révész 1979). *Assume that conditions* (6.3.1)–(6.3.7) *are fulfilled. Then*

(6.3.9) $\quad \lim_{n\to\infty} P\{k_n^{1/2}\sigma_n^{-1} \sup_{0\leq x\leq 1} |r_n(x)-r(x)| \leq a(n/k_n, y)\} = \exp(-2e^{-y})$,

$$(-\infty < y < +\infty).$$

As to our second definition of an empirical regression function (cf., e.g., Nadaraya 1964, 1965), let $K(x)$ $(-\infty < x < +\infty)$ be an arbitrary density function and $\{a_n\}$ be a decreasing sequence of positive numbers tending to 0. We define

$$\psi_n(x) = \frac{\sum_{i=1}^n Y_i K\left(\frac{x-X_i}{a_n}\right)}{\sum_{i=1}^n K\left(\frac{x-X_i}{a_n}\right)}.$$

Here we only consider the special case of this definition when K is the uniform law. Hence we let

$$e(x) = \begin{cases} 1 & \text{if } |x| \leq \tfrac{1}{2} \\ 0 & \text{otherwise} \end{cases}$$

and

$$\varrho_n(x) = \frac{\sum_{i=1}^n Y_i e\left(\frac{x-X_i}{a_n}\right)}{\sum_{i=1}^n e\left(\frac{x-X_i}{a_n}\right)}.$$

Theorem 6.3.3 (Révész 1979). *Assume that conditions* (6.3.1)–(6.3.6) *are fulfilled. Also assume*

(6.3.10) $\qquad f(x)$ *is differentiable with* $|f'(x)| \leq K < \infty$,

(6.3.11) $\qquad a_n^3 n \log n \to 0, \quad (na_n)^{-1}\log n \to 0 \quad (n\to\infty).$

Then we have

(6.3.12) $\quad \lim_{n\to\infty} P\{(na_n)^{1/2}\sigma_n^{-1} \sup_{0\leq x\leq 1} (f_n(x))^{1/2}|\varrho_n(x)-r(x)| \leq a(a_n^{-1}, y)\}$

$$= \exp(-2e^{-y}) \quad (-\infty < y < +\infty),$$

where

$$f_n(x) = (na_n)^{-1} \sum_{i=1}^n e\left(\frac{x-X_i}{a_n}\right).$$

Remark 6.3.1. Using a hystogram-type definition of the empirical regression function, Major (1973) proved a result, similar to the above ones.

We note that the role of a_n in Theorem 6.3.3 is that of k_n/n in Theorem 6.3.1.

For further reference we introduce the following notations:

$$Z_i^{(n)} = Y_i^{(n)} - r(X_i^{(n)}) \quad (i=1,2,\ldots,n; \; n=1,2,\ldots)$$

and

$$\bar{r}_n(x) = \begin{cases} k_n^{-1} \sum_{j=1}^{k_n} r(X_j^{(n)}) & \text{if } x \leq X_{k_n}^{(n)}, \\ k_n^{-1} \sum_{j=i-k_n/2}^{i+k_n/2-1} r(X_j^{(n)}) & \text{if } X_i^{(n)} \leq x < X_{i+1}^{(n)} \\ & (i = k_n/2, k_n/2+1, \ldots, (n+1)-k_n/2), \\ k_n^{-1} \sum_{j=n-k_n+1}^{n} r(X_j^{(n)}) & \text{if } x \geq X_{n+2-k_n}^{(n)}. \end{cases}$$

First a few simple lemmas.

Lemma 6.3.1.

$$\varlimsup_{n \to \infty} \frac{n}{k_n} \sup_{1 \leq i \leq n-k_n} (X_{i+k_n}^{(n)} - X_i^{(n)}) \leq 2/\varepsilon \quad a.s.$$

where $\varepsilon > 0$ is defined in (6.3.3).

The proof of this lemma is trivial.

Applying the above lemma as well as conditions (6.3.2) and (6.3.7), by Taylor's expansion one gets

Lemma 6.3.2.

$$\lim_{n \to \infty} (k_n \log n)^{1/2} \sup_{0 \leq x \leq 1} |\bar{r}_n(x) - r(x)| \overset{a.s.}{=} 0.$$

By a simple transformation Theorem 1.5.5 implies

Lemma 6.3.3. For $-\infty < y < +\infty$, we have

$$\lim_{n \to \infty} P\{k_n^{-1/2} \max_{0 \leq j \leq n} |W(j+k_n) - W(j)| \leq a(n/k_n, y)\} = \exp(-2e^{-y}).$$

Lemma 6.3.4. Under the conditions of Theorem 6.3.1 we have

(6.3.13) $\lim_{n \to \infty} P\{k_n^{-1/2} \sigma^{-1} \sup_{0 \leq x \leq 1} |r_n(x) - \bar{r}_n(x)| \leq a(n/k_n, y)\} = \exp(-2e^{-y}),$

where $-\infty < y < +\infty$.

Proof. By Theorem 4.4.1 one can construct a sequence of Wiener processes $\{W_n(x), x \geqq 0\}_{n=1}^{\infty}$ for which

$$(6.3.14) \qquad \overline{\lim_{n \to \infty}} (\log n)^{-1} \max_{1 \leqq j \leqq n} \left| \sigma^{-1} \sum_{i=1}^{j} Z_j^{(n)} - W_n(j) \right| \leqq C \quad \text{a.s.,}$$

where C is a constant depending only on the distribution G of (6.3.4). Since

$$r_n(x) - \bar{r}_n(x) = \begin{cases} k_n^{-1} \sum_{j=1}^{k_n} Z_j^{(n)} & \text{if } x < X_k^{(n)}, \\ k_n^{-1} \sum_{j=i-k_n/2}^{i+k_n/2-1} Z_j^{(n)} & \text{if } X_i^{(n)} \leqq x < X_{i+1}^{(n)} \\ & (i = k_n/2, k_n/2+1, \ldots, (n+1)-k_n/2), \\ k_n^{-1} \sum_{j=n-k_n+1}^{n} Z_j^{(n)} & \text{if } x \geqq X_{n+2-k_n}^{(n)}, \end{cases}$$

(6.3.14), Lemma 6.3.3 and condition (6.3.7) imply (6.3.13).

Proof of Theorem 6.3.1. Lemmas 6.3.2 and 6.3.4 immediately imply (6.3.8).

Proof of Theorem 6.3.2. Theorem 2.4.4 implies

$$P\left\{ \left| \frac{\sigma_n^2}{2} - 1 \right| \geqq 4n^{-1/2} \log n \right\} \leqq n^{-2}.$$

Now (6.3.9) follows from Theorem 6.3.1 and the above line.

For the sake of the proof of Theorem 6.3.3 we need some further notation and lemmas.

Let $i = i(x)$ resp. $j = j(x)$ be the smallest resp. largest integer for which

$$e\left(\frac{x - X_i^{(n)}}{a_n} \right) = 1 \quad \text{resp.} \quad e\left(\frac{x - X_j^{(n)}}{a_n} \right) = 1.$$

Let $j(x) - i(x) = v_n(i) = n a_n f_n(x)$. Put also

$$\bar{\varrho}_n(x) = \frac{\sum_{i=1}^{n} r(X_i) e\left(\frac{x - X_i}{a_n} \right)}{\sum_{i=1}^{n} e\left(\frac{x - X_i}{a_n} \right)}.$$

Then we have

Lemma 6.3.5.
$$\lim_{n\to\infty} (na_n \log n)^{1/2} \sup_{0\leq x\leq 1} (f_n(x))^{1/2} |\bar{\varrho}_n(x) - r(x)| \stackrel{a.s.}{=} 0.$$

The proof is trivial by Taylor's expansion and condition (6.3.11). Applying our previous notations we have.

Lemma 6.3.6. *With F and f as in* (6.3.3)
$$\lim_{n\to\infty} P\left\{\max_{1\leq i\leq n} \left(na_n f\left(\operatorname{inv} F\left(\frac{i}{n}\right)\right)\right)^{-1/2} \left|W\left(i + na_n f\left(\operatorname{inv} F\left(\frac{i}{n}\right)\right)\right) - W(i)\right|\right.$$
$$\leq a(a_n^{-1}, y)\} = \exp(-2e^{-y}) \quad (-\infty < y < +\infty).$$

Proof. This lemma can be obtained by applying Lemma 6.3.3 for small intervals $i \in \left(l\frac{n}{d}, (l+1)\frac{n}{d}\right)$ $(l=0, 1, 2, \ldots, d-1$ and d is big enough) where $f\left(\operatorname{inv} F\left(\frac{i}{n}\right)\right)$ can be viewed as a constant.

The following simple lemma is a consequence of Theorem 6.2.5.

Lemma 6.3.7.
$$\lim_{n\to\infty} \frac{(na_n)^{1/2}}{\log n} \sup_x |f_n(x) - f(x)| \stackrel{a.s.}{=} 0.$$

Lemma 6.3.8. *Let $\{v_n(i), i=1, 2, \ldots, n; n=1, 2, \ldots\}$ be as in the proof of Theorem 6.3.2. Then*
$$\lim_{n\to\infty} P\{\max_{1\leq i\leq n} (v_n(i))^{-1/2} |W(i+v_n(i)) - W(i)| \leq a(a_n^{-1}, y)\}$$
$$= \exp(-2e^{-y}) \quad (-\infty < y < +\infty),$$
provided that $\{v_n(i), i=1, 2, \ldots, n; n=1, 2, \ldots\}$ and $\{W(x), x\geq 0\}$ are independent.

Proof. This lemma easily follows from Lemmas 6.3.6 and 6.3.7.

Proof of Theorem 6.3.3. Applying again the approximation (6.3.14), one gets (6.3.12) by Lemmas 6.3.5 and 6.3.8.

6.4. Empirical characteristic functions

In the previous sections of this chapter we have seen that empirical density and regression functions can be defined several reasonably plausible ways. Empirical characteristic functions have a most natural definition. It seems to have appeared first in Cramér's book (1946) and then in

Parzen (1962). A first related weak convergence result was proved by Kent (1975), whose work was inspired by Kendall (1974). A systematic study of its limiting behaviour was initiated by Feuerverger and Mureika (1977) and carried out by S. Csörgő (1980).

Let $X_1, X_2, \ldots$ be i.i.d.r.v. with distribution function $F(x)$ and characteristic function $c(t) = \int_{-\infty}^{+\infty} e^{itx} dF(x)$. Let $F_n(x)$ be the empirical distribution function based on the sample $X_1, X_2, \ldots, X_n$. The empirical characteristic function $c_n(t)$ of a random sample is defined as

(6.4.1) $\quad c_n(t) = n^{-1} \sum_{k=1}^{n} e^{itX_k} = \int_{-\infty}^{+\infty} e^{itx} dF_n(x), \quad -\infty < t < +\infty.$

First we prove a Glivenko–Cantelli type theorem.

Theorem 6.4.1 (Feuerverger, Mureika 1977, S. Csörgő 1980). *For arbitrary F we have*

(6.4.2) $\quad \sup_{T_n^{(1)} \leq t \leq T_n^{(2)}} |c_n(t) - c(t)| \xrightarrow{a.s.} 0 \quad as \quad n \to \infty,$

where $T_n = |T_{(n)}^2| \vee |T_{(n)}^1| = o((n/\log \log n)^{1/2})$.

Proof. Let $\quad \Delta_n = \sup_{T_n^{(1)} \leq t \leq T_n^{(2)}} |c_n(t) - c(t)|.$

Let $0 < \varepsilon < 1$, and choose $K > 0$ so that $F(-K), 1 - F(K) < \varepsilon/6$. For (random) large enough n we have by the Glivenko–Cantelli theorem a.s. that $F_n(-K), 1 - F_n(K) > \varepsilon/6$, and hence also $|F_n(\pm K) - F(\pm K)| < \varepsilon/6$. For still larger (if necessary) n, with probability 1,

$$\Delta_n \leq \varepsilon + \sup_{T_n^{(1)} \leq t \leq T_n^{(2)}} \left| -it \int_{-K}^{K} (F_n(x) - F(x)) \exp(itx) \, dx \right|$$

$$= \varepsilon + 2KT_n \sup_{-\infty < x < \infty} |F_n(x) - F(x)|$$

$$\leq \varepsilon + 3KT_n (n^{-1} \log \log n)^{1/2},$$

by the log log law for the empirical process (cf. Theorem 5.1.1).

Next we define the *empirical characteristic process* $Y_n(t)$ by

(6.4.3) $\quad Y_n(t) = n^{1/2}(c_n(t) - c(t)) = \int_{-\infty}^{+\infty} e^{itx} dn^{1/2}(F_n(x) - F(x))$

$$= \int_{-\infty}^{+\infty} e^{itx} d\beta_n(x), \quad -\infty < t < +\infty.$$

It is natural to replace $\beta_n(x)$ in (6.4.3) by its approximating process $B_n(F(x))$ of Theorem 4.4.1, and compare the Gaussian process

$$(6.4.4) \quad Z_n(t) = \int_{-\infty}^{+\infty} e^{itx} d_x B_n(F(x)) = \int_0^1 e^{it(\operatorname{inv} F(y))} d_y B_n(y)$$

$$\stackrel{\mathscr{D}}{=} \int_0^1 e^{it(\operatorname{inv} F(y))} d_y B(y) = Y(t), \quad -\infty < t < +\infty,$$

where $\overline{Y(t)} = Y(-t)$, $EY(t) = 0$ and $EY(t)Y(s) = c(t+s) - c(t)c(s)$, to that of (6.4.3). Clearly we hope that Theorem 4.4.1 should guarantee some closeness of these processes. Actually this problem of nearness turns out to be quite complicated, and $Y_n(t)$ is going to be close to $Z_n(t)$ only if F satisfies some moment conditions. The mentioned difficulty is due to the fact that, in general, the function $e^{it(\operatorname{inv} F(y))}$ does not have a bounded variation. This is demonstrated by the proof of the following:

Theorem 6.4.2. *Assume that F has a bounded support, say $F(a) = 0$, $F(b) = 1$ ($-\infty < a < b < +\infty$), and is arbitrary otherwise. Then, for any $-\infty < T_1 < T_2 < +\infty$ we have*

$$(6.4.5) \quad \sup_{T_1 \leq t \leq T_2} |Y_n(t) - Z_n(t)| \stackrel{a.s.}{=} O(n^{-1/2} \log n).$$

Proof. Clearly we have (cf. Theorem 4.4.1 and Remark 4.4.3)

$$\sup_{T_1 \leq t \leq T_2} |Y_n(t) - Z_n(t)| = \sup_{T_1 \leq t \leq T_2} \left| \int_a^b e^{itx} d_x (\beta_n(x) - B_n(F(x))) \right|$$

$$\leq \sup_{T_1 \leq t \leq T_2} \int_a^b |\beta_n(x) - B_n(F(x))| d_x e^{itx}$$

$$\leq O(n^{-1/2} \log n) \sup_{T_1 \leq t \leq T_2} \operatorname*{var}_{a \leq x \leq b} e^{itx}$$

with probability one. Hence the constant of $O(n^{-1/2} \log n)$ in (6.4.5) depends on the total variation of e^{itx}, i.e., it depends on a, b and T_1, T_2.

The above proof shows that if the support of F is infinite, then Theorem 4.4.1 is not enough to prove a (6.4.5)-type result. Indeed, in order to describe the behaviour of the process $Z_n(t)$ vis-a-vis the process $Y_n(t)$, in addition to Theorem 4.4.1, one also needs a careful investigation of the influence of the tail properties of F. The best available positive result of this nature is

Theorem 6.4.3 (S. Csörgő 1980). Let $-\infty < T_1 < T_2 < \infty$ and let $h(x)$ be a continuous function on $(0, \infty)$ such that

(6.4.6) $$\frac{h(x)}{x^\alpha} \nearrow \infty, \quad \text{as} \quad x \to \infty,$$

with some positive α. Assume that

(6.4.7) $\quad h(x)F(-x) = O(1), \quad h(x)(1-F(x)) = O(1) \quad \text{as} \quad x \to \infty.$

Then there exist for each n a Brownian bridge $B_n(\cdot)$ and a Kiefer process $K(\cdot,\cdot)$ such that for the processes

$$\left\{ Z_n(t) = \int_{-\infty}^{\infty} e^{itx} \, dB_n(F(x)); \, T_1 \leq t \leq T_2 \right\},$$

$$\left\{ K_n(t) = \int_{-\infty}^{\infty} e^{itx} \, d(n^{-1/2} K(F(x), n)); \, T_1 \leq t \leq T_2 \right\}$$

one has

(6.4.8) $\quad P\{\sup_{T_1 \leq t \leq T_2} |Y_n(t) - Z_n(t)| > C_1 r_1(n)\} \leq L_1 n^{-(1+\delta)},$

(6.4.9) $\quad P\{\sup_{T_1 \leq t \leq T_2} |Y_n(t) - K_n(t)| > C_2 r_2(n)\} \leq L_2 n^{-(1+\delta)},$

where $\delta > 0$ is arbitrary large, and the constants $0 < C_1$, C_2 depend only on δ, F, T_1, T_2, while L_1, L_2 on $T_2 - T_1$. The rate-functions $r_k(x)$, $k=1, 2$, are defined as

(6.4.10) $\quad r_k(x) = u_k(x) x^{-1/2} (\log x)^k,$

where $u_k(x)$ is a function, whose inverse $\operatorname{inv} u_k(x)$, for large enough x, is defined by

(6.4.11) $$\frac{\operatorname{inv} u_k(x)}{(\log(\operatorname{inv} u_k(x)))^{2k-1}} = h(x)x^2.$$

From (6.4.8) and (6.4.9) it follows that

(6.4.12) $\quad \Delta_n^{(1)} = \sup_{T_1 \leq t \leq T_2} |Y_n(t) - Z_n(t)| \stackrel{a.s.}{=} O(r_1(n)),$

(6.4.13) $\quad \Delta_n^{(2)} = \sup_{T_1 \leq t \leq T_2} |Y_n(t) - K_n(t)| \stackrel{a.s.}{=} O(r_2(n)).$

Applying (6.4.12) or (6.4.13), it is easy to see that (with F as in Theorem 6.4.3) the distribution of appropriate functionals of $\{Y_n(t); T_1 \leq t \leq T_2\}$ converge to those of $\{Y(t); T_1 \leq t \leq T_2\}$ of (6.4.4), since for each n

$$\{Z_n(t); T_1 \leq t \leq T_2\} \stackrel{\mathscr{D}}{=} \{K_n(t); T_1 \leq t \leq T_2\} \stackrel{\mathscr{D}}{=} \{Y(t); T_1 \leq t \leq T_2\}.$$

Let $C^2 = C^2(T_1, T_2) = C(T_1, T_2) \times C(T_1, T_2)$, $-\infty < T_1 < T_2 < +\infty$, where $C(T_1, T_2)$ is the Banach space of continuous functions on $[T_1, T_2]$ endowed with the supremum norm.

Corollary 6.4.1 (S. Csörgő 1980). *Consider the following functionals on C^2:*

$$\psi_1(u) = \int_{T_1}^{T_2} |u(t)|^2 \, dG(t),$$

$$\psi_2(u) = \int_{T_1}^{T_2} (\operatorname{Re} u(t))^2 \, dG(t),$$

$$\psi_3(u) = \int_{T_1}^{T_2} \operatorname{Im} u(t))^2 \, dG(t),$$

where G is some distribution function with support $[T_1, T_2]$. Also, let $\psi_4(u)$ be an arbitrary real-valued functional, for which the Lipschitz condition

$$|\psi_4(u) - \psi_4(v)| \leq L \sup_{T_1 \leq t \leq T_2} |u(t) - v(t)|, \quad u, v \in C^2$$

holds with some positive constant L. Suppose that $\psi_k(Y)$ has the density function $f_k(x)$, $k=1, 2, 3, 4$, with respect to the Lebesgue measure. Then, under the conditions of Theorem 6.4.3,

(6.4.14)
$$\sup_{-\infty < x < \infty} |P\{\psi_k(Y_n) < x\} - P\{\psi_k(Y) < x\}| = O(r_1(n)), \quad k = 1, 2, 3, 4,$$

provided that the functions $f_4(x)$, $x^{1/2} f_k(x)$, $k=1, 2, 3$, are bounded.

The proof of this Corollary is similar to that of Corollary 5.4.3.

Corollary 6.4.2 (S. Csörgő 1980). *If $h(x) = x^a$ in (6.4.7), with some positive a, then for $r_1(n)$ of (6.4.8), (6.4.12) and (6.4.14) one has*

$$r_1(n) \approx n^{-\frac{a}{2a+4}} (\log n)^{\frac{a+1}{a+2}},$$

and for $r_2(n)$ of (6.4.9), (6.4.13) one has

$$r_2(n) \approx n^{-\frac{a}{2a+4}} (\log n)^{\frac{2a+1}{a+2}}.$$

Specifically, if $\int_{-\infty}^{\infty} |x|^a dF(x) < \infty$ for arbitrary large a, then $r_1(n) \approx n^{-1/2} \log n$, $r_2(n) \approx n^{-1/2} (\log n)^2$, the respective rate-functions of Theorems 4.4.1 and 4.4.3.

S. Csörgő (1980) also showed that the left hand sides of (6.4.12) and (6.4.13) cannot converge to zero a.s. if the supremum is extended to an infinite interval. However, it can be extended to an interval $[T_n^{(1)}, T_n^{(2)}]$ whose end points tend to infinity at a moderate rate. Studying also the sequence of stochastic processes

$$\{(2 \log \log n)^{-1/2} K_n(t);\ T_1 \leq t \leq T_2\},$$

he proves via Theorem 6.4.3 the following

Theorem 6.4.4 (S. Csörgő 1980). *Let F be as in Theorem 6.4.3. Then the sequence*

$$\{(2 \log \log n)^{-1/2} Y_n(t);\ t \in [T_1, T_2]\}$$

is a.s. relatively compact in $C^2[T_1, T_2]$, and the set of its limit points is

$$\mathscr{G}(F) = \left\{g(t) = \int_{-\infty}^{\infty} \exp(itx)\, df(F(x)), t \in [T_1, T_2]; f \in \mathscr{F}\right\},$$

where $\mathscr{F}$ is the set of those absolutely continuous functions f of $C(0, 1)$ for which we have

$$f(0) = f(1) = 0 \quad \text{and} \quad \int_0^1 (f'(y))^2\, dy \leq 1.$$

As mentioned already, a Theorem 6.4.3-type statement cannot be proved without some conditions which make the tails of F behave regularly. Now for $m = 2, 3, \ldots$ and $\varepsilon > 0$, put

(6.4.15) $$g_m(x) = (\log x)\left(\prod_{k=2}^{m} \log_k x\right)^2,$$

where $\log_j$ denotes the j times iterated logarithm and $\prod_{k=2}^{1}$ is understood as 1. Then we have

Theorem 6.4.5 (S. Csörgő 1980). *For each $m = 2, 3, \ldots$ there exists a distribution function F such that $g_m(x) F(-x) = O(1)$, as $x \to \infty$, and any version of the process $Y(t) = \int_{-\infty}^{\infty} \exp(itx)\, dB(F(x))$ is almost surely discontinuous on each finite interval.*

Hence, if $Y_n(\cdot)$ is defined with an F as in Theorem 6.4.5, then it cannot converge weakly in C^2 to $Y(\cdot)$.

The proof of Theorem 6.4.5 hinges on constructing a discrete distribution function F so that the resulting process $Y(\cdot)$ becomes a discontinuous random Fourier series. On the other hand, a necessary and sufficient condition for the sample-continuity of $Y(\cdot)$ is also given by S. Csörgő (1980) in terms of the behaviour of the characteristic function $c(t)$ around the origin. This condition is, of course, a necessary one for weak convergence of $Y_n(\cdot)$ to $Y(\cdot)$ and his conjecture that it is also sufficient for the latter was subsequently proved to be true by Marcus (1980).

Supplementary remarks

Section 6.1. A number of further empirical density function definitions are used in the literature. Most of them (like, e.g., the definitions by spline functions) are special cases of our Definition 5. Here we mention one which is not a special case of Definition 5.

Definition 6

$$f_n(x) = b_n \left(n(X^{(n)}_{i+[b_n/2]} - X^{(n)}_{i-[b_n/2]}) \right)^{-1} \quad \text{if} \quad X^{(n)}_i \leq x < X^{(n)}_{i+1}.$$

This definition is closely related to Definition 2. Only the length of the "window" here depends on x and the sample. The window will be narrow where the sample is dense. This definition appears to be natural, however its mathematical treatment is quite complicated. Some results are obtained by Tusnády (1974) and Csörgő, Révész (1980c).

The idea of using the method of strong approximation in the investigation of the properties of empirical density functions is due to M. Rosenblatt (1971) (cf. also Bickel–Rosenblatt (1973)). These papers also study the limit distribution of the maximal deviation via evaluating the limit distribution of the supremum of $\Gamma_n(x)$. Our Theorem 6.1.1 gives a framework for the evaluation of the limit distribution of the maximal deviation in more general cases than those of Rosenblatt, provided one could only evaluate the limit distribution of the supremum of Γ_n in these more general situations. The latter, however, seems to be quite difficult to do. Case (b) of Theorem 6.1.5 is essentially due to Bickel and Rosenblatt (1973).

The proof of Theorem 6.1.2 is based on the Karhunen–Loève expansion of a Gaussian process (cf. Karhunen 1946, 1947). This theory was not covered at all in this book. For details we refer to Yeh ((1973), p. 283) and here we give a short hint of it. Let $\{G(x); A < x < B\}$ be a Gaussian process with

$$EG(x) = 0, \quad EG(x)G(y) = R(x, y)$$

and assume that
$$\int_A^B \int_A^B R^2(x, y)\, dx\, dy < \infty.$$

Then we can consider the integral operator
$$\mathcal{R}f = \int_A^B R(x, y) f(y)\, dy \colon L^2(A, B) \to L^2(A, B).$$

Let $\varphi_k(u)$ resp. λ_k be the sequence of eigenfunctions resp. eigenvalues of the operator $\mathcal{R}$. It is well known (cf. Riesz, Sz.-Nagy 1955, p. 243) that $\{\varphi_k(x);\ A < x < B,\ k = 1, 2, \ldots\}$ is an orthonormal sequence,
$$\int_A^B R(x, x)\, dx = \sum_{k=1}^{\infty} \lambda_k \quad \text{and} \quad \int_A^B \int_A^B R^2(x, y)\, dx\, dy = \sum_{k=1}^{\infty} \lambda_k^2.$$

The theory of Karhunen–Loève expansion says that the Gaussian process $G(x)$ can be written in the form
$$G(x) = \sum_{k=1}^{\infty} \lambda_k^{1/2} \varphi_k(x)\, N_k,$$
where $\{N_k\}$ is a sequence of independent $\mathcal{N}(0, 1)$ r.v. and the nature of convergence of the above series is meant to be
$$\int_A^B \left(\sum_{k=1}^{m} \lambda_k^{1/2} \varphi_k(x) N_k - G(x) \right)^2 dx \xrightarrow{\text{a.s.}} 0 \quad \text{as} \quad m \to \infty.$$

This expansion of $G(x)$ immediately implies that
$$\int_A^B G^2(x)\, dx = \sum_{k=1}^{\infty} \lambda_k N_k^2,$$
$$E \int_A^B G^2(x)\, dx = \sum_{k=1}^{\infty} \lambda_k = \int_A^B R(x, x)\, dx,$$
$$E\left(\int_A^B G^2(x)\, dx \right)^2 = E\left\{ \sum_{k=1}^{\infty} \lambda_k^2 N_k^4 + 2 \sum_{1 \le j < k < \infty} \lambda_j \lambda_k N_j^2 N_k^2 \right\}$$
$$= 3 \sum_{k=1}^{\infty} \lambda_k^2 + 2 \sum_{1 \le j < k < \infty} \lambda_j \lambda_k$$
and
$$\operatorname{Var} \int_A^B G^2(x)\, dx = 2 \sum_{k=1}^{\infty} \lambda_k^2.$$

Section 6.3. For the consistency of a general class of non-parametric regression estimators we refer to Stone (1977).

7. Random Limit Theorems via Strong Invariance Principles

7.0. Introduction and some historical remarks

One of the earliest papers dealing with random sum limit theorems was published by Kac (1949). He introduced Poisson random-size samples for the sake of making the problem of weak convergence of the empirical process easier. His idea of Poissonization turned out to be very useful also later on, when the need for tackling the problem of strong invariance principles for multivariate empirical processes became apparent after Kiefer's 1972 paper (cf. e.g. Wichura 1973, Csörgő, Révész 1975b, 1975d, Révész 1976).

Let N_λ be a Poisson r.v. with mean λ and, for every $\lambda > 0$, let N_λ, $U_1, U_2, \ldots$ be independent r.v., where for $i = 1, 2, \ldots$ the U_i are $U(0, 1)$ r.v.

Kac (1949) defines his modified empirical distribution function by the formula

$$(7.0.1) \qquad E_\lambda^*(y) = \lambda^{-1} \sum_{i=1}^{N_\lambda} I_{(0, y]}(U_i), \quad 0 \le y \le 1,$$

where the sum is taken to be zero if $N_\lambda = 0$, and his modified empirical process by

$$(7.0.2) \qquad K_\lambda(y) = \sqrt{\lambda}(E_\lambda^*(y) - y), \quad 0 \le y \le 1, \quad \lambda \ge 0,$$

and notes that, for every fixed λ, $\{K_\lambda(y); 0 \le y \le 1\}$ is an independent increment process and $EK_\lambda(y) = 0$, $EK_\lambda^2(y) = y$. Moreover if $\lambda \to \infty$, then, for every fixed y, $K_\lambda(y)$ tends in distribution to a $\mathcal{N}(0, y)$ r.v. He then writes (cf. Kac (1949), pp. 253–254): "Thus it is natural to expect that the limiting properties of the process $K_\lambda(y)$ will be those of a Gaussian process $X(y)$ ($X(0) = 0$) with independent increments and such that $EX(y) = 0$, $EX^2(y) = y$. These are characteristics of a Wiener process." Kac does not actually prove

$$(7.0.3) \qquad K_\lambda(\cdot) \xrightarrow{\mathscr{D}} W(\cdot), \quad \lambda \to \infty.$$

Hewever, he proves rigorously that

(7.0.4) $$\sup_{0\leq y\leq 1} |K_\lambda(y)| \xrightarrow{\mathscr{D}} \sup_{0\leq y\leq 1} |W(y)|,$$

using the fact that $K_\lambda(y)$ is an independent increment process. With proving (7.0.4) however, Kac came pretty close to proving also

(7.0.5) $$\sup_{0\leq y\leq 1} |\alpha_n(y)| \xrightarrow{\mathscr{D}} \sup_{0<y<1} |B(y)|, \quad \text{as} \quad n\to\infty,$$

the very question Doob (1949) posed and argued.

Just to see how $\alpha_n(y)$ and $K_\lambda(y)$ hang together, we may write

$$K_\lambda(y) = \sqrt{\frac{N_\lambda}{\lambda}} \sqrt{N_\lambda}(E_{N_\lambda}(y)-y) + y(N_\lambda-\lambda)/\sqrt{\lambda}$$

$$= \sqrt{\frac{N_\lambda}{\lambda}} \alpha_{N_\lambda}(y) + y(N_\lambda-\lambda)/\sqrt{\lambda}, \quad 0\leq y\leq 1.$$

Since N_λ is independent of $\alpha_n(y)$, by (4.2.2) we have $\alpha_{N_\lambda}(\cdot) \xrightarrow{\mathscr{D}} B(\cdot)$ as $\lambda\to\infty$, and, since $N_\lambda/\lambda \xrightarrow{P} 1$ as $\lambda\to\infty$, we also have

(7.0.6) $$\sqrt{\frac{N_\lambda}{\lambda}} \alpha_{N_\lambda}(\cdot) \xrightarrow{\mathscr{D}} B(\cdot), \quad \lambda\to\infty.$$

Also, $y(N_\lambda-\lambda)/\sqrt{\lambda} \xrightarrow{\mathscr{D}} y\tilde{W}(1)$, where $\tilde{W}(1)$ is a standard normal r.v. and, since by assumption N_λ is independent of the sequence $\{U_i\}$, $\tilde{W}(1)$ is independent of the Brownian bridge $B(\cdot)$ of (7.0.6). Hence $W(y) = = B(y) + y\tilde{W}(1)$ is again a standard Wiener process, and so we have $K_\lambda(\cdot) \xrightarrow{\mathscr{D}} W(\cdot)$. Viewing and getting (7.0.3) this way, also throws further light on why Kac's idea works for a direct verification of (7.0.4).

We could, of course, also argue the other way around, saying that $K_\lambda(\cdot) \xrightarrow{\mathscr{D}} W(\cdot)$, $N_\lambda/\lambda \xrightarrow{P} 1$ and $y(N_\lambda-\lambda)/\sqrt{\lambda} \xrightarrow{\mathscr{D}} y\tilde{W}(1)$, and then conclude that $\alpha_{N_\lambda}(\cdot) \xrightarrow{\mathscr{D}} B(\cdot)$. Another simple step now takes us to the statement $\alpha_n(\cdot) \xrightarrow{\mathscr{D}} B(\cdot)$ as $n\to\infty$.

The example of the process $K_\lambda(y)$, the quoted further papers which use the idea of Poissonization and, indeed, also the very stopping time theorem of Skorohod itself, show that the notion of randomly stopped processes has played a significant role in the development of our view of invariance. One of the aims of this chapter is to call attention to the fact that the now extensive theory and methodology of strong invariance principles can, in turn, be applied to studying similar and weak convergence properties of randomly selected sequences. We demonstrate this in Section 7.3 for partial sums of r.v. and, in Section 7.4, for empirical and quantile processes. In Section 7.2 we review some almost sure and in-probability convergence statements for randomly selected sequences, which will then be applied in Sections 7.3 and 7.4.

This whole chapter is based on a recent paper by M. Csörgő, S. Csörgő, Fischler and Révész (1975), where a somewhat more extensive bibliography than that of this chapter can be also found (cf. also Karlsson and Szász (1974)).

7.1. Laws of large numbers for randomly selected sequences

Let $\{Z_n\}$ be a sequence of random variables and let $\{v_n\}$ be a sequence of positive integer valued random variables defined on the same probability space. We have

Theorem 7.1.1. (a) Let r and s be real numbers, $r \geq 1$ and $s > r+2$. Assume that for any $\varepsilon > 0$ we have

$$\sum_{n=1}^{\infty} n^{s-2} P\{|Z_n| > \varepsilon\} < \infty \quad \text{and} \quad \sum_{n=1}^{\infty} n^{r-2} P\left\{\left|\frac{v_n}{n} - v\right| > \varepsilon\right\} < \infty,$$

where v is a positive random variable with $P\{a \leq v \leq b\} = 1$ for some $0 < a < b < \infty$. Then

$$\sum_{n=1}^{\infty} n^{r-2} P\{|Z_{v_n}| > \varepsilon\} < \infty.$$

(b) $Z_n \xrightarrow{\text{a.s.}} 0$ and $v_n \xrightarrow{\text{a.s.}} \infty$ imply $Z_{v_n} \xrightarrow{\text{a.s.}} 0$.

(c) $Z_n \xrightarrow{\text{a.s.}} 0$ and $v_n \xrightarrow{P} \infty$ imply $Z_{v_n} \xrightarrow{P} 0$.

(d) $Z_n \xrightarrow{P} 0$ and $v_n \xrightarrow{P} \infty$ do not necessarily imply $Z_{v_n} \xrightarrow{P} 0$.

(e) $Z_n \xrightarrow{P} 0$ and $v_n/f(n) \xrightarrow{P} v$, where v is an arbitrary positive r.v. and $f(n) \nearrow \infty$, imply $Z_{v_n} \xrightarrow{P} 0$, provided that $Z_n = \sum_{i=1}^{n} X_i/n$, where $X_1, X_2, \ldots$ are independent r.v.

Proof. The statements (b), (c), (d) and (e) are well known and immediate. The proof of (b) can, for example, be found in Csörgő (1968) while that of (c) in Csörgő–Fischler (1970), where we also deal with (d). The statement (e) was proved by Mogyoródi (1965). Also Révész ((1968), Theorem 10.2) proves (e) via replacing the condition $v_n/f(n) \xrightarrow{P} v$ by the weaker one: for every $\varepsilon > 0$ there exist $0 < a = a(\varepsilon) < b = b(\varepsilon) < \infty$ and $n_0(\varepsilon)$ such that

$P\{af(n) < v_n < bf(n)\} \geq 1-\varepsilon$, whenever $n > n_0(\varepsilon)$, where $f(n) \nearrow \infty$. Hence we prove only (a) here, which, roughly speaking, says that the total convergence of $\{Z_n\}$ and $\{v_n/n - v\}$ implies that of $\{Z_{v_n}\}$.

Clearly, for any n and $\varepsilon > 0$, and for a, b as in (a), we have

$$P\{|Z_{v_n}| > \varepsilon\} \leq P\{|Z_{v_n}| > \varepsilon, \; n(a-\varepsilon) \leq v_n \leq n(b+\varepsilon)\} + P\left\{\left|\frac{v_n}{n} - v\right| > \varepsilon\right\}$$

$$\leq P\{\sup_{m \geq n(a-\varepsilon)} |Z_m| > \varepsilon\} + P\left\{\left|\frac{v_n}{n} - v\right| > \varepsilon\right\}$$

$$\leq \sum_{m=[n(a-\varepsilon)]}^{\infty} P\{|Z_m| > \varepsilon\} + P\left\{\left|\frac{v_n}{n} - v\right| > \varepsilon\right\}.$$

Given our assumptions in (a), we have $P\{|Z_m| > \varepsilon\} = (1/m)^{s-2} o(1)$ for $s > r+2$, $r \geq 1$. Let θ be an arbitrary positive number. There exists then $m_0 = m_0(\varepsilon, \theta)$ so that, whenever $m \geq m_0$, we have

$$\sum_{m=[n(a-\varepsilon)]}^{\infty} P\{|Z_m| > \varepsilon\} \leq \sum_{m=[n(a-\varepsilon)]}^{\infty} \frac{\theta}{m^{s-2}}, \quad s > r+2, \quad r \geq 1.$$

Whence, in order to prove (a), it suffices to consider the following problem of convergence:

$$\theta \sum_{n=1}^{\infty} n^{r-2} \sum_{m=[n(a-\varepsilon)]}^{\infty} \frac{1}{m^{s-2}}$$

$$\leq \theta \sum_{n=1}^{\infty} n^{r-2} \left\{ \frac{1}{[n(a-\varepsilon)]^{s-3}} + \int_{[n(a-\varepsilon)]}^{\infty} \frac{1}{x^{s-2}} dx \right\}$$

$$\leq \frac{\theta}{(a-\varepsilon)^{s-3}} \sum_{n=1}^{\infty} \frac{n^{r-2}}{n^{s-3}} \left(1 - \frac{1}{s-3}\right).$$

The latter sum is finite for every $r \geq 1$, provided we choose $s > r+2$ as stipulated in our assumptions. This also completes the proof of (a) of Theorem 7.1.1.

Example 7.1.1. Let X_k be a sequence of i.i.d.r.v. with $EX_1 = 0$ and let $S_n = \sum_{k=1}^{n} X_k$, $S_0 \equiv 0$. Let $\{v_n\}$ be a sequence of Poisson r.v. with $Ev_n = n$ ($n = 1, 2, \ldots$). Then we know (cf. Katz (1963)) that the statements

(i) $\qquad\qquad EX_1 = 0, \quad E|X_1|^r < \infty, \quad r \geq 1,$

and

(ii) $\qquad\qquad \sum_{n=1}^{\infty} n^{r-2} P\{|S_n| > n\varepsilon\} < \infty \quad \text{for any} \quad \varepsilon > 0$

are equivalent (the equivalence of (i) and (ii) was first proved by Spitzer (1956) for $r=1$ and, for $r=2$, by Erdős (1949, 1950)). Now $v_n \overset{\mathscr{D}}{=} \sum_{i=1}^{n} Y_i$, where the Y_i are i.i.d. Poisson r.v. with $EY_i=1$. Whence, for any $r \geq 1$,

$$\sum_{n=1}^{\infty} n^{r-2} P\left\{\left|\frac{v_n}{n}-1\right|>\varepsilon\right\}<\infty, \quad \varepsilon>0.$$

Consequently,

(7.1.1) $$\sum_{n=1}^{\infty} n^{r-2} P\{|S_{v_n}|>v_n \varepsilon\}<\infty,$$

provided that, in addition to $EX_1=0$, we also assume $E|X_1|^s<\infty$ for $s>r+2, r \geq 1$.

We also note that a special case of $r=2$ of (a) of Theorem 7.1.1 was proved by Szynal (1973) for a sequence of quantiles of a random-size sample. Szynal (1972) proved also (7.1.1) for $r=2$. Example 7.1.1 shows that (a) of Theorem 7.2.1 always implies an Erdős–Spitzer–Katz type statement for randomly selected partial sums of i.i.d.r.v., provided the summands have at least $3+\delta$ ($\delta>0$) moments.

Example 7.1.2. It is clear from the proof of (a) of Theorem 7.1.1 that $P\{|Z_{v_n}|>\varepsilon\}$ converges to zero exponentially fast, provided $P\{|Z_n|>\varepsilon\}$ and $P\left\{\left|\frac{v_n}{n}-v\right|>\varepsilon\right\}$, with v as in (a), do so themselves. In the latter case we have

$$\sum_{n=1}^{\infty} g(n) P\{|Z_{v_n}|>\varepsilon\}<\infty, \quad \varepsilon>0,$$

for any polynomial function $g(n)$ of n. An example for this situation is provided by letting Z_n of (a) be $Z_n=n^{-1/2} \sup_{0 \leq y \leq 1}|\alpha_n(y)|$, where $\alpha_n(y)$ is the empirical process. Then, by (4.1.6), $P\{|Z_n|>\varepsilon\}$ converges to zero exponentially fast, and so does also $P\{|Z_{v_n}|>\varepsilon\}$, provided that $P\left\{\left|\frac{v_n}{n}-v\right|>\varepsilon\right\}$ does the same (e.g. if $v_n \overset{\mathscr{D}}{=} \sum_{i=1}^{n} Y_i$ is as in Example 7.1.1).

7.2. Invariance (strong and weak) principles for random-sum limit theorems

While it is true that in this book we deal only with the problem of strong approximation of partial sums and empirical processes of i.i.d.r.v., it will be seen that the main idea of this section and also that of the next one simply amount to saying that whatever processes one might have succeeded in strongly approximating by appropriate Gaussian processes, then weak convergence properties of certain randomized versions of the latter will be also inherited by the former. In order to be able to fully utilize also this latter type of inheritance for partial sums, we are going to deviate now somewhat from our i.i.d. setting. Philipp and Stout (1975) (cf. also Berkes and Philipp (1979)) developed methods to prove almost sure invariance principles for sums of weakly dependent (e.g. strongly mixing, lacunary trigonometric, Gaussian, asymptotic martingale difference sequences and certain Markov processes) random variables. In order to be able to "summarize" their results in a single statement, here we are not going to make the notion of weak dependence precise but will simply call all those sequences of random variables weakly dependent for which Philipp and Stout (1975), and Berkes and Philipp (1979) proved relation (7.2.1) underneath. In this setting then, we have:

Proposition 7.2.1. *Let $X_1, X_2,$ be a sequence of weakly dependent r.v. with $EX_i = 0$ and $E|X_i|^{2+\delta} < \infty$ for some $\delta > 0$, and $i = 1, 2, \ldots$. Assume also that $\lim_{n \to \infty} n^{-1} ES_n^2 = 1$. Then, possibly under further conditions which, in turn, would depend on any given specific notion of weak dependence, there exists a Wiener process $\{W(t); 0 \leq t < \infty\}$ such that, with $S_t = \sum_{i \leq t} X_i$, we have*

(7.2.1) $$|S_t - W(t)| \stackrel{a.s.}{=} O(t^{1/2 - \eta}),$$

where η is a positive number, depending only on the sequence $\{X_i\}$.

Combining first (a) of Theorem 7.1.1 with Theorem 2.6.7, we get

Theorem 7.2.1. *Let $\{v_n\}$ be as in (a) of Theorem 7.1.1, i.e.,*

$$\sum_{n=1}^{\infty} n^{r-2} P\left\{\left|\frac{v_n}{n} - v\right| > \varepsilon\right\} < \infty$$

for some $r \geq 1$ and any $\varepsilon > 0$, where v is such a positive r.v. that for some

$0 < a < b < \infty$ we have $P\{a < v < b\} = 1$. Further let $X_1, X_2, \ldots$ be i.i.d.r.v. with $EX_1 = 0$, $EX_1^2 = 1$. Assume also that $E|X_1|^p < \infty$ for some $p > 2r+4$ (≥ 6). Then, if the sequences of r.v. $\{v_n\}$ and $\{X_n\}$ are defined on the same probability space, there exists a Wiener process $\{W(t); 0 \leq t < \infty\}$ such that, for any $\varepsilon > 0$, we have

$$(7.2.2) \qquad \sum_{n=1}^{\infty} n^{r-2} P\left\{\left|\sup_{0 \leq t \leq 1} S_{[v_n t]} - W(v_n t)\right| > \varepsilon \sqrt{v_n}\right\} < \infty.$$

Proof. Letting now $p/2 > s > r+2$ and $Z_n = \sup_{0 \leq t \leq 1} |S_{[nt]} - W(nt)|/\sqrt{n}$, it follows by Theorem 2.6.7 that

$$\sum_{n=1}^{\infty} n^{s-2} P\{|Z_n| > \varepsilon\} < \frac{C_p}{\varepsilon^p} \sum_{n=1}^{\infty} \frac{O(n)}{n^{p/2-s+2}} < \infty,$$

and (a) of Theorem 7.1.1, in turn, implies (7.2.2).

Replacing the conditions of Theorem 7.2.1 on $\{v_n\}$ by $v_n \xrightarrow{\text{a.s.}} \infty$ as $n \to \infty$, then, by (2.6.4) and (b) of Theorem 7.1.1, instead of (7.2.2), we get

$$(7.2.3) \qquad v_n^{-1/p} \sup_{0 \leq t \leq 1} |S_{[v_n t]} - W(v_n t)| \xrightarrow{\text{a.s.}} 0, \quad p > 2,$$

and, if we only assume that $v_n \xrightarrow{P} \infty$ as $n \to \infty$, then, by (2.6.4) and (c) of Theorem 7.1.1,

$$(7.2.4) \qquad v_n^{-1/p} \sup_{0 \leq t \leq 1} |S_{[v_n t]} - W(v_n t)| \xrightarrow{P} 0, \quad p > 2.$$

Since $p > 2$, (7.2.3) implies

$$(7.2.5) \qquad v_n^{-1/2} \sup_{0 \leq t \leq 1} |S_{[v_n t]} - W(v_n t)| \xrightarrow{\text{a.s.}} 0 \quad \text{as} \quad n \to \infty,$$

and (7.2.4) implies

$$(7.2.6) \qquad v_n^{-1/2} \sup_{0 \leq t \leq 1} |S_{v_n t} - W(v_n t)| \xrightarrow{P} 0 \quad \text{as} \quad n \to \infty.$$

Remark 7.2.1. In the sequel we are going to study the question of how a (7.2.4)-type statement should imply weak convergence of a (properly normed) sequence of processes $\{S_{[v_n t]}; 0 \leq t \leq 1\}$. We should also note that (b) of Theorem 7.1.1, (7.2.1) and $v_n \xrightarrow{\text{a.s.}} \infty$ ($n \to \infty$) imply (7.2.5), and (c) of Theorem 7.1.1, (7.2.1) and $v_n \xrightarrow{P} \infty$ ($n \to \infty$) imply (7.2.6), with $S_{[v_n t]}$ in the general terms of Proposition 7.2.1 both times.

As just mentioned, our aim now is to study how our results so far should imply weak convergence for randomly selected partial sum type processes.

A general formulation of the latter problem can be described as follows. Let $\{S_k\}$ be a sequence of partial sums of any r.v., and let $\{v_n\}$ be a sequence of positive integer valued r.v. on the same probability space. Assuming that $n^{-1/2} S_{[n \cdot]} \xrightarrow{\mathscr{D}} W(\cdot)$, where $\{W(t); 0 \leq t \leq 1\}$ is a standard Wiener process, then for what kind of r.v. is it going to be also true that $v_n^{-1/2} S_{[v_n \cdot]} \xrightarrow{\mathscr{D}} W(\cdot)$, where it is also assumed that the sequence $\{v_n\}$ satisfies the

Condition (cf. (e) of Theorem 7.1.1)

(7.2.7) $\quad v_n/f(n) \xrightarrow{P} v, \quad P\{v > 0\} = 1, \quad f(n) \nearrow \infty \quad \text{as} \quad n \to \infty.$

Such a general formulation of our problem was initiated by Billingsley (1962), and general methods for its solution were given, for example, by M. Csörgő and S. Csörgő (1973), Fischler (1976) and Rootzén (1974). A common property of the latter four papers is that, in order to solve the above problem of weak convergence, they work directly with the random-sequence of stochastic processes $\{v_n^{-1/2} S_{[v_n t]}; 0 \leq t \leq 1\}$. Our thesis now amounts to saying that the above invariance principles enable one to work with the generally simpler random sequence of stochastic processes $\{v_n^{-1/2} W(v_n t); 0 \leq t \leq 1\}$ in order to obtain a weak convergence statement for the latter and then to translate the thus obtained weak convergence to that of $\{v_n^{-1/2} S_{[v_n t]}; 0 \leq t \leq 1\}$.

As to the Condition (7.2.7), we note that stonger conditions like those of Theorem 7.2.1 on $\{v_n\}$ are only assumed for the sake of having rate of convergence statements like, e.g., that of (7.2.2) in mind. In general, the least we must assume is that $v_n \xrightarrow{P} \infty$ as $n \to \infty$. On the other hand, the latter condition is not enough (cf. pp. 143–144 in Billingsley 1968) if we do not wish to assume anything re. the independence of $\{v_n\}$ and $\{S_n\}$. While the latter independence assumption can be very helpful on occasions (cf., e.g., Section 0 of this chapter), we do not wish to, and, indeed, we cannot assume it in general. For this reason we must, therefore, postulate something about the way v_n goes to infinity in probability as compared to a sequence of numbers $f(n) \nearrow \infty$. Since an example of Rényi (1960) shows that it is also not enough to require only the convergence in distribution of $v_n/f(n)$ to a positive r.v. v, we can, therefore conclude that (7.2.7) is indeed the most general and meaningful condition to assume for our problem at hand (cf. also Aldous 1978).

Theorem 7.2.2. Let $\{X_i\}$ be such a sequence of r.v. for which (7.2.1) of Proposition 7.2.1 holds, and let $\{v_n\}$ be a sequence of positive integer valued r.v. defined on the probability space of the latter. Assume also that $\{v_n\}$ satisfies Condition (7.2.7). Then

(7.2.8) $$v_n^{-1/2} W(v_n \cdot) \xrightarrow{\mathscr{D}} W(\cdot),$$

and whence,

(7.2.9) $$v_n^{-1/2} S_{[v_n \cdot]} \xrightarrow{\mathscr{D}} W(\cdot).$$

The proof of this theorem is based on three lemmas.

Lemma 7.2.1. Under the conditions of Theorem 7.2.2 we have

(7.2.10) $$\sup_{0 \leq t \leq 1} \left| \frac{W(v_n t)}{\sqrt{v_n}} - \frac{W(f(n)vt)}{\sqrt{f(n)v}} \right| \xrightarrow{P} 0 \quad as \quad n \to \infty.$$

Proof. $\frac{v_n}{f(n)} = v + \varepsilon_n$. Then by Condition (7.2.7) we have that for any $\varepsilon > 0$ there exist $0 < \delta < a < b < \infty$ and an integer n_0 such that

(7.2.11) $$P\{a \leq v \leq b\} \geq 1 - \varepsilon, \quad P\{|\varepsilon_n| \leq \delta\} \geq 1 - \varepsilon,$$

whenever $n \geq n_0$.

Now on the set $\{a \leq v \leq b\} \cap \{|\varepsilon_n| \leq \delta\}$ we have

$$\sup_{0 \leq t \leq 1} \left| \frac{W(v_n t)}{\sqrt{v_n}} - \frac{W(f(n)vt)}{\sqrt{f(n)v}} \right|$$

$$= \sup_{0 \leq t \leq 1} \left| \frac{W(f(n)vt + \varepsilon_n f(n)t)}{\sqrt{vf(n)}} \cdot \sqrt{\frac{v}{v + \varepsilon_n}} - \frac{W(f(n)vt)}{\sqrt{vf(n)}} \right|$$

$$\leq \sup_{0 \leq t \leq 1} \left| \frac{W(f(n)vt + \varepsilon_n f(n)t)}{\sqrt{vf(n)}} - \frac{W(f(n)vt)}{\sqrt{vf(n)}} \right|$$

$$+ \frac{|\varepsilon_n|}{a - \delta} \sup_{0 \leq t \leq 1} \frac{|W(f(n)vt + \varepsilon_n f(n)t)|}{\sqrt{vf(n)}}$$

$$\leq \sup_{0 \leq x \leq f(n)b} \sup_{0 \leq s \leq \delta f(n)} \frac{|W(x+s) - W(x)|}{\sqrt{af(n)}}$$

$$+ \frac{|\varepsilon_n|}{a - \delta} \sup_{0 \leq x \leq f(n)(b+\delta)} \frac{|W(x)|}{\sqrt{af(n)}}$$

$$= I_1(n) + \frac{|\varepsilon_n|}{a - \delta} I_2(n).$$

By (7.2.11) and the above inequalities, we get for any $\gamma > 0$,

$$P\left\{\sup_{0 \leq t \leq 1}\left|\frac{W(v_n t)}{\sqrt{v_n}} - \frac{W(f(n)vt)}{\sqrt{f(n)v}}\right| > \gamma\right\}$$

$$\leq 2\varepsilon + P\left\{I_1(n) > \frac{\gamma}{2}\right\} + P\left\{\frac{|\varepsilon_n|}{a-\delta} I_2(n) > \frac{\gamma}{2}\right\},$$

where $P\left\{I_1(n) > \frac{\gamma}{2}\right\} \to 0$ by Lemma 1.2.1, and $P\left\{\frac{|\varepsilon_n|}{a-\delta} I_2(n) > \frac{\gamma}{2}\right\} \to 0$ since $\varepsilon_n \xrightarrow{P} 0$. Hence (7.2.10) is now proved.

Lemma 7.2.2. *Let A be any event of positive probability. Then*

(7.2.12) $\left\{\dfrac{W(nt)}{\sqrt{n}}; 0 \leq t \leq 1 | A\right\} \xrightarrow{\mathscr{D}} \{W(t); 0 \leq t \leq 1\},$

where (7.2.12) means that the sequence of conditional probability measures generated by the process $\dfrac{W(nt)}{\sqrt{n}}$ given A, converges to the Wiener measure.

The proof of this lemma is quite simple. Here we mention only that, for example, the proof of Rényi's mixing theorem (cf. Theorem 2.1.3) can be adopted also in this situation. Another possibility is to check that the conditional finite dimensional distributions of $W(nt)/\sqrt{n}$ converge to the finite dimensional distributions of W and then (7.2.12) follows from continuity of the latter.

Now Lemma 7.2.2 implies

Lemma 7.2.3. *Let v be any positive r.v. Then we have*

(7.2.13) $\left\{\dfrac{W(nvt)}{\sqrt{nv}}; 0 \leq t \leq 1\right\} \xrightarrow{\mathscr{D}} \{W(t); 0 \leq t \leq 1\}.$

Proof of Theorem 7.2.2. (7.2.8) follows by Lemmas 7.2.1 and 7.2.3 and (7.2.9) is a consequence of (7.2.6) (cf. Remark 7.2.1) and (7.2.8).

Remark 7.2.2. Naturally (7.2.9) implies that for every continuous functional $h: C(0, 1) \to C(0, 1)$ we have

(7.2.14) $h\left(\dfrac{S_{[v_n t]}}{\sqrt{v_n}}\right) \xrightarrow{\mathscr{D}} h(W(t)).$

We have also

(7.2.15) $\left|h\left(\dfrac{S_{[v_n t]}}{\sqrt{f(n)}}\right) - h\left(\dfrac{W(v_n t)}{\sqrt{f(n)}}\right)\right| \xrightarrow{P} 0,$

provided that h satisfies also the Lipschitz condition of (5.4.11). Let us now consider the functional $h(x(t))=x(1)$. Then by Lemma 7.2.2, and given the conditions of Theorem 7.2.2, we have

$$(7.2.16) \qquad \lim_{n\to\infty} P\{S_{v_n} \le x f^{1/2}(n)\} = \int_0^\infty \Phi\left(\frac{x}{\sqrt{y}}\right) dP\{v \le y\},$$
$$-\infty < x < +\infty.$$

The latter result is proved for an i.i.d. sequence $\{X_i\}$ by Wittenberg (1964) with $f(n)=n$ and, independently, also by Mogyoródi (1966) in the above form. In their case only two moments are needed for (7.2.16) with i.i.d.r.v. Whence their result does not follow from the more general but $2+\delta$ moments-setting of our (7.2.16). Indeed in many important theorems concerning weak convergence of partial sums of r.v. one assumes only the existence of the second moments of the summands (e.g. Donsker's theorem; cf. Theorem 0.1). Hence, results like the randomized version of Donsker's theorem (cf. Theorem 17.2 in Billingsley 1968 and also the predecessor of the latter, namely the random sum central limit theorem of Mogyoródi 1962 and Blum, Hanson and Rosenblatt 1963), and also the just mentioned Wittenberg–Mogyoródi theorem as well as S. Csörgő's (1974) random versions of the Erdős–Kac (1946) theorems do not follow from our Theorem 7.2.2. However, in case of two moments only, it is again possible to prove a Theorem 7.2.2 type statement for partial sums of i.i.d.r.v. This result will imply also the just mentioned ones. The latter program is feasible on account of Theorem S.2.2.1.

Corollary 7.2.1 (Horváth 1978). Let $X_1, X_2, \ldots$ and $Y_1, Y_2, \ldots$ be r.v. as in Theorem S.2.2.1, and let $\{v_n\}$ be a sequence of positive integer valued r.v. defined on the probability space of the latter. Assume also that $\{v_n\}$ satisfies Condition (7.2.7). Then

$$(7.2.17) \qquad v_n^{1/2} \sup_{0 \le t \le 1} |S_{[v_n t]} - T_{[v_n t]}| \xrightarrow{P} 0.$$

Also,

$$(7.2.18) \qquad v_n^{1/2} T_{[v_n \cdot]} \xrightarrow{\mathscr{D}} W(\cdot),$$

and, whence,

$$(7.2.19) \qquad v_n^{1/2} S_{[v_n \cdot]} \xrightarrow{\mathscr{D}} W(\cdot),$$

where T_n is as in Theorem S.2.2.1.

Proof. Condition (7.2.7) and Theorem S.2.2.1 imply (7.2.17) by (c) of Theorem 7.1.1. The rest of the proof is similar to that of Theorem 7.2.2. For details we refer to Horváth (1978).

Remark 7.2.3. We note that in case of two moments only, a direct approach like that of M. Csörgő and S. Csörgő (1973) and especially that of Aldous (1978) for example, leads to more general results than the ones we can deduce from Corollary 7.2.1. We will now see that a more natural application of our method is to empirical processes with random size samples.

7.3. Invariance (strong and weak) principles for random size empirical processes

Our first theorem is a parallel of Theorem 7.2.1.

Theorem 7.3.1. *Let $\alpha_n(y)$ and $K(y, n)$ be as in Theorem 4.4.3 and let $\{v_n\}$ be a sequence of positive integer valued random variables such that for some number $r \geq 1$ we have*

$$\sum_{n=1}^{\infty} n^{r-2} P\left\{\left|\frac{v_n}{n} - v\right| > \varepsilon\right\} < \infty$$

for any $\varepsilon > 0$, where v is a positive random variable with $P\{a \leq v \leq b\} = 1$ for some $0 < a < b < \infty$. Then

(7.3.1) $$\sum_{n=1}^{\infty} n^{r-2} P\left\{\sup_{0 \leq y \leq 1} |\alpha_{v_n}(y) - v_n^{-1/2} K(y, v_n)| > \varepsilon\right\} < \infty.$$

Proof. For an arbitrary $\varepsilon > 0$, let $x = \frac{\varepsilon}{2} \sqrt{n}/\log n$ in (4.4.22). Then, by the latter inequality, we have (for n large)

$$P_n(\varepsilon) = P\left\{\sup_{0 \leq y \leq 1} |\alpha_n(y) - K(y, n)/n^{1/2}| > \varepsilon\right\} \leq L e^{-\lambda \frac{\varepsilon}{2}\sqrt{n}/\log n}.$$

Whence, for any polynomial function $g(n)$ of n, we have

$$\sum_{n=1}^{\infty} g(n) P_n(\varepsilon) < \infty.$$

The latter combined with (a) of Theorem 7.1.1 now gives (7.3.1).

Theorem 7.3.2. Let $\alpha_n(y)$, $u_n(y)$, $q_n(y)$, $K(y, n)$ be as in Theorems 4.4.3, 4.5.3 and 4.5.7 respectively, and let $\{v_n\}$ be a sequence of positive integer valued random variables, and assume that $v_n \xrightarrow{P} \infty$ as $n \to \infty$. Then

$$(7.3.2) \qquad \sup_{0 \leq y \leq 1} |\alpha_{v_n}(y) - v_n^{-1/2} K(y, v_n)| \xrightarrow{P} 0,$$

$$(7.3.3) \qquad \sup_{0 \leq y \leq 1} |u_{v_n}(y) - v_n^{-1/2} K(y, v_n)| \xrightarrow{P} 0,$$

$$(7.3.4) \qquad \sup_{0 < y < 1} |f(F^{-1}(y)) q_{v_n}(y) - v_n^{-1/2} K(y, v_n)| \xrightarrow{P} 0.$$

We note that (7.3.2) also holds true with $y = F(x)$ for $\alpha_n(F(x)) = \beta_n(x)$. Further, if instead of $v_n \xrightarrow{P} \infty$ we have that $v_n \xrightarrow{a.s.} \infty$, then (7.3.2)–(7.3.4) also hold true with probability one.

Proof of Theorem 7.3.2. Combining the respective statements (b) and (c) of Theorem 7.1.1 with (4.4.23), (4.5.8) and (4.5.25), the above statements follow.

Theorem 7.3.3. *If v_n of Theorem 7.3.2 also satisfies the condition (7.2.7), then*

$$(7.3.5) \qquad \alpha_{v_n}(\cdot) \xrightarrow{\mathscr{D}} B(\cdot),$$

$$(7.3.6) \qquad f(F^{-1}(\cdot)) q_{v_n}(\cdot) \xrightarrow{\mathscr{D}} B(\cdot).$$

Proof. First we note that (7.2.7) implies that $v_n \xrightarrow{P} \infty$ as $n \to \infty$. It follows then from (7.3.2) and (7.3.4) that, in order to prove (7.3.5) and (7.3.6), it suffices to show that

$$(7.3.7) \qquad v_n^{-1/2} K(\cdot, v_n) \xrightarrow{\mathscr{D}} B(\cdot).$$

Now the proof of (7.3.7) can be done along the lines of that of (7.2.8).

Remark 7.3.1. The first paper on the random sample size empirical process α_{v_n} was written by Pyke (1968). He proved (7.3.5) under the assumption that $v_n/n \xrightarrow{P} 1$. The above method of proof also extends to empirical processes defined in terms of multivariate random variables (cf. M. Csörgő, S. Csörgő, Fischler and Révész 1975). As to random sum limit theorems, one of the first papers was that of Anscombe (1952) (cf. also Doeblin 1938, 1940).

References

ALDOUS, D J.
  (1978) Weak convergence of randomly indexed sequences of random variables. *Math. Proc. Cambridge Philos. Soc.* **83** 117–126.
ALI, Mir M.–CHAN, L. K.
  (1964) On Gupta's estimates of the parameters of the normal distribution. *Biometrika* **51** 498–501.
ANDERSON, T. W.–DARLING, D. A.
  (1952) Asymptotic theory of certain "goodness of fit" criteria based on stochastic processes. *Ann. Math. Statist.* **23** 193–212.
ANSCOMBE, F. J.
  (1952) Large-sample theory of sequential estimation. *Proc Cambridge Philos. Soc.* **48** 600–607.
BAHADUR, R. R.
  (1966) A note on quantiles in large samples. *Ann. Math. Statist.* **37** 577–580.
BÁRTFAI, P.
  (1966) Die Bestimmung der zu einem wiederkehrenden Prozess gehörenden Verteilungsfunktion aus den mit Fehlern behafteten Daten einer Einzigen Relation. *Studia Sci. Math. Hung.* **1** 161–168.
  (1977) Connections between the convex analysis and the theory of large deviations. Technical report, Math. Inst. Hung. Acad. Sci. Budapest.
BARTLETT, M. S.
  (1949) Some evalutionary stochastic processes. *J. Roy. Statist. Soc. Ser. B.* **11** 211–229.
BAXTER, G.
  (1956) A strong limit theorem for Gaussian processes. *Proc. Amer. Math. Soc.* **7** 522–527.
BERKES, I.
  (1972) A remark to the law of the iterated logarithm. *Studia Sci. Math. Hung.* **7** 189–197.
BERKES, I.–PHILIPP, W.
  (1979) Approximation theorems for independent and weakly dependent random vectors. *Ann. Probability* **7** 29–54.
BICKEL, P. J.–ROSENBLATT, M.
  (1973) On some global measures of the deviations of density function estimates. *Ann. Statist,* **1** 1071–1095.
BICKEL, P. J.
  (1974) Edgeworth expansions in nonparametric statistics. *Ann. Statist.* **2** 1–20.

BILLINGSLEY, P.
(1962) Limit theorems for randomly selected partial sums. *Ann. Math. Statist.* **33** 85–92.
(1968) *Convergence of Probability Measures.* J. Wiley, New York.

BLUM, J.–HANSON, D.–ROSENBLATT, J.
(1963) On the central limit theorem for the sum of a random number of independent random variables. *Z. Wahrscheinlichkeitstheorie verw. Gebiete.* **1** 389–393.

BOCHNER, S.
(1955) *Harmonic analysis and the theory of probability.* Univ. of California Pr. Berkeley–Los Angeles.

BOLTHAUSEN, E.
(1978) On the speed of convergence in Strassen's law of the iterated logarithm. *Ann. Probability* **6** 668–672.

BOOK, S. A.
(1976) Large deviation probabilities and the Erdős–Rényi law of large numbers. *Canad. J. Statist.* **4** 185–210.

BOOK, S. A.–SHORE, T. R.
(1978) On large intervals in the Csörgő–Révész theorem on increments of a Wiener process. *Z. Wahrscheinlichkeitstheorie verw. Gebiete* **46** 1–11.

BOROVKOV, A. A.
(1973) On the speed of convergence in the invariance principle. *Teor. Ver. i ee Prim.* **18** 217–234.

BREIMAN, L.
(1967) On the tail behaviour of sums of independent random variables. *Z. Wahrscheinlichkeitstheorie und Verw. Gebiete* **9** 20–25.
(1968) *Probability.* Addison-Wesley. Reading, Mass.

BRILLINGER, D. L.
(1969) An asymptotic representation of the sample distribution function. *Bull. Amer. Math. Soc.* **75** 545–547.

BURKE, M. D.–CSÖRGŐ, M.
(1976) Weak approximations of the empirical process when parameters are estimated. In *Empirical Distributions and Processes* (P. Gaenssler and P. Révész, Eds.). *Lecture Notes in Mathematics* **566** 1–16, Springer-Verlag, Berlin.
(1976b) Strong approximations of the k-sample quantile and multivariate empirical process. *Carleton Mathematical Lecture Note.* No. 16.

BURKE, M. D.–CSÖRGŐ, M.–CSÖRGŐ, S.–RÉVÉSZ, P.
(1979) Approximations of the empirical process when parameters are estimated. *Ann. Probability* **7** 790–810.

CANTELLI, F. P.
(1917) Sulla probabilitá come limita della frequenza. *Rend. Accad. Lincei.* **26** 39.

ČENCOV, N. N.
(1962) Evaluation of an unknown density from observations. *Soviet Math.* **3** 1559–1569.
(1956) Wiener random fields depending on several parameters. *Dokl. Akad. Nauk. SSSR* **106** 607–609.

CHAN, A. H. C.
(1977) *On the increments of Multiparameter Gaussian Processes.* Ph. D. Thesis, Carleton University.

CHAN, A. H. C.–CSÖRGŐ, M.–RÉVÉSZ, P.
(1978) Strassen type limit points for moving averages of a Wiener process. *Canad. J. Statist.* **6** 57–75.

CHERNOFF, H.
(1952) A measure of asymptotic efficiency for tests of a hypothesis based on sums of observations. *Ann. Math. Statist.* **23** 493–507.

CHUNG, K. L.
(1948) On the maximum partial sums of sequences of independent random variables. *Trans. Amer. Math. Soc.* **64** 205–233.
(1949) An estimate concerning the Kolmogorov limit distribution. *Trans. Amer. Math. Soc.* **67** 36–50.

CHUNG, K. L.–ERDŐS, P.–SIRAO, T.
(1959) On the Lipschitz condition for Brownian motion. *J. of Math. Soc. Japan* **11** 263–274.

CRAMÉR, H.
(1946) *Mathematical methods of statistics.* Princeton Univ. Press, Princeton, N. J.
(1962) *Random variables and probability distributions.* Cambridge Univ. Press, London.

CSÁKI, E.
(1968) An iterated logarithm law for semimartingales and its application to empirical distribution function. *Studia Sci. Math. Hung.* **3** 287–292.
(1975) Some notes on the law of iterated logarithm for empirical distribution function. In *Coll. Math. Soc. J. Bolyai* **11**, *Limit Theorems of Probability Theory* (P. Révész, Ed.), Keszthely, Hungary 1974, North Holland, Amsterdam–London, 47–58.
(1977) The law of iterated logarithm for normalized empirical distribution function. *Z. Wahrscheinlichkeitstheorie Verw. Gebiete* **38** 147–167.
(1978) On the lower limits of maxima and minima of Wiener process and partial sums. *Z. Wahrscheinlichkeitstheorie verw. Gebiete* **43** 205–222.

CSÁKI, E.–RÉVÉSZ, P.
(1979) How big must be the increments of a Wiener Process? *Acta Math. Acad. Sci. Hung.* **33** 37–49.

CSÖRGŐ, M.
(1966) Some k-sample Kolmogorov–Smirnov–Rényi type theorems for empirical distribution functions. *Acta Math. Acad. Sci. Hung.* **17** 325–334.
(1967) A new proof of some results of Rényi and the asymptotic distribution of the range of his Kolmogorov–Smirnov type random variables. *Can. J. Math.* **19** 550–558.
(1968) On the strong law of large numbers and the central limit theorem for martingales. *Trans. Amer. Math. Soc.* **131** 259–275.

CSÖRGŐ, M.–CSÖRGŐ, S.
(1973) On weak convergence of randomly selected partial sums. *Acta. Sci. Math. (Szeged)* **34** 53–60.

CSÖRGŐ, M.–CSÖRGŐ, S.–FISCHLER, R.–RÉVÉSZ, P.
(1975) Random limit theorems via strong invariance principles (in Hungarian). *Matematikai Lapok* **26** 39–66.

CSÖRGŐ, M.–FISCHLER, R.
(1970) Departure from independence: the strong law, standard and random-sum central limit theorems. *Acta Math. Acad. Sci. Hung.* **21** 105–114.

CSÖRGŐ, M.–KOMLÓS, J.–MAJOR, P.–RÉVÉSZ, P.–TUSNÁDY, G.
(1977) On the empirical process when parameters are estimated. In *Trans. Seventh Prague Conference* 1974, Academia, Prague, 87–97.

CSÖRGŐ, M.–RÉVÉSZ, P.
(1975) A new method to prove Strassen type laws of invariance principle, I. *Z. Wahrscheinlichkeitstheorie verw. Gebiete* **31** 255–260.
(1975b) A new method to prove Strassen type laws of invariance principle, II. *Z. Wahrscheinlichkeitstheorie verw. Gebiete* **31** 261–269.
(1975c) Some notes on the empirical distribution function and the quantile process. In *Coll. Math. Soc. J. Bolyai* **11**, *Limit Theorems of Probability Theory* (P. Révész, Ed.), Keszthely, Hungary 1974, North Holland, Amsterdam–London, 59–71.
(1975d) A strong approximation of the multivariate empirical process. *Studia Sci. Math. Hung.* **10** 427–434.
(1978) How big are the increments of a multi-parameter Wiener process? *Z. Wahrscheinlichkeitstheorie verw. Gebiete* **42** 1–12.
(1978b) Strong approximations of the quantile process. *Ann. Statist.* **6** 882–894.
(1979a) How small are the increments of a Wiener process? *Stochastic Processes Appl.* **8** 119–129.
(1979b) How big are the increments of a Wiener process? *Ann. Probability* **7** 731–737.
(1979) On the standardized quantile process. In *Optimizing Methods in Statistics, Procedings of an International Conference* (I. S. Rustagi, Ed.), Bombay India, December 1977, Academic Press, New York, 125–140.
(1980) Quadratic ... ity tests. *Carleton Mathematical Series*. No. 162.
(1980a) Quantile processes and sums of weighted spacings for composite goodnes-of-fit, II. *Carleton Mathematical Lecture Note*.
(1980b) An estimation of the quantile function via density estimation. *Carleton Mathematical Series*.
(1980c) An invariance principle for N.N. empirical density functions. *Carleton Mathematical Lecture Note*.

CSÖRGŐ, M.–STEINEBACH, J.
(1980) Improved Erdős–Rényi and strong approximation laws for increments of partial sums. *Carleton Mathematical Series*. No. 166.

CSÖRGŐ, S.
(1974) On the limit distributions of sequences of random variables with random indices. *Acta Math. Acad. Sci. Hung.* **25** 227–232.
(1976) On an asymptotic expansion for the Mises ω^2 statistic. *Acta Sci. Math. (Szeged)*. **38** 45–67.
(1979) Erdős–Rényi laws. *Ann. Statist.* **7** 772–787.
(1980) Limit behaviour of the empirical characteristic function. *Ann. Probability* **8**.

CSÖRGŐ, S.–STACHÓ, L.
(1979) A step toward an asymptotic expansion for the Cramér–von Mises statistic. In *Coll. Math. Soc. J. Bolyai*. **21**, *Analytic Function Methods in Probability Theory* (B. Gyires, Ed.), Debrecen, Hungary 1977, North Holland, Amsterdam–London, 53–65.

DARLING, D. A.
(1955) The Cramér–Smirnov test in the parametric case. *Ann. Math. Statist.* **26** 1–20.

DARLING, D. A.–ERDŐS, P.
(1956) A limit theorem for the maximum of normalized sums of independent random variables, *Duke Math. J.* **23** 143–145.

DEO, C. M.
(1977) A note on increments of a Wiener process. *Technical Report.* University of Ottawa.

DeWET, T.–VENTER, J. H.
(1972) Asymptotic distributions of certain tests criteria of normality. *S. Afr. Statist. J.* **6** 135–149.

DOEBLIN, W.
(1938) Sur de problémes de M. Kolmogoroff concernant le chaines dénombrables. *Bull. Soc. Math. France* **66** 210–220.

(1940) Éléments d'une théorie générale des chaines simples constantes de Markov. *Ann. Sci. École Norm. Sup.* **57** 61–111.

DONSKER, M.
(1951) An invariance principle for certain probability limit theorems. *Four papers on probability. Mem. Amer. Math. Soc. No.* **6**.

(1952) Justification and extension of Doob's heuristic approach to the Kolmogorov–Smirnov theorems. *Ann. Math. Statist.* **23** 277–283.

DONSKER, M –VARADHAN, S. S. R.
(1977) On laws of iterated logarithm for local times. *Comm. Pure Appl. Math.* **30** 707–753.

DOOB, J. L.
(1949) Heuristic approach to the Kolmogorov–Smirnov theorems. *Ann. Math. Statist.* **20** 393–403.

(1953) *Stochastic Processes.* J. Wiley, New York.

DUDLEY, R. M.
(1968) Distances of probability measures and random variables. *Ann. Math. Statist.* **39** 1563–1572.

(1973) Sample functions of the Gaussian process. *Ann. Probability* **1** 66–103.

DURBIN, J.
(1973a) Weak convergence of the sample distribution function when parameters are estimated. *Ann. Statist.* **1** 279–290.

(1973b) *Distribution Theory for Tests based on the Sample Distribution function. Regional Conference Series on Appl. Math.* **9** S.I.A.M., Philadelphia.

(1976) Kolmogorov–Smirnov tests when parameters are estimated. In *Empirical Distributions and Processes* (P. Gaenssler and P. Révész, Ed.). *Lecture Notes in Mathematics* **566** 33–44, Springer-Verlag, Berlin.

DVORETZKY, A.–KIEFER, J.–WOLFOWITZ, J.
(1956) Asymptotic minimax character of the sample distribution function and of the multinomial estimator. *Ann. Math. Statist.* **27** 642–669.

DVORETZKY, A.
(1963) On the oscillation of the Brownian motion process. *Israel J. Math.* **1** 212–214.

EICKER, F.
(1976) The asymptotic distribution of the supremum of the standardized empirical distribution function. *Preprint No. 76/7 of the Dept. Statist.*, Dortmund University

(1979) The asymptotic distribution of the suprema of standardized empirical process. 7 116–138.

EINSTEIN, A.
(1905) On the movement of small particles suspended in a stationary liquid demended by the molecular-kinetic theory of heat. *Ann. Physik* **17** 549–560.

EPANECHNIKOV, V. A.
(1969) Nonparametric estimates of a multivariate probability density. *Theor. Probability Appl.* **14** 153–158.

ERDŐS, P.
(1949) On a theorem of Hsu and Robbins. *Ann. Math. Statist.* **20** 286–291.
(1950) Remark on my paper "On a theorem of Hsu and Robbins." *Ann. Math. Statist.* **21** 138.

ERDŐS, P.–KAC, M.
(1946) On certain limit theorems of the theory of probability. *Bull. Amer. Math. Soc.* **52** 292–302.
(1947) On the number of positive sums of independent random variables. *Bull. Amer. Math. Soc.* **53** 1011–1020.

ERDŐS, P.–RÉNYI, A.
(1970) On a new law of large numbers. *J. Analyse Math.* **23** 103–111.

ERDŐS, P.–RÉVÉSZ, P.
(1976) On the length of the longest head-run. In *Coll. Math. Soc. J. Bolyai* **16**, *Topics in Information Theory*, Keszthely, Hungary 1975, North Holland, Amsterdam.

FELLER, W.
(1943) The general form of the so-called law of the iterated logarithm. *Trans. Amer. Math. Soc.* **54** 373–402.
(1948) On the Kolmogorov–Smirnov limit theorems for empirical distributions. *Ann. Math. Statist.* **19** 177–189.
(1968) *An Introduction to Probability Theory and Its Applications*, Volume I: Third Edition. J. Wiley, New York.
(1966) *An Introduction to Probability Theory and Its Applications*, Volume II. J. Wiley, New York.

FEUERVERGER, A.–MUREIKA, R. A.
(1977) The empirical characteristic function and its applications. *Ann. Statist.* **5** 88–97.

FINKELSTEIN, H.
(1971) The law of iterated logarithm for empirical distributions. *Ann. Math. Statist.* **42** 607–615.

FISCHLER, R.
(1976) Convergence faible avec indices aléatoires. *Ann. Inst. Henri Poincaré, B: Calcul des Probabilités et Statistique* **12** 391–399.

FISHER, R. A.
(1929) Tests of significance in harmonic analysis. *Proc. Roy. Soc. (A)* **125** 54–59. Reprinted 1950 as Paper No. 16 of R. A. Fisher's *Contributions to Mathematical Statistics*, J. Wiley, New York; Chapman & Hall, London.

FÖLDES, A.–RÉVÉSZ, P.
(1974) A general method for density estimation. *Studia Sci. Math. Hung.* **9** 81–92.

GAENSSLER, P.–STUTE, W.
(1979) Empirical processes: a survey of results for independent and identically distributed random variables. *Ann. Probability* **7** 193–243.

GARLING, D. J. H.
(1976) Functional central limit theorems in Banach spaces. *Ann. Probability* **4** 600–611.

GLIVENKO, V. I.
(1933) Sulla determinazione empirica delle leggi di probabilitá. *Giorn. Ist. Ital. Attuari.* **4** 92–99.

GNEDENKO, B. V.–KOROLYUK, V. S.–SKOROHOD, A. V.
(1960) Asymptotic expansions in probability theory. In *Proc. Fourth Berkeley Symp. Math. Statist. Prob.* **2** 153–169, Univ. California Press.

GÖTZE, F.
(1979) Asymptotic expansions for bivariate von Mises functionals. *Preprints in Statistics. No. 49. Univ of Cologne.*

GUIBAS, L. J.–ODLYZKO, A. M.
(1980) Long repetitive patterns in random sequences. *Z. Wahrscheinlichkeitstheorie verw. Gebiete.*

HARTER, H. L.
(1961) Expected values of normal order statistics. *Biometrika* **48** 151–165.

HARTMAN, P.–WINTNER, A.
(1941) On the law of iterated logarithm. *Amer. J. Math.* **63** 169–176.

HIRSCH, W. M.
(1965) A strong law for the maximum cumulative sum of independent random variables. *Comm. Pure Appl. Math.* **18** 109–217.

HOFFMAN–JØRGENSEN, J.–PISIER, G.
(1976) The law of large numbers and the central limit theorem in Banach spaces. *Ann. Probability* **4** 587–599.

HORVÁTH, L.
(1978) A note on random limit theorems via strong invariance principles. *JATE Student Mathematical Workshop Notes.*

IBRAGIMOV, I. A.–HAS'MINSKIĬ, R. Z.
(1972) Asymptotic behaviour of statistical estimators in the smooth case I. Study of the likelihood ratio. *Theor. Probability Appl.* **17** 445–462.
(1973a) Asymptotic behaviour of some statistical estimators II. Limit theorems for the a posteriori density and Bayes' estimators. *Theory Probability Appl.* **18** 76–91.
(1973b) On the approximation of statistical estimators by sums of independent variables. *Soviet Math. Dokl.* **14** 883–887.

ITÔ, K.
(1960) *Probability Processes*, Volume I (In Japanese; Russian Translation by ИЛ, M., 1960). Tokyo, Iwanami Shoten Publishers.

ITÔ, K.–NISIO, M.
(1968) On the oscillation functions of Gaussian processes. *Math. Scand.* **22** 209–223.

JAESCHKE, D.
(1975) Über die Grenzverteilung des Maximums der normierten empirischen Verteilungsfunktion. Dissertation, Dortmund University.

(1976) The asymptotic distribution of the supremum of the standardized empirical distribution function on subintervals. *Preprint No. 76/11 of the Dept. Statist.*, Dortmund University.

(1979) The asymptotic distribution of the supremum of the standardized empirical distribution function on subintervals *Ann. Statist.* **7** 108–115.

JAMES, B. R.

(1975) A functional law of the iterated logarithm for weighted empirical distributions. *Ann. Probability* **3** 762–772.

JAIN, N. C.–PRUITT, W. E.

(1976) The other law of the iterated logarithm. *Ann. Probability* **3** 1046–1049.

KAC, M.

(1946) On the average of a certain Wiener functional and a related limit theorem in calculus of probability. *Trans. Amer. Math. Soc.* **59** 404–414.

(1949) On deviations between theoretical and empirical distributions. *Proc. Nat. Acad. Sci. USA.* **35** 252–257.

(1951) On some connections between probability theory and differential and integral equations. In *Second Berkeley Symp. on Probability and Statistics.* 180–215. Univ. California Press, Berkeley–Los Angeles.

KAC, M.–KIEFER, J.–WOLFOWITZ, J.

(1955) On tests of normality and other tests on goodness of fit based on distance methods. *Ann. Math. Statist.* **26** 189–211.

KARHUNEN, K.

(1946) Zur Spektraltheorie stochastischen Processe. *Ann. Acad. Sci. Fennicae Ser A.* I. Math. Phys. No. **34** 7.

(1947) Über lineare Methoden in der Wahrscheinlichkeitsrechnung. *Ann. Acad. Sci. Fennicae Ser. A.* I. Math. Phys. No. **37** 79.

KARLSSON, J E.–SZÁSZ, D. E.

(1974) *Random limit theorems. A bibliography.* Math. Inst. Hung. Acad. Sci. Budapest.

KATZ, M. L.

(1963) The probability in the tail of a distribution. *Ann. Math. Statist.* **34** 312–318.

KENDALL, D. G.

(1974) Hunting quanta. *Phil. Trans. Roy. Soc. London A* **276** 231–266. Second edition:

(1977) *Proc. Symp. Honour Jerzy Neyman,* Warsawa 1974, Polish Acad. Sci. 111–159.

KENT, J. T.

(1975) A weak convergence theorem for the empirical characteristic function. *J. Appl. Prob.* **12** 515–523.

KHINCHINE, A.

(1933) *Asymptotische Gesetze der Wahrscheinlichkeitsrechnung.* Springer, Berlin.

KIEFER, J.

(1959) K-sample analogues of the Kolmogorov–Smirnov and Cramér–von Mises tests. *Ann. Math. Statist.* **30** 420–447.

(1969a) On the deviations in the Skorohod–Strassen approximation scheme. *Z. Wahrscheinlichkeitstheorie verw. Gebiete* **13** 321–332.

(1969b) Old and new methods for studying order statistics and sample quantiles. In *Proc. First International Conf. Nonparametric Inference.* 349–357. Cambridge Univ. Press. 1970. London.

(1970) Deviations between the sample quantile process and the sample D. F. *Nonparametric Techniques in Stat. Inference* (M. L. Puri, Ed.) 299–319, Cambridge Univ. Press. London.

(1972) Skorohod Embedding of Multivariate RV's and the sample DF. *Z. Wahrscheinlichkeitstheorie verw. Gebiete* 24 1–35.

KLASS, M. J.

(1976) Forward a universal law of the iterated logarithm I. *Z. Wahrscheinlichkeitstheorie verw. Gebiete* 36 165–178.

(1977) Forward a universal law of the iterated logarithm II. *Z. Wahrscheinlichkeitstheorie verw. Gebiete* 39 151–165.

KOLMOGOROV, A. N.

(1931) Eine Veralgemeinerung des Laplace–Liapunoffschen Satzes. *Izv. Akad. Nauk SSSR, Ser. Fiz-Mat.* 959–962.

(1933) Sulla determinazione empirica di une legge di distribuzione. *Giorn. Inst. Ital. Attuari* 4 83–91.

(1933a) Über die Grenzwertsätze der Wahrscheinlichkeitsrechnung. *Izv. Akad. Nauk SSSR, Ser. Fiz-Mat.* 363–372.

KOMLÓS, J.–MAJOR, P.–TUSNÁDY, G.

(1975) An approximation of partial sums of independent R.V.'s and the sample DF. I. *Z. Wahrscheinlichkeitstheorie verw. Gebiete* 32 111–131.

(1975a) Weak convergence and embedding. In *Coll. Math. Soc. J. Bolyai* 11, *Limit Theorems of Probability Theory* (P. Révész, Ed.), Keszthely, Hungary 1974, North Holland, Amsterdam–London, 149–165.

(1976) An approximation of partial sums of independent R.V.'s and the sample DF. II. *Z. Wahrscheinlichkeitstheorie verw. Gebiete* 34 33–58.

KOMLÓS, J.–TUSNÁDY, G.

(1975) On sequences of "pure heads". *Ann. Probability* 3 608–617.

KUELBS, J.

(1973) The invariance principle for Banach space valued random variables. *J. Multivariate Anal.* 3 161–172.

KUIPER, N. H.

(1960) Tests concerning random points on a circle. *Proc. Nederl. Akad. Wetensch. Indag. Math.* (A). 63 38–47.

LAI, T. L.

(1973) On Strassen Type Laws of the Iterated Logarithm for Delayed Averages of the Wiener Process. *Bull. Inst. Math. Acad. Sinica* 1 29–39.

(1974) Limit theorems for delayed sums. *Ann. Probability* 2 432–440.

LÉVY, P.

(1937) *Théorie de l'addition des variables aléatoires.* Gauthier-Villars, Paris.

(1940) Le Mouvement brownien plan. *Amer. J. Math.* 62 487–550.

(1948) *Procesus stochastique et mouvement Brownien.* Gauthier-Villars, Paris.

LOÉVE, M.

(1963) *Probability Theory,* Third Edition, Van Nostrand–Reinhold, Princeton, N. J.

MAJOR, P.

(1973) On non-parametric estimation of the regression function. *Studia Sci. Math. Hungar.* 8 347–361.

(1976a) The approximation of partial sums of independent r.v.'s. *Z. Wahrscheinlichkeitstheorie verw. Gebiete* **35** 213–220.

(1976b) Approximation of partial sums of i.i.d.r.v.'s when the summands have only two moments. *Z. Wahrscheinlichkeitstheorie verw. Gebiete* **35** 221–230.

(1979) An improvement of Strassen's invariance principle. *Ann. Probability* **7** 55–61.

MARCUS, M. B.

(1980) Weak convergence of the empirical characteristic function. *Ann. Probability* **8**.

MICHEL, R.

(1974) Results on probabilities of moderate deviation. *Ann. Probability* **2** 349–353.

MOGUL'SKIĬ, A. A.

(1974) Small deviations in a space of trajectories. *Theor. Probability Appl.* **19** 726–736.

(1977) Laws of iterated logarithm on function spaces (in Russian). In *Abstracts of communications* **2** 44–47, Second Vilnius Conference on Probability Theory and Mathematical Statistics, Vilnius 1977.

MOGYORÓDI, J.

(1962) A central limit theorem for the sum of a random number of random variables. *Publ. Math. Inst. Hung. Acad. Sci.* **7** 409–424.

(1965) On the law of large numbers for the sum of a random number of independent random variables. *Annales Univ. Sci. Budapest de Eötvös nom. Sect. Math.* **8** 33–38.

(1966) A remark on stable sequences of random variables and a limit distribution theorem for the sum of independent random variables. *Acta Math. Acad. Sci. Hung.* **17** 401–409.

MÜLLER, D. W.

(1970) On Glivenko–Cantelli convergence. *Z. Wahrscheinlichkeitstheorie verw. Gebiete* **16** 195–210.

NADARAYA, E. A.

(1964) On estimating regression. *Theory Probability Appl.* **9** 141–142.

(1965) On non-parametric estimates of density functions and regression. *Theory Probability Appl.* **10** 199–203.

OREY, S.–TAYLOR, S. J.

(1974) How often on a Brownian path does the law of iterated logarithm fail? *Proc. London Math. Soc.* **28** 174–192.

ORLOV, A. I.

(1974) A speed of convergence for the distribution of the Mises–Smirnov statistic (in Russian). *Teorija Verojatn. Primen.* **19** 765–786.

PALEY, R. E. A. C.–WIENER, N.

(1934) *Fourier transforms in the complex domain.* Amer. Math. Soc. Colloq. Publ. **19** Amer. Math. Soc., Providence, R. I.

PARANJAPE, S. R.–PARK, C.

(1973) Laws of iterated logarithm of multiparameter Wiener process. *J. Multivariate Analysis* **3** 132–136.

PARK, W. J.

(1974) On Strassen's version of the law of the iterated logarithm for the two-parameter Gaussian process. *J. Multivariate Analysis* **4** 479–485.

PARTHASARATHY, K. R.

(1967) *Probability measures on metric spaces.* Academic Press, New York.

PARZEN, E.
(1962) On estimation of a probability density function and mode. *Ann. Math. Statit.s* **33** 1065–1076.

PETROV, V. V.
(1975) *Sums of Independent Random Variables.* Springer, Berlin–Heidelberg–New York.

PHILIPP, W.–STOUT, W.
(1975) Almost sure invariance principles for partial sums of weakly dependent random variables. *Mem. Amer. Math. Soc. No.* **161**.

PROHOROV, YU. V.
(1956) The convergence of random processes and limit theorems in probability theory. *Teorija Verojatn. Primen.* **1** 177–238.

PRUITT, W. E.–OREY, S.
(1973) Sample functions of the N-parameter Wiener process. *Ann. Probability* **1** 138–163.

PYKE, R.
(1968) The weak convergence of the empirical process with random sample size. *Proc. Cambridge Philos. Soc.* **64** 155–160.

QUALLS, C.–WATANABE, H.
(1972) Asymptotic properties of Gaussian processes. *Ann. Math. Statist.* **43** 580–596.

RAO, K. C.
(1972) The Kolmogorov, Cramér–von Mises Chisquare statistics for goodness of fit tests in the parametric case. Abstract: *Bulletin, Inst. Math. Statist.* **1** 87.

RÉNYI, A.
(1953) On the theory of order statistics. *Acta Math. Acad. Sci. Hung.* **4** 191–232.
(1958) On mixing sequences of sets. *Acta Math. Acad. Sci. Hung.* **9** 215–228.
(1960) On the central limit theorem for the sum of a random number of independent random variables. *Acta Math. Acad. Sci. Hung.* **11** 97–102.
(1970) *Probability theory.* North Holland, Amsterdam and Akadémiai Kiadó, Budapest.

RÉVÉSZ, P.
(1968) *The laws of large numbers.* Academic Press, New York.
(1972) On empirical density function. *Periodica Math. Hung.* **2** 85–110.
(1976) On strong approximation of the multidimensional empirical process. *Ann. Probability* **4** 729–743.
(1978) Strong theorems on coin tossing. In *Proc. Int. Congress of Mathematicians.* 749–754. Helsinki.
(1979) On the nonparametric estimation of the regression function. *Problems of Control and Information Theory* **8** 297–302.
(1979a) A note to the Chung–Erdős–Sirao theorem. In *Asymptotic Theory of Statistical Tests and Estimation.* 147–158. Academic Press, New York.
(1979b) A generalization of Strassen's functional law of iterated logarithm. *Z. Wahrscheinlichkeitstheorie verw. Gebiete* **50** 257–264.

RIESZ, F.–SZ.–NAGY. B.
(1955) *Functional Analysis.* Frederick Ungar, New York.

ROOTZÉN, H.
(1974) Some properties of convergence in distribution of sums and maxima of dependent random variables. *Z. Wahrscheinlichkeitstheorie verw. Gebiete* **29** 295–307.

ROSENBLATT, M.
(1956) Remarks on some nonparametric estimates of density function. *Ann. Math. Statist.* **27** 832–837.
(1971) Curve estimates. *Ann. Math. Statist.* **42** 1815–1842.

RUBIN, H.–SETHURAMAN, J.
(1965) Probabilities of moderate deviations, *Sankhya,* Ser. A, **27** 325–346.

SARAHAN, A. E.–GREENBERG, B. G.
(1956) Estimation of location and scale parameters by order statistics from singly and doubly censored samples, I. *Ann. Math. Statist.* **27** 427–451.

SCHWARTZ, S. C.
(1967) Estimation of probability density by an orthogonal series. *Ann. Math. Statist.* **38** 1261–1265.

SHAPIRO, S. S.–FRANCIA, R. S.
(1972) An approximate analysis of variance tests for normality. *J. Amer. Statist. Assoc.* **67** 215–216.

SHAPIRO, S. S.–WILK, M. B.
(1965) An analysis of variance test for normality (complete samples). *Biometrika* **52** 591–611.

SHORACK, G. R.
(1972a) Functions of order statistics. *Ann. Math. Statist.* **43** 412–427.
(1972b) Convergence of quantile and spacings processes with applications. *Ann. Math. Statist.* **43** 1400–1411.

SKOROHOD, A. V.
(1956) Limit theorems for random processes. *Teorija Verojatn. Primen.* **1** 289–319.
(1961) *Studies in the Theory of Random Processes.* Addison–Wesley. Reading, Mass.

SMIRNOV, N. V.
(1937) On the distribution of the von Mises ω^2-criterion (in Russian). *Matem Sbornik.* **5** 973–993.
(1939) On the estimation of the discrepancy between empirical curves of distribution for two independent samples. *Bull. Math. de l' Université de Moscou* **2** (Fasc. 2).
(1944) Approximate laws of distribution of random variables from empirical data (in Russian). *Uspehi Mat. Nauk.* **10** 179–206.

SPITZER, F.
(1956) The probability in the tail of a distribution. *Ann. Math. Statist.* **34** 312–318.

STEINEBACH, J.
(1979) *Erdős–Rényi-Zuwächse bei Erneuerungsprozessen und Partialsummen auf Gittern.* Habilitationsschrift, Univ. of Düsseldorf.

STONE, Ch. J.
(1977) Consistent nonparametric regression. *Ann. Statist.* **5** 595–645.

STOUT, W. F.
(1974) *Almost sure convergence.* Academic Press, New York.

STRASSEN, V.
(1964) An invariance principle for the law of the iterated logarithm. *Z. Wahrscheinlichkeitstheorie verw. Gebiete* **3** 211–226.
(1965a) The existence of probability measures with given marginals. *Ann. Math. Statist.* **36** 423–439.

(1965b) Almost sure behaviour of sums of independent random variables and martingales. In *Proc. Fifth Berkeley Symp. Math. Statist. Prob.* **2** 315–344.

(1966) A converse to the law of the iterated logarithm. *Z. Wahrscheinlichkeitstheorie verw. Gebiete* **4** 265–268.

SZYNAL, D.

(1972) On almost complete convergence for the sum of a random number of independent random variables. *Bull. Acad. Sci. Polon. Ser. Math. Astronom. Phys.* **20** 571–574.

(1973) On the limit behaviour of a sequence of quantiles of a sample with a random number of items. *Applications Matematicae* **13** 321–327.

TAYLOR, S. J.

(1974) Regularity of irregulartities on a Brownian path. *Ann. Inst. Fourier*, Grenoble **24** 195–203.

TUSNÁDY, G.

(1974) On testing density functions. *Periodica Math. Hung.* **5** 161–169.

(1977) Strong invariance principles. In *Recent Developments in Statistics* (J. R. Barra et al., Ed.) 289–300. North-Holland, Amsterdam.

(1977b) *A study of statistical hypotheses*. Dissertation, The Hungarian Academy of Sciences, Budapest.

VAN RYZIN, I.

(1966) Bayes risk consistency of classification procedures using density estimation. *Sankhya, Ser. A.* **28** 261–270.

WICHURA, M. J.

(1970) On the construction of almost uniformly convergent random variables with given weakly convergent image laws. *Ann. Math. Statist.* **41** 284–291.

(1973) Some Strassen-type laws of the iterated logarithm for multiparameter stochastic processes with independent increments. *Ann. Probability* **1** 272–296.

WITTENBERG, H.

(1964) Limit distributions of random sums of independent random variables. *Z. Wahrscheinlichkeitstheorie verw. Gebiete* **3** 7–18.

YEH, I.

(1960) Wiener measure in a space of functions of two variables. *Trans. Amer. Math. Soc.* **95** 433–450.

(1973) *Stochastic processes and the Wiener integral*. Marcel Dekker, New York.

ZIMMERMANN, G.

(1972) Some sample function properties of the two-parameter Gaussian process. *Ann. Math. Statist.* **43** 1235–1246.

ZYGMUND, A.

(1968) *Trigonometric Series*, Volumes I, II. Cambridge Univ. Press, London.

Author Index

Aldous, D. J. 257, 261
Ali, M. M. 204
Anderson, T. W. 43, 164, 168
Anscombe, F. J. 262

Bahadur, R. R. 160
Bártfai, P. 96, 101, 113, 114
Bartlett, M. S. 154
Baxter, G. 83
Berkes, I. 123, 255
Bickel, P. J. 166, 248
Billingsley, P. 13, 43, 257, 260
Blum, J. 260
Bochner, S. 224
Bolthausen, E. 124
Book, S. A. 35, 113, 123
Borovkov, A. A. 95
Breiman, L. 88, 89, 93, 94, 95, 97, 108
Brillinger, D. L. 16, 130
Brown, R. 21
Burke, M. D. 191, 201, 217, 218

Cantelli, F. P. 128
Čencov, N. N. 86, 221
Chan, A. H. C. 78, 85, 87
Chan, L. K. 204
Chernoff, H. 99
Chung, K. L. 35, 48, 82, 85, 122, 124, 132, 157
Cramér, H. 102, 242
Csáki, E. 35, 85, 86, 125, 126, 147, 148, 153, 159, 217
Csörgő, M. 30, 44, 47, 63, 85, 101, 102, 114, 123, 132, 144, 146, 148, 149, 153, 165, 173, 182, 191, 201, 205, 217, 218, 248, 250, 252, 257, 261, 262
Csörgő, S. 113, 123, 166, 167, 168, 191, 201, 217, 218, 243, 245, 246, 247, 248, 252, 257, 260, 261, 262

Darling, D. A. 43, 56, 164, 168, 188
Deo, C. M. 35
DeWet, T. 180, 181, 204
Doeblin, W. 262
Donsker, M. 12, 13, 16, 85, 88, 129, 130, 154
Doob, J. L. 15, 16, 43, 129, 251
Dudley, R. M. 15, 83
Durbin, J. 188, 191, 192, 200, 202, 218
Dvoretzky, A. 85, 128

Eicker, F. 169, 170
Einstein, A. 21
Epanechnikov, V. A. 221
Erdős, P. 11, 12, 15, 31, 56, 82, 97, 98, 101, 113, 122, 254, 260

Feller, W. 23, 113, 118, 123, 124, 129
Feuerverger, A. 243
Finkelstein, H. 77, 86, 157
Fischler, R. A. 252, 257, 262
Fisher, R. A. 216
Földes, A. 221
Francia, R. S. 204

Gaenssler, P. 18
Garling, D. J. H. 19
Glivenko, V. J. 128

Gnedenko, B. V. 166
Götze, F. 167
Greenberg, B. G. 203
Guibas, L. I. 123

Hanson, D. 260
Harter, H. L. 204
Hartman, P. 15, 119
Hasminskiĭ, R. Z. 217, 218
Hirsch, W. M. 35, 124, 125
Hoffman-Jørgensen, J. 19
Horváth, L. 260, 261

Ibragimov, I. A. 217, 218
Ito, K. 55, 82

Jaeschke, D. 169, 170
James, B. R. 217
Jain, N. C. 122

Kac, M. 11, 12, 15, 122, 154, 188, 250, 251, 260
Karhunen, K. 248
Karlsson, J. 252
Katz, M. L. 253, 254
Kendall, D. G. 154, 243
Kent, J. T. 243
Khinchine, A. 13
Kiefer, J. 17, 87, 95, 101, 128, 132, 160, 161, 182, 188, 250
Klass, M. J. 123
Kolmogorov, A. N. 13, 128, 129
Komlós, J. 107, 108, 110, 112, 113, 122, 132, 133, 140, 141, 154, 165, 217, 218
Korolyuk, V. S. 166
Kuelbs, J. 19
Kuiper, N. H. 44, 164

Lai, T. L. 83, 118
Lévy, P. 26, 36, 40, 43, 44, 82, 83, 84, 123
Loéve, M. 94

Major, P. 93, 107, 108, 110, 112, 132, 133, 140, 141, 154, 165, 217, 218, 239
Marcus, M. B. 248
Michel, R. 103, 106

Mogul'skiĭ, A. A. 121, 159
Mogyoródi, J. 252, 260
Mureika, R. A. 243
Müller, D. W. 17, 87

Nadaraya, E. A. 239
Nisio, M. 55

Odlyzko, A. M. 123
Orey, S. 41, 61, 66, 74, 82
Orlov, A. I. 166

Paley, R. E. A. C. 55
Paranjape, S. R. 61
Park, C. 61
Park, W. J. 61, 62
Parthasarathy, K. R. 13
Parzen, E. 221, 224, 243
Pisier, G. 19
Petrov, V. V. 18
Philipp, W. 19, 255
Prohorov, Yu. V. 13, 16
Pruitt, W. E. 61, 66, 74, 82, 122
Pyke, R. 262

Qualls, C. 44, 82, 86

Rao, K. C. 218
Rényi, A. 17, 31, 91, 95, 97, 98, 101, 113, 143, 164, 165, 257
Révész, P. 30, 35, 44, 47, 63, 82, 85, 101, 102, 113, 114, 123, 132, 144, 146, 148, 149, 153, 173, 191, 201, 205, 217, 218, 221, 224, 230, 231, 238, 239, 248, 250, 252, 262
Riesz, F. 36, 84, 249
Rootzén, H. 257
Rosenblatt, I. 260
Rosenblatt, M. 221, 248
Rubin, H. 103

Sarahan, A. E. 203
Schwartz, S. C. 221
Sethuraman, J. 103
Shapiro, S. S. 203, 204
Shorack, G. R. 155
Shore, T. R. 35
Sirao, T. 82

Skorohod, A. V. 13, 16, 88, 89, 166
Smirnov, N. V. 43, 128, 129, 132, 157
Spitzer, F. 254
Stacho, L. 167
Steinebach, I. 123
Stone, Ch. J. 249
Stout, W. F. 18, 19, 255
Strassen, V. 14, 15, 36, 37, 40, 89, 91, 94, 95, 101, 114, 119, 123, 130
Stute, W. 18
Szász, D. E. 252
Sz.-Nagy, B. 36, 84, 249
Szynal, D. 254

Taylor, S. J. 41, 82, 85
Tusnády, G. 107, 108, 110, 112, 113, 122, 132, 133, 140, 141, 154, 165, 217, 218, 248

van Ryzin, I. 221
Varadhan, S. S. P. 85
Varga, T. 97, 120
Venter, J. H. 180, 181, 204

Watanabe, H. 44, 82, 86
Wichura, M. J. 15, 61, 75, 250
Wiener, N. 55
Wilk, M. B. 203
Wintner, H. 15, 119
Wittenberg, H. 260
Wolfowitz, J. 128, 188

Yeh, I. 86, 226, 248

Zimmermann, G. 61
Zygmund, A. 225

Subject Index

BROWNIAN BRIDGE
  continuity modulus 42
  covariance function 41
  definition 41
  distribution of functionals 43, 44
  infinite series representation 55
  law of iterated logarithm 42

CHERNOFF THEOREM 99

DOOB TRANSFORMATION 42

DONSKER THEOREMS 13, 30

EMPIRICAL CHARACTERISTIC
FUNCTION
  approximation by Gaussian processes 244, 245
  definition 243
  law of iterated logarithm 247
  strong theorems 243, 247

EMPIRICAL DENSITY FUNCTION
  approximation by Gaussian processes 224
  definitions 220, 221, 222
  limit theorems 225, 228,
  strong theorems 230, 237

EMPIRICAL DISTRIBUTION
FUNCTION 127
  (see also empirical process)

EMPIRICAL PROCESS
  approximation by Brownian bridge 129, 130, 133
  approximation by Kiefer process 132, 141
  definition 128
  distribution of functionals 128, 163, 170
  law of iterated logarithm 148, 156, 157, 159
  random limit theorems 261

EMPIRICAL REGRESSION
FUNCTION
  definition 238, 239
  limit theorems 238, 239

FISCHER INFORMATION MATRIX 217

GLIVENKO—CANTELLI THEOREM 128

KARHUNEN—LOÉVE EXPANSION 249

KIEFER PROCESS
  continuity modulus 81, 87
  covariance function 80
  definition 80
  increments 87
  law of iterated logarithm 80, 81

ORSTEIN—UHLENBECK PROCESS
  covariance function 56
  definition 55
  distribution of functionals 56, 57

PARTIAL SUM PROCESS OF I.I.D.R.V.
  approximation by Wiener process 92, 93, 101, 106, 112

distribution of functionals 122
Erdős—Rényi law 97, 98, 109, 113, 122, 123
increments 117, 120
law of iterated logarithm 119, 121, 123
random limit theorems 258

QUADRATIC QUANTILE TESTS 178, 202

QUANTILE FUNCTION 143
(see also quantile process)

(SAMPLE) QUANTILE PROCESS
approximation by Brownian bridge 144, 153
approximation by Kiefer process 146, 153
approximation by uniform quantile process 149
definition 143
distribution of functionals 143, 171
law of iterated logarithm 148, 162

SKOROHOD (EMBEDDING SCHEME) STOPPING TIME 127

STOCHASTIC GEYSER 95

STRASSEN STRONG INVARIANCE PRINCIPLE 91

UNIFORM (SAMPLE) QUANTILE PROCESS 143

Summary of Notations and Abbrevations

This list includes only symbols used systematically throughout the book in some special way.

Probability Space, Random Variables, Expectation

$(\Omega, \mathscr{A}, P)$ is a probability space, with P a probability measure on a measurable space $(\Omega, \mathscr{A})$. Events (elements of $\mathscr{A}$) are usually denoted by $A, B, \ldots$ etc.; ω is the generic element of Ω. Random variable(s) (r.v.) are usually denoted by $X, Y, \ldots$ etc., $\{X_i\}$ is a sequence of r.v. and for independent identically distributed r.v. we write: i.i.d.r.v. EX denotes the expected value of the r.v. X. $\text{Var}\, X$ is the variance, while $\text{var}\, f(\,.\,)$ is the total variation of the function $f(\,.\,)$. $P\{X \leq x\}$ denotes the probability of the event $\{\omega : X(\omega) \leq x\}$. I_A is the indicator function of the event A and card A is the cardinality of A.

Distributions, Densities

$F, G, \ldots$ etc. usually stand for distribution functions, and $f, g, \ldots$ etc. for density functions of r.v. $\mathscr{N}(\mu, \sigma^2)$ stands for the normal family of distribution functions with mean $\mu \in R^1$ and variance $\sigma^2 > 0$, where R^k is Euclidean k-space ($k \geq 1$). We frequently write $X \in \mathscr{N}(\mu, \sigma^2)$ or $F \in \mathscr{N}(\mu, \sigma^2)$, both having the same meaning and for $F \in \mathscr{N}(0, 1)$ we use the notation Φ with density function φ. $\mathscr{U}(a, b)$, a and $b \in R^1$, stand for the family of uniform distributions on (a, b); $\text{Exp}\,(a, b), a \in R^1, b > 0$ denote the exponential family of distributions $b^{-1} \exp(-b^{-1}(x-a))$, $x \geq a$. $\mathscr{B}(n, p)$ is the binomial family with parameters $n \geq 1$, $0 < p < 1$. The inverse of a distribution function F is denoted by inv F. The derivative of a function f is usually denoted by f', f'' is the second derivative of f, etc.

Convergence and Equality in Distribution Notions

$\mathscr{P}$ stands for equality in distribution, $\xrightarrow{\mathscr{P}}$ means convergence in distribution, $\xrightarrow{P}$ is convergence in probability, and $\stackrel{a.s.}{=}$ resp. $\xrightarrow{a.s.}$ stands for a.s. (almost sure) equality resp. convergence of r.v. The symbols $o(.)$, $O(.)$ are used in the usual, Landau sense. When the corresponding relations hold in probability, then we frequently write $o_P(.) O_P(.)$. The symbol $\approx$ stands for asymptotic equality.

Special Stochastic Processes

$W(t) = \{W(t); t \geq 0\}$ stands for the standard Wiener process, $B(t) = \{B(t); 0 \leq t \leq 1\}$ is the Brownian bridge, $W(s, t) = \{W(s, t); s, t \geq 0\}$ is two parameter Wiener process, $K(s, t) = \{K(s, t); 0 \leq s \leq 1; t \geq 0\}$ is Kiefer process. When emphasizing sample path properties of a stochastic process $X(t)$, then we sometimes write $X(t, \omega)$.

Special Metric Spaces

$C = C(0, 1)$ is the space of continuous functions $x(.)$ on $[0, 1]$ with the uniform metric $\varrho(x, y) = \sup_t |x(t) - y(t)|$. $D = D(0, 1)$ is the space of functions $x(.)$ on $[0, 1]$, having points of discontinuity of the first kind only, endowed with the Skorohod topology.

Special Notation for Section 5.7

The transpose of a vector V is V^t. The norm $\|.\|$ on R^p is defined by $\|(y_1, y_2, \ldots, y_p)\| = \max_{1 \leq i \leq p} |y_i|$. For a function $g(x; \theta)$, where $\theta = (\theta_1, \theta_2, \ldots, \theta_p) \in R^p$, $\nabla_\theta g(x; \theta_0)$ denotes the vector of parial derivatives

$$((\partial/\partial\theta_1)g(x; \theta), \ldots, (\partial/\partial\theta_p)g(x; \theta))$$

evaluated at $\theta = \theta_0$. Similarly, $\nabla_\theta^2 g(x; \theta_0)$ denotes the vector

$$((\partial^2/\partial\theta_1^2)g(x; \theta), \ldots, (\partial^2/\partial\theta_p^2)g(x; \theta))$$

evaluated at $\theta = \theta_0$. The matrix $[(\partial^2/\partial\theta_i\partial\theta_j)g(x; \theta)]_{i,j}$ is denoted by $g''_{\theta\theta}(x; \theta)$. For a matrix or vector $V = (v_{ij})$, $|V|$ denotes the matrix $(|v_{ij}|)$, and $\int V$ stands for $(\int v_{ij})$, while V^δ is meant to be (v_{ij}^δ).